Optical Remote Sensing of Air Pollution

A series devoted to the publication of courses and educational seminars given at the Joint Research Centre, Ispra Establishment, as part of its education and training programme.

Published for the Commission of the European Communities, Directorate-General Information Market and Innovation.

Volumes already published

– APPLICATIONS OF MASS SPECTROMETRY TO TRACE ANALYSIS
  Edited by S. Facchetti

– ANALYTICAL TECHNIQUES FOR HEAVY METALS IN BIOLOGICAL FLUIDS
  Edited by S. Facchetti

The publisher will accept continuation orders for this series which may be cancelled at any time and which provide for automatic billing and shipping of each title in the series upon publication. Please write for details.

Optical Remote Sensing of Air Pollution

Lectures of a course held at the Joint Research Centre, Ispra (Italy)
12–15 April 1983

Edited by

P. Camagni and S. Sandroni

Joint Research Centre, Ispra, Italy

Published for the Commission of the European Communities
by
ELSEVIER Amsterdam – Oxford – New York – Tokyo 1984

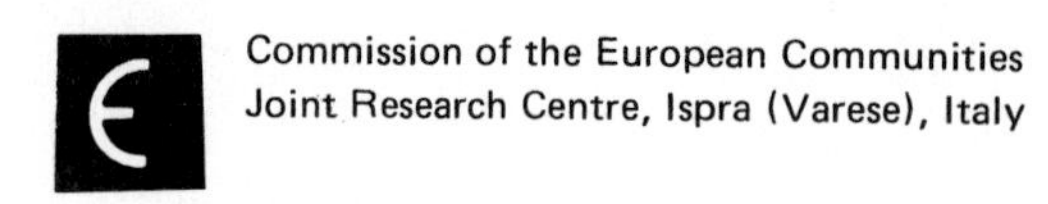

Commission of the European Communities
Joint Research Centre, Ispra (Varese), Italy

Publication arranged by:
Directorate-General Information Market and Innovation
Luxembourg

Published under licence by:
Elsevier Science Publishers B.V.
Molenwerf 1,
P.O. Box 211, 1000 AE Amsterdam,
The Netherlands

Distributors for the United States and Canada:
Elsevier Science Publishing Company Inc.
52, Vanderbilt Avenue
New York, N.Y. 10017

EUR 8754 EN

Library of Congress Cataloging in Publication Data
Main entry under title:

Optical remote sensing of air pollution.

Includes index.
1. Air--Pollution--Measurement--Optical methods--Addresses, essays, lectures. 2. Air--Pollution--Remote sensing--Addresses, essays, lectures. I. Camagni, P. II. Sandroni, S. III. Commission of the European Communities. Joint Research Center. Ispra Establishment.
TD890.O68 1983 628.5'3'0287 84-6052
ISBN 0-444-42343-5

Legal Notice

Neither the Commission of the European Communities nor any person acting on behalf of the Commission is responsible for the use which might be made of the following information.

ISBN 0-444-42343-5

Printed in The Netherlands

Foreword

Environmental research is charged nowadays with larger and larger commitments, owing to an increasing conflict between the needs for energy production and the preservation of environmental quality. Face to the growth of energy-related pollution, one of the major tasks for research resides in the continuous improvement of methods for the monitoring of pollutant behaviour. To this end classical and remote sensing techniques have been developed which combine multi-disciplinary competences in optics, atmospheric physics, meteorology, physical-chemistry and instrumentation.

Further promotion and exploitation of such means require a permanent scientific effort and multi-sided cooperation. In this frame the Commission of the European Communities has sponsored a research and development programme which, besides financing the studies of national laboratories (indirect action) and of the JRC (direct action), promotes a constant interaction and joint exercise on problems of environmental concern. Raising panels and workshops on special topics related to these matters is an integral part of this activity, which is also recommended by our assisting Advisory Committees.

Aim of the present Course, held in April 1983, at the JRC-Ispra in the frame of the Education and Training Programme, was to give an updated overview of the present status and applications of optical remote sensing in the study of atmospheric pollution. This is a topic of central interest, marked in recent years by the successful development of various specialized techniques and by a growing demonstration of their use for field monitoring of atmospheric phenomena. In the light of the experience accumulated in Western Europe from past exercises and collaborations, it was thought that the time was appropriate to make an assessment of specific achievements and to discuss the potentials for future exploitation. These aspects are dealt with in a series of 18 lectures by qualified experts of the various fields.

In our intention the practical goal of this Course was to provide people committed to environmental research, at various levels, with the following information:

- principles of optical remote sensing, including a general discussion of optical phenomena in the atmosphere and the conceptual basis of monitoring techniques;
- present development of specific techniques (Lidars, Cospec, radiometry, fluorosensing); their limits and perspectives contributions in the study of pollutant properties or pollutant behaviour in the troposphere;
- methodological and practical problems in field exercise; combination and intercomparison between different techniques; monitoring of complex situations and large-scale phenomena (e.g. areal dispersion and fall-out).

We are indebted to the lecturers for their qualified contributions and to all the participants for their active interest, which helped, we hope, to make the Course successful.

JRC Ispra, June 1983 P. Camagni and S. Sandroni

LIST OF CONTRIBUTORS

H. Quenzel	Meteorologisches Institut der Universität, München (West Germany)
M.M. Millán	Laboratorios de Ensayos e Investigacion Industrial J. Terrontegui, Bilbao (Spain)
R.E.W. Pettifer	Meteorological Office, Wokingham, Berkshire (England)
R.H. Varey	Central Electricity Research Laboratories, Leatherhead, Surrey (England)
D.J. Brassington	Central Electricity Research Laboratories, Leatherhead, Surrey (England)
R. Capitini	Commissariat Energie Atomique, CEN, Saclay (France)
D. Renaut	Météorologie Nationale, Magny-les-Hameaux (France)
E. Joos	Electricité de France, Chatou (France)
P. Camagni	Commission of the European Communities, Joint Research Centre, Electronics Division, Ispra (Italy)
D. Anfossi	Istituto di Cosmogeofisica del CNR, Torino (Italy)
S. Sandroni	Commission of the European Communities, Joint Research Centre, Chemistry Division, Ispra (Italy)
G. Restelli	Commission of the European Communities, Joint Research Centre, Electronics Division, Ispra (Italy)
C. Tomasi	Istituto FISBAT, CNR, Bologna (Italy)
N. Omenetto	Commission of the European Communities, Joint Research Centre, Chemistry Division, Ispra (Italy)
R. Zellner and **J. Haegele**	Institut für Physikalische Chemie der Universität, Göttingen (West Germany)
D. Onderdelinden	Rijksinstituut voor de Volksgezondheid, Bilthoven (The Netherlands)

CONTENTS

Optical Remote Sensing of Air Pollution,
Lectures of a course held at the Joint Research Centre, Ispra, Italy, 12—15 April 1983,
P. Camagni and S. Sandroni (Eds). 1—25

SCATTERING, ABSORPTION, EMISSION AND RADIATIVE TRANSFER IN THE ATMOSPHERE

H. QUENZEL

1 QUANTITIES FOR DESCRIBING THE RADIATION FIELD

Remote Sensing of atmospheric properties is based on the modification of the electromagnetic field by constituents of the atmosphere. To measure the radiation field requires its quantitative description, and hence, the definition of suitable quantities. There are several suitable quantities which are somewhat redundant. In this paper only fundamental quantities needed for a quantitative description of the interaction between radiation and atmospheric matter are given (Table 1).

1.1 WAVELENGTH, FREQUENCY, WAVENUMBER

There are interaction processes between radiation and matter which can be understood by imagining that the radiation has the properties of waves (e.g. interference) and others which require imagining that the radiation consists of corpuscles (photons) (e.g. photovoltaic effect). Consequently either the wave or the corpuscular imagery will be chosen, all according to which image makes it easier to interpret each particular problem in this paper. The energy E of a photon is

$$E = h \cdot \nu \quad (1)$$

where h = Planck's constant = $6.626 \cdot 10^{-34}$ J·s and $\nu = c/\lambda$ = frequency, with λ being the wavelength of the radiation as a wave and c being the velocity of light; c = $2.998 \cdot 10^{8}$ m/s. In addition to wavelength and frequency the wavenumber $1/\lambda$ is often used.

1.2 AMPLITUDE, PHASE, POLARIZATION AND ENERGY

In electrodynamics the radiation is described by the amplitude, phase and the position of the plane of oscillation. Radiation is said to be unpolarized when its wavetrains are distributed equally over all possible planes of oscillation, and is said to be 100 % linearly polarized when all the wavetrains oscillate in just one plane. Polarization as a property of radiation will not be treated further here, because remote sensing techniques that take account

TABLE 1.
RADIOMETRIC QUANTITIES

Quantity	Symbol	Dimension	Unit
Wavelength	λ	length	μm
Frequency	ν	1/time	$Hz = s^{-1}$
Wavenumber	k, ν^*	1/length	cm^{-1}
Radiance	L	power/(area·solid angle)	$W \cdot m^{-2} sr^{-1}$
Spectral radiance	L_λ	power/(area·solid angle ·length)	$W \cdot m^{-2} \cdot sr^{-1} \cdot \mu m^{-1}$
Radiant flux density		power/area	$W \cdot m^{-2}$
as irradiance	E	power/area	$W \cdot m^{-2}$
as exitance	M	power/area	$W \cdot m^{-2}$
Spectral radiant flux density		power/(area·length)	$W \cdot m^{-2} \cdot \mu m^{-1}$
as spectral irradiance	E_λ	power/(area·length)	$W \cdot m^{-2} \cdot \mu m^{-1}$
as spectral exitance	M_λ	power/(area·length)	$W \cdot m^{-2} \cdot \mu m^{-1}$
Scattering coefficient	$\alpha_{s\lambda}$	1/length	m^{-1}
Absorption coefficient	$\alpha_{a\lambda}$	1/length	m^{-1}
Extinction coefficient	$\alpha_{e\lambda} = \alpha_{s\lambda} + \alpha_{a\lambda}$	1/length	m^{-1}
Scattering function			
absolute	$f(\theta)$	1/(solid angle·length)	$sr^{-1} \cdot m^{-1}$
normalized	$f'(\theta)$	1/solid angle	sr^{-1}

of polarization are still under development. Nor does the phase need to be discussed because it does not contain any information that would be of use in remote sensing. Instead remote sensing uses the square of the amplitude of the wave which is proportional to the energy of the radiation.

1.3 RADIANCE

The radiation field which is to be measured, is anisotropic, that is radiation of different strengths arrives at the detector from different directions. To describe it one uses the radiance L, that is the energy per time emerging from a small solid angle and penetrating an area. The dimension of radiance is power per unit area and solid angle and so the units are, for instance, $W.m^{-2}.sr^{-1}$.

In general, remote sensors do not integrate over all wavelengths but measure radiation only in small wavelength intervals.

The descriptive monochromatic quantity is called spectral radiance L_λ and has the dimensions power per unit area and solid angle and length. So the units are, for instance, $W.m^{-2}.sr^{-1}.\mu m^{-1}$.

1.4 RADIANT FLUX DENSITY

The total radiation from a hemisphere through an area is a radiant flux density and is called irradiance E. The total radiation from an area into a hemisphere is also a radiant flux density and is called exitance M. Both are obtained by integrating the radiance over the hemisphere and therefore have the dimension power per unit area, and have units of e.g. $W.m^{-2}$. The corresponding monochromatic quantities are called spectral irradiance E_λ and spectral exitance M_λ. (The quantities that have been introduced, describe the local state of the radiation field independently of whether there is a radiating or an irradiated surface or not and whether a detector is present or not).

2 RADIATION SOURCES

2.1 SUN, EARTH

The radiation field of the atmosphere contains photons that come from the sun and also those that are emitted by the earth (from the surface and the atmosphere). The sun radiates somewhat like a black body with a temperature of 6000 K while the earth's surface is like one at about 300 K. Fig. 2.1 shows, for different temperatures, the radiant flux density emitted by a blackbody plotted against the wavelength (Planck's law). As the temperature drops, the maximum of the curves is shifted towards longer wavelengths, and the radiant flux density drops in general. Despite this, nearly all of the photons in the longwave region of the atmosphere's radiation field are emitted by the earth.

For instance, Fig. 2.2 shows the radiant flux density, arising from a fictitious earth that radiates in the longwave region like a black body with a temperature of 300 K, and that reflects the radiation from the sun in the shortwave region with a reflectance of $\rho = 0.1$. The curves in Fig. 2.2 give an

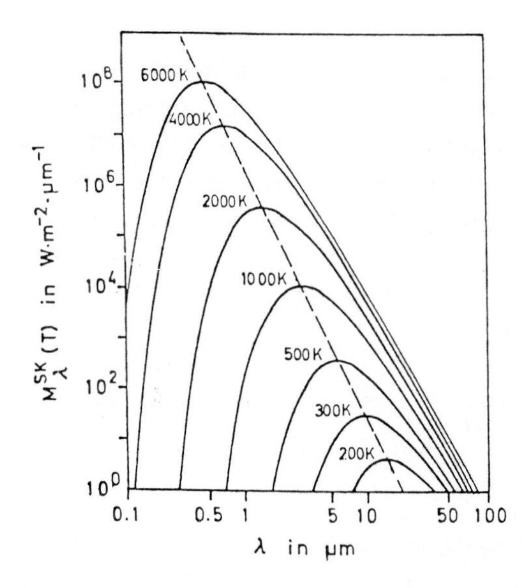

Fig. 2.1. Planck curves of blackbody radiation

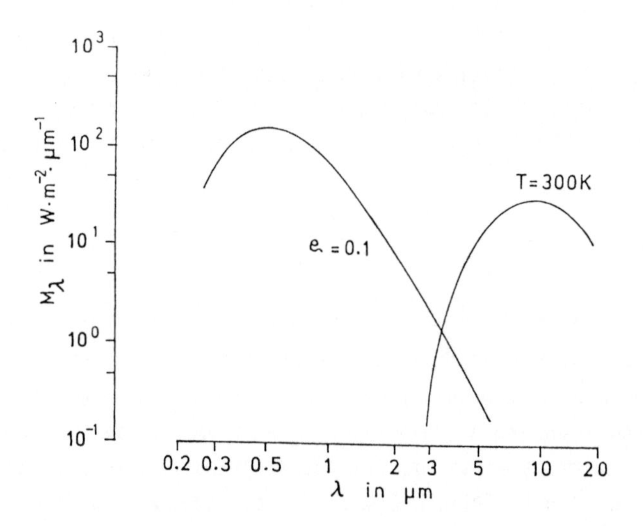

Fig. 2.2. Comparison of exitances due to reflected solar irradiance and thermal emission

idea of the radiant power arriving at an orbiting radiometer observing the earth with a solid angle of view of 2π . The curves in Fig. 2.2 differ from Planck's law (Fig. 2.1) because the sun irradiates the earth with a solid angle of only $6.8 \cdot 10^{-5}$ sr, but the reflected solar radiation is distributed over the whole solid angle of 2π. This results in a minimum in the radiation field at about 3.5 μm, so that the radiation field is divided, de facto, into a shortwave (solar) and a longwave (terrestrial) region. The terrestrial wavelength region includes the microwave region towards its longwave end. Because different sensing techniques are needed, a further distinction is drawn between optical and microwave regions.

2.2 REFLECTION AT THE EARTH'S SURFACE

Some of the radiation that arrives at the earth's surface is reflected. For remote sensing the atmosphere, the earth's surface can be treated as a known boundary condition of the radiation field, which is why the reflection will be treated here in the section 'Radiation Sources'.

The angular dependence of the reflection is described by the reflection function γ. The reflection of radiant flux densities is described by the reflectance ρ which is also called albedo A (proportion of reflected to incoming radiant flux density). The reflection properties are, of course, dependent on the wavelength and vary according to the type of surface.

In the solar spectral region, the solid surface of the earth reflects anisotropically but it is nevertheless more similar to isotropic reflection (diffuse reflection, Lambert reflection) than it is to specular reflection. Nearly specular reflection takes place on water surfaces and, to a lesser extent, on ice.

The values of the albedo in the solar spectral region are about 6 % for water surfaces and they are between 10 % and 30 % for land surfaces, while snow gives rise to values as high as 80 %. It should be pointed out that great spectral differences occur over the solar region.

In the terrestrial spectral region, surfaces also reflect anisotropically. However, because the irradiating field is mostly atmospheric radiation and therefore nearly isotropic, the reflected radiation field is also isotropic. But if sources with a small beamwidth are used in remote sensing, in particular lasers, the anisotropy of the reflection has to be fully taken into account. Reflectances of surfaces in the terrestrial spectral region typically lie between 0 and 15 %, if angular averages are taken. Single pairs of angles of incidence and of reflection may show reflectances up to 100 % i.e. in Reststrahlen bands of quartz sand surfaces. One minus reflectance is the emittance which is azimuthally isotropic as long as one can assume rotational symmetry of the surface. This is the case as long as no significant linear texture is present. The values of the emittance lie, according to the reflectances, typically between 85 % and 100 %.

So the earth can be treated like a blackbody with an appropriate emittance ≤ 1. Because the radiation field leaving the earth in thus nearly isotropic, the input to a radiometer with a narrow field-of-view (radiance) will be similar to that given in Fig. 2.2.

Obviously radiation leaving the earth is modified by the atmosphere according to its wavelength but this does not affect the separation into solar and terrestrial spectral regions, or the order of magnitude of the radiant power at a detector.

2.3 ARTIFICIAL RADIATION SOURCES

For remote sensing one can, of course, also use radiation from non-natural sources. In the solar and the IR-terrestrial spectral regions only lasers are suitable in practice. A laser emits a very high spectral radiance because its output by stimulated emission is concentrated in a very small solid angle in the order of a milliradiant. However, to date no satellite-borne lasers have been operated for remote sensing.

In the microwave region, however, radar methods have already been successfully used, even from satellites.

3 INTERACTION BETWEEN RADIATION AND MATTER

The fundamental interaction between radiation and matter consists of a photon transferring its energy to an atom or a molecule and thus being removed from the radiation field. The energy of a photon raises an electron to a higher energy level or in molecules or solids it can set up higher rotational or vibrational states.

This increase in energy of the substance can be released in several different ways (see Fig. 3.1).

One possibility is for the activated molecule to collide with another molecule, and to drop back into a lower energy state; the energy thus freed becomes kinetic energy of the molecules. This corresponds to warming the gas. This interaction is called absorption. In this case the photon is permanently lost from the radiation field; the radiation is attenuated by absorption.

A second possibility for releasing the energy increase is the spontaneous transition (in about 10^{-9}s) into the original state by emitting a photon which is identical to the absorbed one except for its direction of propagation. This interaction is called scattering, more precisely, elastic scattering. In this process, the photon remains part of the radiation field, but the direct beam is attenuated.

A third possibility is that the activated molecule releases its energy spontaneously but in two steps. This results in the release of two photons with different, lower energies which add up to the energy of the absorbed photon. The direct beam is attenuated, the original photon is no longer part of the radiation field as it has been replaced by two photons at longer wavelengths. This interaction is called Raman-scattering. For the purpose of remote sensing by means of Raman-scattering it is not the attenuation of the direct beam which is used but the photons which are released by this process.

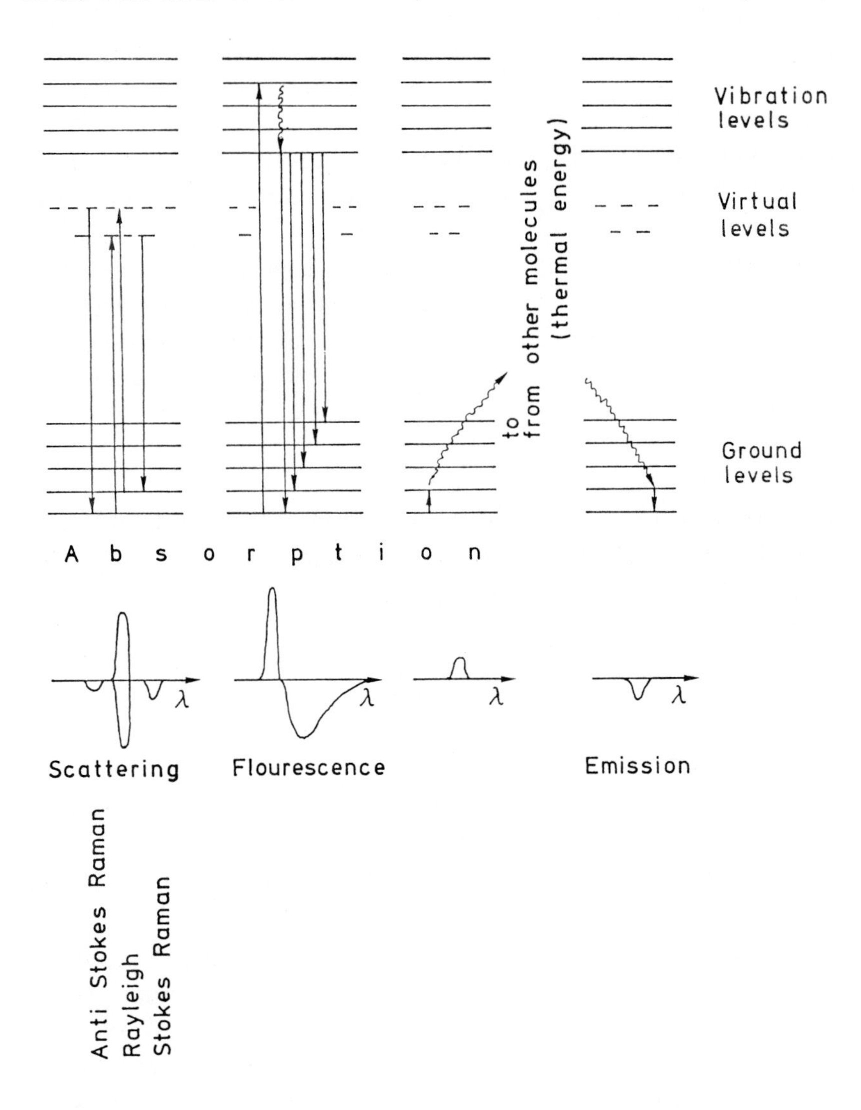

Fig. 3.1. Energy term scheme for interaction processes

The fourth possibility is that the energy is not released spontaneously, but after a relaxation time of about 10^{-9} to 10^{-4} s, depending on the type of molecule, and on the temperature and pressure. If the transition from the activated state to the ground state takes only one step, this interaction is called resonance fluorescense, but if it takes two or more steps, creating two or more photons then it is just called fluorescence. The latter scattering processes are called inelastic scattering processes.

Relaxation times of seconds to hours also occur and are called phosphorescence but they have no practical application to remote sensing.

The term extinction does not refer to an additional interaction process but rather it refers to the attenuation of the radiation caused by the combination of the scattering and absorption processes.

The interaction that have been described are also valid when the radiation strikes a solid object. The scattering at larger solid objects is called reflection.

A further fundamental interaction process consists in kinetic energy of the molecule (thermal energy) being converted into electromagnetic energy (created photons). In this process molecules are activated by collisions with each other, and the activation energy is emitted as photons. This interaction process is therefore called emission.

4 LAW OF EXTINCTION

A plane wave that has radiance L_λ at place 0 is attenuated by extinction (scattering and absorption) over a path ds by the amount dL_λ (Fig. 4.1).

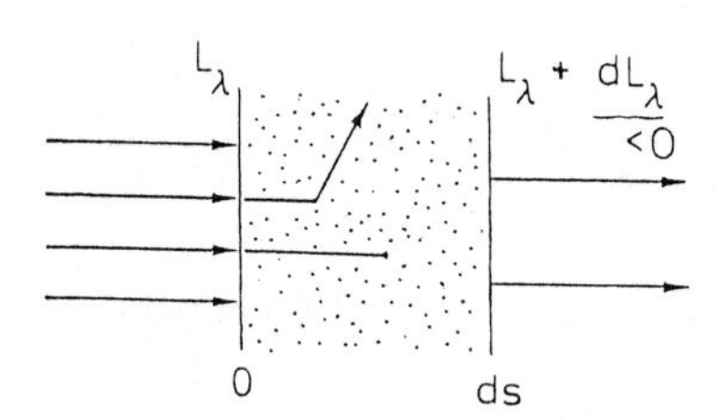

Fig. 4.1. Law of extinction

Under the conditions that are to be found in the atmosphere, the attenuation is proportional to the incoming radiance, L_λ, to the pathlength, ds, and in the case of a gas to the density, ρ. The properties of the substance (different substances cause differing amounts of extinction) are described by the proportionality factor $\varkappa_\lambda$. The factor $\varkappa_\lambda$ can be expressed by quantities used in electrodynamics, namely the dielectric constant, the permeability and the conductivity. So, the law of extinction expressed in differential form is

$$dL_\lambda = -\varkappa_\lambda \cdot \rho \cdot L_\lambda \cdot ds \tag{2}$$

Changing to a finite path s is done by integrating

$$\int_{L_{o\lambda}}^{L_\lambda} \frac{dL_\lambda}{L_\lambda} = -\int_0^s \varkappa_\lambda \cdot \rho \cdot ds \tag{3}$$

$$\ln L_\lambda - \ln L_{o\lambda} = \ln L_\lambda / L_{o\lambda} = -\varkappa_\lambda \cdot \rho \cdot s \tag{4}$$

The outcome of this derivation is the law of extinction in integral form

$$L_\lambda = L_{o\lambda} \cdot \exp(-\varkappa_\lambda \cdot \rho \cdot s) \tag{5}$$

So the term exp $(-\varkappa_\lambda \cdot \rho \cdot s)$ describes the transmission T_λ of a layer of thickness s:

$$T_\lambda = \exp(-\varkappa_\lambda \cdot \rho \cdot s) \tag{6}$$

The law of extinction is also called the Lambert or Bouguer-Lambert or Lambert-Beer law depending on how the factors in the exponent are combined.

Because ρ has the dimension mass/volume, s has the dimension length and the exponent $\varkappa_\lambda \cdot \rho \cdot s$ must be dimensionless, the mass extinction coefficient $\varkappa_\lambda$ has the dimension volume/(mass·length) and so the units are e.g. $cm^3/(g \cdot cm) = cm^2/g$.

If the density ρ is replaced by the particle number N per unit volume in cm^{-3} then $\varkappa_\lambda$ is replaced by the absorption cross-section σ_λ with the dimension of an area e.g. in cm^2 i.e.

$$\varkappa_\lambda \cdot \rho = \sigma_\lambda \cdot N \text{ in } cm^{-1} \tag{7}$$

The product $\varkappa_\lambda \cdot \rho$ is called optical density (not optical depth) or extinction coefficient and is often written as α_λ and has the dimension 1/length and so has the units of e.g. cm^{-1}. The whole exponent $\varkappa_\lambda \cdot \rho \cdot s$ is called optical depth of a layer with the length s. If the layer is the vertical column of the total atmosphere, the optical depth $\varkappa_\lambda \cdot \rho \cdot s$ is called $a_{E\lambda}$. This is the terminologie I want to use.

5 EXTINCTION IN THE ATMOSPHERE

5.1 COMPONENTS OF THE EXTINCTION

There are different substances in the atmosphere that cause extinction and so contribute to the spectral optical depth of a layer or the whole atmosphere. These substances are: air molecules, aerosol particles, water vapour, ozone and minor constituents. The minor constituents are the substances this course is focussed on.

From the derivation of the law of extinction it follows that the total optical depth of the atmosphere $a_{E\lambda}$ is the sum of the optical depths of its individual components:

$$a_{E\lambda} = \overbrace{a_{RS\lambda} + \underbrace{a_{DS\lambda}}_{} }^{= a_{S\lambda}} \!\!\!\!\!\! \quad \overbrace{\underbrace{\;+ a_{DA\lambda}}_{= a_{DE\lambda}} + a_{gasA\lambda}}^{= a_{A\lambda}} \tag{8}$$

($a_{S\lambda} = a_{RS\lambda} + a_{DS\lambda}$; $a_{A\lambda} = a_{DA\lambda} + a_{gasA\lambda}$; $a_{DE\lambda} = a_{DS\lambda} + a_{DA\lambda}$)

$a_{RS\lambda}$ is the optical depth of the O_2 and N_2 air molecules which scatter in the spectral region of interest but do not absorb except in a weak O_2-band. The extinction due to haze, $a_{DE\lambda}$, is separated into two components, scattering, $a_{DS\lambda}$, and absorption, $a_{DA\lambda}$. For ozone and the other trace gases (e.g. H_2O, CO_2) the scattering they cause is negligible compared to $a_{RS\lambda}$ but their absorption is described by $a_{gasA\lambda}$.

Ignoring its components, the optical depth of the atmosphere can be considered to be the sum of a scattering part $a_{S\lambda}$ and an absorption part $a_{A\lambda}$.

Another quantity that is important for the description of the radiation field is the ratio of the optical depth due to absorption by a substance to the total optical depth due to that substance. This quantity is called absorption fraction k_λ. The scattering fraction $\omega_{o\lambda} = 1-k_\lambda$ is usually called "albedo of single scattering".

For instance for an aerosol, $1-k_\lambda = \omega_{o\lambda} = a_{DS\lambda}/a_{DE\lambda}$. For a nonabsorbing substance, $\omega_{o\lambda} = 1$ and $k_\lambda = 0$.

5.2 RAYLEIGH SCATTERING

In 1871 Lord Rayleigh solved the Maxwell equations for the case of a plane electromagnetic wave which penetrates a medium in which dielectric spheres are statistically distributed and whose sizes are small compared to the wavelength. The radius of an air molecule is about 10^{-4} µm so it is small compared to the wavelength. Therefore the scattering at air molecules can be described by Rayleigh scattering. Taking the wavelength dependence of the refractive index of air into account, the result is approximately $a_{RS\lambda} \sim \lambda^{-4.09}$ in the solar spectral region.

5.3 MIE EXTINCTION

In 1908 Gustav Mie solved the Maxwell equations for the case of a particle whose radius r is of the same order of magnitude as the wavelength of the radiation. The radii of the usual aerosol particles extend from 0.1 to 10 µm and so their size is comparable with that of wavelengths in the solar spectral region. So the Mie theory describes the scattering and absorption properties of the aerosol particles. The spectral behaviour of the aerosol optical depth $a_{DE\lambda}$ depends on the aerosol size distribution dN/dr.

As a rough approximation, the size distribution of the aerosol particles can be described after Junge by

$$dN/dr \sim r^{-(\nu^*+1)} \quad \text{where } \nu^* = 3.3 \tag{9}$$

From Mie theory it then follows that $a_{DE\lambda} \sim \lambda^{-1.3}$.

The wavelength exponent of 1.3 was also arrived at by Angstrom as the average of many optical measurements. However the aerosol size distributions are not, in fact, simple power law distributions but usually consists of three log normal distributions, with different aerosol types showing different relative strengths and distributions of the three modes. (Fig. 5.1) Conse-

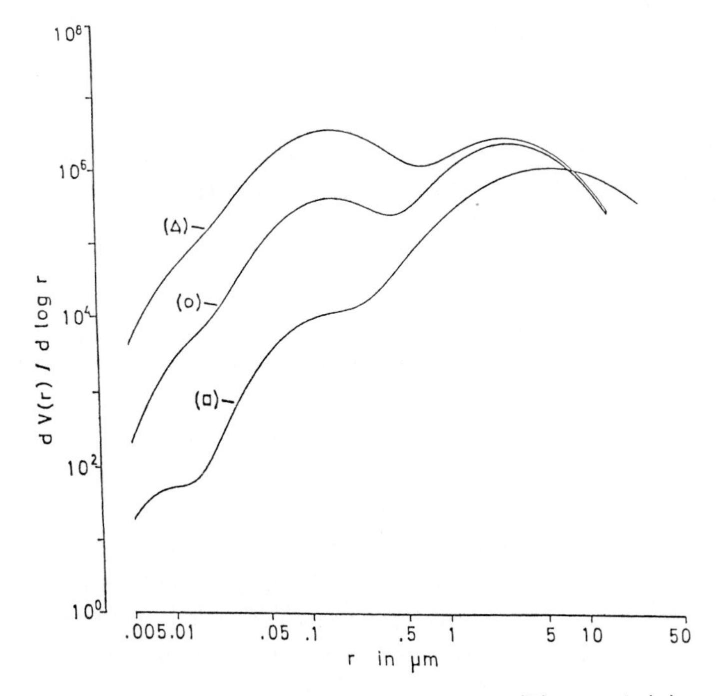

Fig. 5.1. Size distributions for maritime (□), rural (O), and urban (Δ) aerosol

quently the wavelength dependence of $a_{DE\lambda}$ is not a simple power law. When approximating the true distribution by a power law, the exponent can take values between 0 and -2 in the solar spectral region. In narrow spectral intervals values can even be >0 or <-2. In addition the spectral behaviour depends on the refractive index of the aerosol particles, so the exact determination of the spectral behaviour requires knowledge not only of the aerosol size distribution but also of the refractive index.

The size of the imaginary part of the complex refractive index determines the absorption properties of the aerosol particles and hence their albedo of single scattering. Because the refractive index only varies slowly with wavelength, aerosol absorption is continuum absorption and not as with gases, band absorption.

It should be pointed out that clouds can be treated just like aerosol layers.

5.4 GAS ABSORPTION

Gases can absorb radiation by changing their rotational, vibrational or electronic state or any combination of these. Since these changes of state are not continuous but take place between discrete energy levels, that is in quantum jumps, radiation is not absorbed at all wavelengths but in absorption lines which form the absorption bands when combined.

5.4.1 Ozone

Ozone has absorption bands in the UV and the visible spectral region as well as in the IR (Fig. 5.2): Hartley band 0.24-0.3 µm, Huggins band 0.31-0.34 µm. Chappuis band ~0.6 µm, 9.6 µm band.

5.4.2 Oxygen

Oxygen has absorption bands below 0.24 µm which determine the shortwave end of the solar spectrum reaching the earth, and it has another narrow band at 0.76 µm (O_2-A-band).

5.4.3 Water Vapour

Water vapour has absorption bands (Fig. 5.2) in the solar spectral region at 0.7 µm (a-band), 0.8 µm-band, 0.96 µm ($\rho\sigma\tau$ -band), 1.1 µm (ϕ -band), 1.38 µm (ψ -band), 1.9 µm (Ω -band), 2.7 µm (χ -band), 3.2 µm-band and in the terrestrial spectral region the 6.3 µm-band and the rotational bands from 14 µm to about 100 µm. Water vapour has more absorption bands than any other gas in the atmosphere.

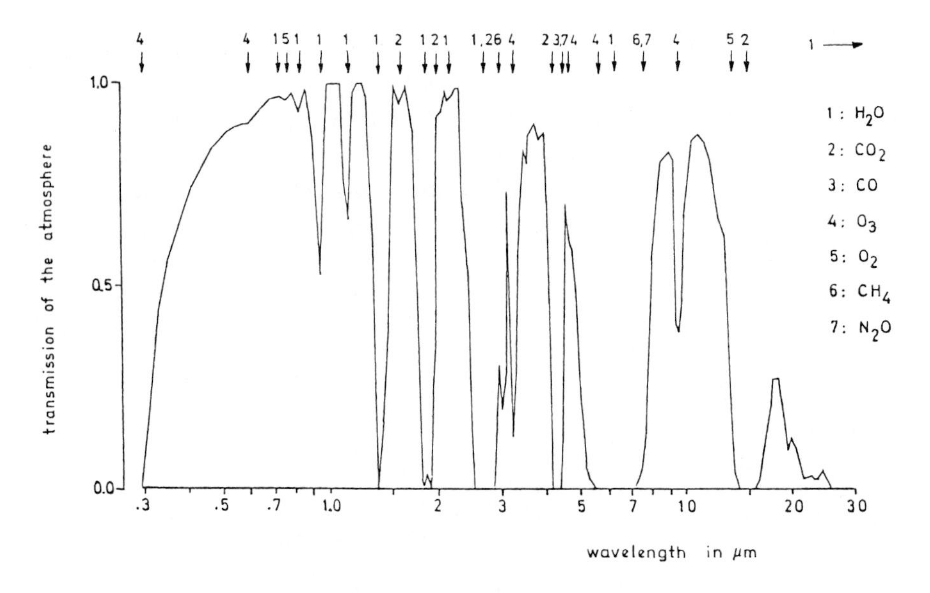

Fig. 5.2. Absorption bands of atmospheric gases

5.4.4 Carbon Dioxide

Carbon dioxide has absorption bands at 1.6 μm, 2.0 μm, 2.7 μm, 4.3 μm and 15 μm and plays the next most important role in the atmosphere's radiation budget after water vapour.

5.4.5 Other Minor Constituents

There are many other minor constituents like nitrous oxides or CFM's which, whether they occur naturally or as a result of man's activities, have many absorption bands of different strengths in the solar spectral region as well as in the infrared. Many of these minor constituents are a source of environmental pollution and, just because they have absorption bands, they can be remotely sensed. A list of the absorption bands of the minor constituents is not given here, firstly because it is very long and would be useless without further explanation, and secondly because some of the gases will be discussed later in the course.

(In any case, here we are only concerned with the basic interactions between radiation and matter, and not with a list of the absorption bands of the minor constituents.) The main reason for having listed the absorption bands was to show that at wavelengths where polluting gases have absorption bands, natural gases may also have absorption bands and may perturb the determination of the polluting gases.

5.5 MODIFICATION OF THE SOLAR SPECTRUM BY ATMOSPHERIC EXTINCTION

Fig. 5.3 shows the contribution to the attenuation of the solar spectrum by ozone (2), air molecules (3), aerosol scattering (4), aerosol absorption (5) and water vapor absorption (6). Curve (1) is the solar spectrum at the top of the atmosphere and curve (6) at the bottom. The spectra (2) to (5) are fictious because the matter interactions numbered (2) to (5) act simultaneously.

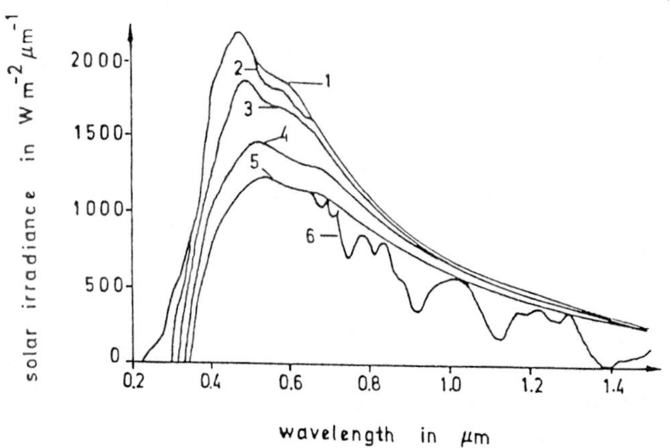

Fig. 5.3. Attenuation of the solar spectral irradiance due to atmospheric absorption and scattering

6 SCATTERING FUNCTIONS

In the interaction between radiation and matter called elastic scattering (Section 3) the photons remain in the radiation field although they travel in a direction different to the one they had before being scattered.

The angular distribution of the scattered radiation (in the photon imagery, the probability with which a photon is scattered into a particular direction) is described with the normalized scattering function f', often simply called phase function, or more precisely, single scattering phase function.

For the scattering particles that are found in the atmosphere (i.e. roughly spherical ones), the scattering function is a function of just one angle, the scattering angle θ, which is the angle between the new and the original direction of travel (Fig. 6.1)

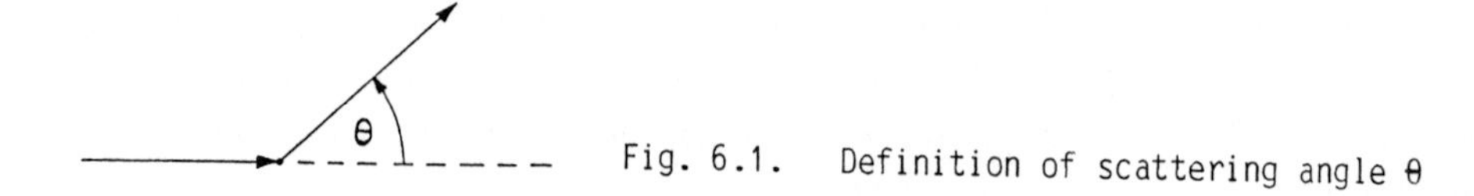

Fig. 6.1. Definition of scattering angle θ

Because the probability that a photon is scattered in some direction is 1 if it is scattered at all, it follows that

$$\int_{\text{sphere}} f'(\Theta)\cdot d\Omega = 1 \qquad (10)$$

The absolute scattering function $f(\theta)$ gives not just the angular distribution but also the strength of the scattered radiation. It is defined by

$$L(\theta) = f(\theta)\cdot E_0 \qquad (11)$$

where E_0 is the irradiance entering the scattering volume from just one direction and $L(\theta)$ is the radiance leaving at a scattering angle θ.

Absolute and normalized scattering functions are linked by the scattering coefficient $\alpha_{S\lambda}$:

$$f(\theta) = \alpha_{S\lambda}\cdot f'(\theta) \qquad (12)$$

The derivation of this relation is beyond the scope of this paper.

The scattering coefficient $\alpha_{S\lambda}$ is thus a measure of how strongly radiation is scattered and so also a measure of the attenuation of the incoming radiation (Section 4).

The unit of the absolute scattering function $f(\theta)$ is $sr^{-1}\cdot m^{-1}$ and that of the normalized scattering function $f'(\theta)$ is sr^{-1}.

7 SCATTERING IN THE ATMOSPHERE

Scattering in the atmosphere is caused by air molecules, aerosol particles and cloud droplets (Section 5.1).

7.1 RAYLEIGH SCATTERING FUNCTION

The scattering function of the air molecules (which are small compared to the wavelength) was derived by Rayleigh from the Maxwell equations. It is

$$f'(\theta) = \frac{3}{16\pi}\,(1+\cos^2\Theta) \qquad (13)$$

The usual graphic representations of the scattering function either use Cartesian or polar coordinates (Fig. 7.1).

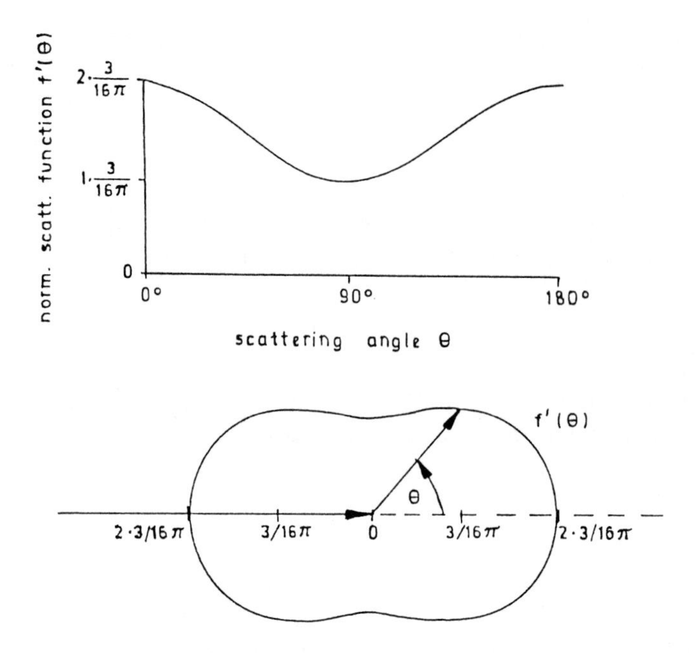

Fig. 7.1. Rayleigh scattering function

7.2 MIE SCATTERING FUNCTION

The scattering functions of particles with radii of the same order of magnitude as the wavelength (aerosol particles and cloud droplets) have been derived by Mie. For the Mie scattering functions there are no analytic formulations as there are for the Rayleigh scattering function. The solutions of the Maxwell equations derived from Mie theory can only be presented as an infinite series.

Their numerical solution is not a problem with current computers. Each aerosol size distribution has its own scattering function. In principle Mie scattering functions show orders of magnitude more forward scattering than backscattering.

As an example of this, Fig. 7.2 shows the scattering function of three aerosol types that more or less cover the variability of the scattering functions which are observed in the atmosphere.

7.3 MIXED SCATTERING FUNCTION

In the atmosphere there are air molecules as well as aerosol particles. The normalized scattering function of their combination is called mixed normalized scattering function $p_\lambda'/4\pi$. The product of the albedo of single scattering of the mixture of components in the atmosphere with $p_\lambda'/4\pi$ is defined as

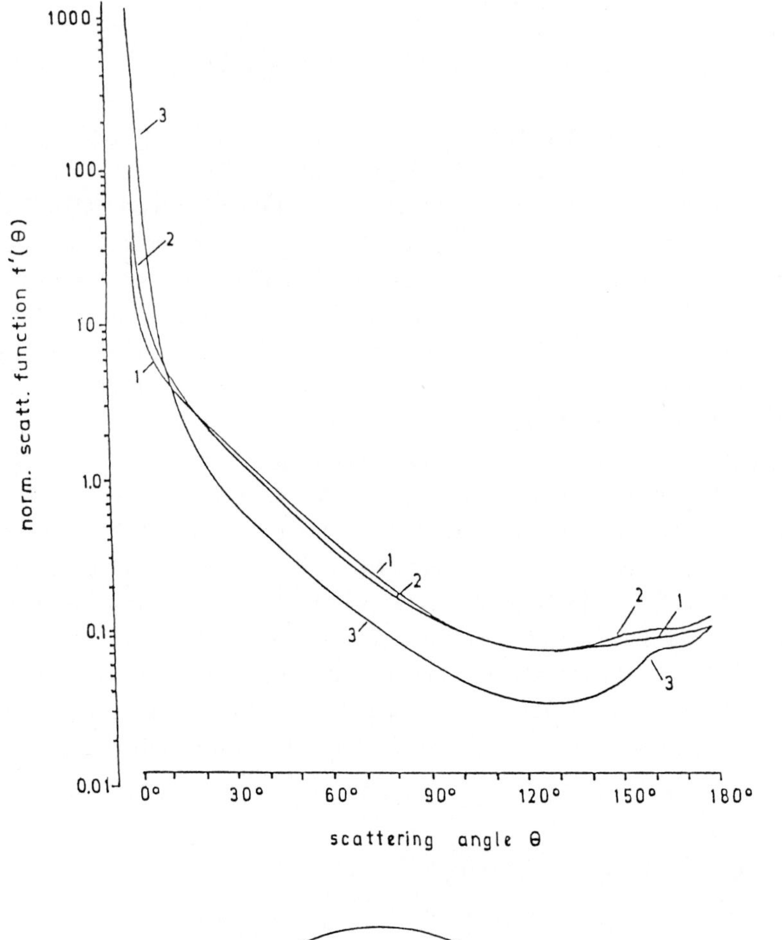

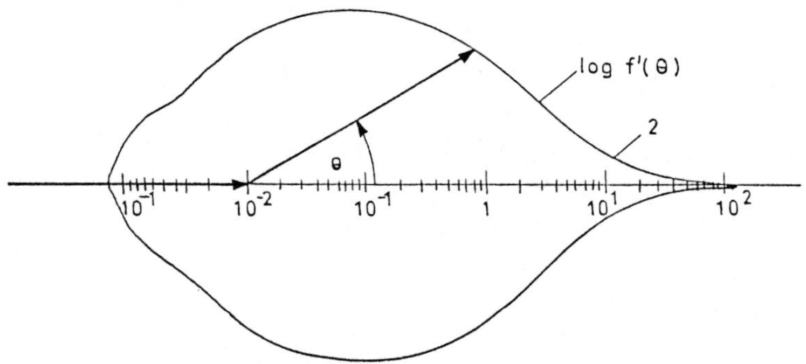

Fig. 7.2. Mie scattering functions for aerosol type 'urban average' (1), for a power law aerosol size distribution with Junge exponent 3 (2), for aerosol type 'marine surface' (3)

$$(1-k_\lambda)\cdot\frac{p_\lambda'}{4\pi}=\frac{a_{RS\lambda}\cdot f_{R\lambda}(\theta)+a_{DS\lambda}\cdot f_{D\lambda}(\theta)}{a_{RS\lambda}+a_{DS\lambda}+a_{DA\lambda}+a_{gasA\lambda}} \tag{14}$$

Little effort is needed to see why this definition is chosen.

8 RADIATIVE TRANSFER EQUATIONS

8.1 CONVERTING FROM SINGLE VOLUMES TO THE WHOLE ATMOSPHERE

Above we have discussed the reduction of the radiation by absorption and scattering at a volume with a relatively small optical depth. But if we want to consider not just single volumes but the radiation field over the whole atmosphere, then we also need to take intensification processes into account. This intensification arises as a result of photons being scattered from neighbouring volumes into the direction of observation or from created photons being emitted into this direction.

Fig. 8.1 presents just a few of the many possible photon paths. The figures indicate the number of scattering processes the photons have undergone before they reach the observer from the direction of observation. Some of the photons which were already traveling in the viewing direction were attenuated by extinction in the volume. Simultaneously, the volume is illuminated by photons which originally had a different direction but have since been scattered in other volumes and now illuminate the volume of interest as part of the scattered radiation field. Additionally the volume is illuminated by the direct solar beam. All these photons will be scattered in this volume into the

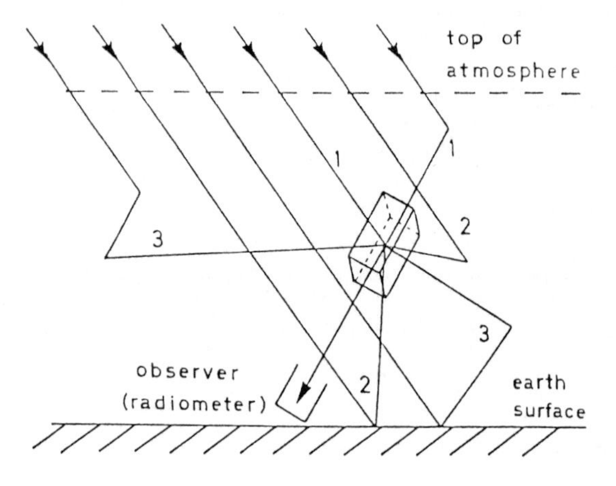

Fig. 8.1. Examples of photon paths in the atmosphere

direction of observation with a probability determined by the absolute scattering function. Finally we must consider the created photons that are emitted from the volume. The complete description of the radiation field as seen by the observer must therefore take account of both attenuation and intensification of the radiation. To obtain the radiation which in fact arrives at the observer, we have to integrate over all volumes along the direction of observation. This qualitative description is formalized in the next section.

8.2 THE RADIATIVE TRANSFER EQUATION IN DIFFERENTIAL FORM

To derive the radiative transfer equation, we consider a volume that is part of the atmosphere. Radiation travelling in the direction of observation through the volume towards the observer (Fig. 8.2) is attenuated in the volume according to the law of extinction. However, as was explained in Section 8.1, some photons which come from radiation illuminating the volume from other directions are scattered into the observation direction, and photons are also created in the volume, some of which are emitted into the observation direction.

The radiative transfer equation (RTE) can thus be written as the law of extinction, but with an additional term that gives the intensification.

$$dL_\lambda = -\alpha_\lambda \cdot ds \cdot L_\lambda + \alpha_\lambda \cdot ds \cdot J_\lambda \tag{15}$$

So the RTE in differential form becomes

$$\frac{dL_\lambda}{\alpha_\lambda \cdot ds} = - L_\lambda + J_\lambda \tag{16}$$

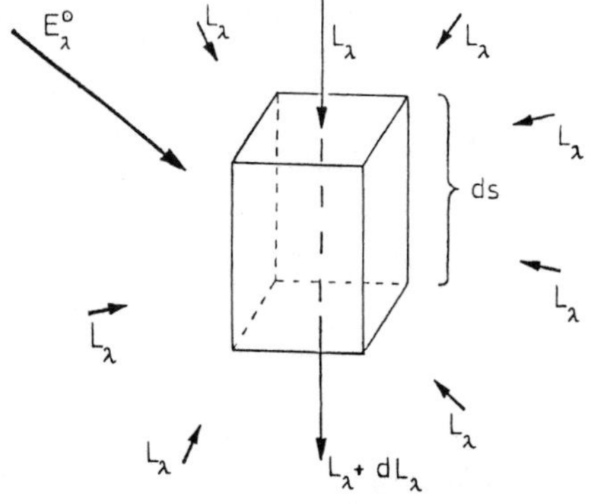

Fig. 8.2. Illustration of radiation intensification by scattering (source functions J^{dif} and J^{dir})

The function describing the intensification J_λ, the source function, consists of three additive terms. J_λ^{dif} describes the new photons in the volume originating from the diffuse radiation surrounding the volume, J_λ^{dir} describes those from the incoming direct solar radiation, and J_λ^{e} describes the emitted ones.

In order to locate the radiation field in the atmosphere, appropriate coordinates will be introduced. As the vertical coordinate we take the optical depth $\tau_\lambda = \int_z^0 \alpha_{E\lambda}(z) \cdot dz$ of the atmosphere from its top, down to height z. At each point in the atmosphere with the horizontal coordinates x,y and the vertical coordinate τ_λ, the direction of the radiation is given by a zenith angle ϑ and an azimuth angle φ. Frequently, instead of the zenith angle ϑ, the quantity $\mu = \cos\vartheta$ is used, combined with the additional convention that the upwelling radiation is expressed by the $+\mu$ coordinate, and the downwelling radiation by the $-\mu$ coordinate, which in both cases yields $0 \leq \mu \leq 1$. The directions of incidence are given as quantities with primes (μ', φ'), the outgoing directions are given without primes (μ, φ). The scattering angle θ is connected to these coordinates by

$$\cos\theta = \mu \cdot \mu' + \sqrt{1-\mu^2} \cdot \sqrt{1-\mu'^2} \cdot \cos(\varphi - \varphi') \tag{17}$$

The pair of coordinates (μ'_0, φ_0') indicates the direction of the incident solar radiation.

The source functions written with these coordinates are as follows:

a) the source function of the photons originating from the illumination of the volume by previously scattered radiation is

$$J_\lambda^{dif} = (1-k_\lambda) \cdot \int_0^{2\pi} \int_{-1}^{+1} \frac{p_\lambda'}{4\pi} (\mu, \varphi; \mu', \varphi') \cdot L_\lambda(\tau_\lambda, \mu', \varphi') \cdot d\mu' \cdot d\varphi' \tag{18}$$

b) the source function of the photons originating from the illumination of the volume by direct solar radiation is

$$J_\lambda^{dir} = (1-k_\lambda) \cdot \frac{p_\lambda'}{4\pi} (\mu, \varphi; -\mu_0', \varphi_0') \cdot E_\lambda^{\odot} \cdot e^{-\tau_\lambda/\mu_0} \tag{19}$$

c) The source function of the photons created in the volume is

$$J_\lambda^{e} = k_\lambda \cdot B_\lambda(T) \tag{20}$$

Now the RTE can be written in the following form

$$\mu \cdot \frac{dL_\lambda}{d\tau_\lambda} = - L_\lambda + J_\lambda^{dif} + J_\lambda^{dir} + J_\lambda^{e} \tag{21}$$

with the source functions as given above.

According to the derivation, we have here a one dimensional model, that is the atmosphere is horizontally homogeneous at each level, and extends out to infinity. For practical purposes a horizontal homogeneity of 30 km is sufficient. We have a plane, not a curved, atmosphere. In practice this means that elevation angles of less than 3° to 5° cannot be properly handled. Of course the RTE could also be written three-dimensionally without adding any new physical thoughts, but it would look more complicated and its numerical treatment would indeed be so. There is only one extraterrestrial source (the sun) and so the strength of the incoming radiation is the same all across the horizontal plane. The scattering behaviour of the particles is axially symmetrical about the direction of the incident beam so that the scattering function only has one independent variable, the scattering angle Θ. The radiative transfer equation only holds for the monochromatic case, that is when at the wavelength in question, the optical effects of the atmosphere can be described by a single optical thickness; then transmission functions are exponential functions. Section 8.3 will show how it is still possible to write the radiative transfer equation for absorption bands in which the optical effects at one wavelength cannot be described by just one optical depth.

Our version of the RTE holds for the radiances without considering the polarization of the radiation. If one wants to take it into account then the RTE must be applied separately to each of the four Stokes parameters which describe the radiation field including the polarization. We have not done so as it would make the presentation unnecessarily tortuous. It should be stressed though, that all the remarks that follow, will hold as if not only the strength of the radiation field had been taken into account, but also the degree and the plane of polarization.

The RTE is a tool which permits the accurate calculation of the complete radiation field in the atmosphere. However, it is an integro-differential equation because the term to the left of the equals sign contains a derivative of the radiation field, the first term to the right of the equals sign contains the radiation field itself and the second term to the right of the equals sign contains the radiation field as an integrand.

It can only be solved numerically and this is only possible if one does not use its differential form but its integral form (introduced in the next section) inclusive of the boundary conditions.

8.3 THE RADIATIVE TRANSFER EQUATION IN INTEGRAL FORM

By formally solving the differential RTE

$$\frac{dL_\lambda(\tau_\lambda,\mu,\varphi)}{d\tau_\lambda} = -L_\lambda(\tau_\lambda,\mu,\varphi) + J_\lambda(\tau_\lambda,\mu,\varphi) \tag{22}$$

and separating the radiance into the upwelling and downwelling components, we get the integral form of the RTE

upwelling:

$$L_\lambda(\tau_\lambda,\mu,\varphi) = L_\lambda(a_{E\lambda},\mu,\varphi)\cdot e^{-\frac{a_{E\lambda}-\tau_\lambda}{\mu}} + \int_{\tau_\lambda}^{a_{E\lambda}} J_\lambda(t_\lambda,\mu,\varphi)\cdot e^{-\frac{t_\lambda-\tau_\lambda}{\mu}}\cdot\frac{dt_\lambda}{\mu} \tag{23}$$

with $0 < \mu \leq 1$

downwelling:

$$L_\lambda(\tau_\lambda,\mu,\varphi) = L_\lambda(0,\mu,\varphi)\cdot e^{-\frac{\tau_\lambda}{\mu}} + \int_0^{\tau_\lambda} J_\lambda(t_\lambda,-\mu,\varphi)\cdot e^{-\frac{\tau_\lambda-t_\lambda}{\mu}}\cdot\frac{dt_\lambda}{\mu} \tag{24}$$

with $0 < \mu \leq 1$

Since we are concerned with a one-dimensional model there is just an upper and a lower boundary condition, and no lateral ones.

At the top of the atmosphere ($\tau_\lambda = 0$) only solar radiation enters, so

$$L_\lambda(0,-\mu,\varphi) = L_\lambda^{\odot}(-\mu_o',\varphi_o')\cdot\delta(-\mu,\mu_o')\cdot\delta(\varphi,\varphi_o') \tag{25}$$

The Dirac delta functions ensure that this term is zero except for the direction of the sun (μ'_0, φ_0').

The bottom of the atmosphere, the earth's surface is described by its angle dependent reflection function and by its thermal emission, which is given by the Planck function with the temperature of the earth's surface $B_\lambda(T_s)$, multiplied by its spectral emittance $\varepsilon_\lambda = k_{\lambda S}$ where $0 \leq \varepsilon_\lambda \leq 1$.

The irradiation of the earth's surface results from the direct solar radiation and the scattered radiation from all over the sky

$$E_\lambda(a_{E\lambda}) = E_\lambda^{\odot}\cdot\mu_o'\cdot e^{-\frac{a_{E\lambda}}{\mu}} + \int_0^{2\pi}\int_0^1 L_\lambda(a_{E\lambda},-\mu',\varphi')\cdot\mu'\cdot d\mu'\cdot d\varphi' \tag{26}$$

and hence

$$L_\lambda(a_{E\lambda},\mu,\varphi) = \gamma_\lambda(\mu,\varphi;\mu_0',\varphi_0')\cdot E_\lambda^{\odot}\cdot\mu_0'\cdot e^{-\frac{a_{E\lambda}}{\mu}} + \int_0^{2\pi}\int_0^{1}\gamma_\lambda(\mu,\varphi;\mu',\varphi')\cdot L_\lambda(a_{E\lambda},-\mu',\varphi')\cdot\mu'\cdot d\mu'\cdot d\varphi' + \varepsilon_\lambda\cdot B_\lambda(T_S) \qquad (27)$$

If one inserts the boundary conditions and the source functions into the integral form of the RTE, it yields the following complete form of the radiative transfer equation:
upwelling radiances:

Term 0 $\quad L_\lambda(\tau_\lambda,\mu,\varphi) =$

Term 1 $$+\int_{\tau_\lambda}^{a_{E\lambda}} E_\lambda^{\odot}\cdot e^{-\frac{\tau_\lambda}{\mu_0'}}\cdot(1-k_\lambda)\cdot\frac{p_\lambda'}{4\pi}(\mu,\varphi;-\mu_0',\varphi_0')\cdot e^{-\frac{t_\lambda-\tau_\lambda}{\mu}}\cdot\frac{dt_\lambda}{\mu}$$

Term 2 $$+\int_{\tau_\lambda}^{a_{E\lambda}}(1-k_\lambda)\int_0^{2\pi}\int_{-1}^{1}\frac{p_\lambda'}{4\pi}(\mu,\varphi;\mu',\varphi')\cdot L_\lambda(\tau_\lambda,\mu',\varphi')\cdot d\mu'\cdot d\varphi'\cdot e^{-\frac{t_\lambda-\tau_\lambda}{\mu}}\cdot\frac{dt_\lambda}{\mu}$$

Term 3 $$+\int_{\tau_\lambda}^{a_{E\lambda}} k_\lambda\cdot B_\lambda(T(t_\lambda))\cdot e^{-\frac{t_\lambda-\tau_\lambda}{\mu}}\cdot\frac{dt_\lambda}{\mu} \qquad (28)$$

Term 4 $$+\varepsilon_\lambda\cdot B_\lambda(T_S)\cdot e^{-\frac{a_{E\lambda}-\tau_\lambda}{\mu}}$$

Term 5 $$+\gamma_\lambda(\mu,\varphi;\mu_0',\varphi_0')\cdot E_\lambda^{\odot}\cdot\mu_0'\cdot e^{-\frac{a_{E\lambda}}{\mu_0'}}\cdot e^{-\frac{a_{E\lambda}-\tau_\lambda}{\mu}}$$

Term 6 $$+\int_0^{2\pi}\int_0^{1}\gamma_\lambda(\mu,\varphi;\mu',\varphi')\cdot L_\lambda(a_{E\lambda},-\mu',\varphi')\cdot\mu'\cdot d\mu'\cdot d\varphi'\cdot e^{-\frac{a_{E\lambda}-\tau_\lambda}{\mu}}$$

with $0<\mu\leq 1$

downwelling radiances:

Term 0 $\quad L_\lambda(\tau_\lambda,-\mu,\varphi) =$

Term 1 $\quad + \int_0^{\tau_\lambda} E_\lambda^{\odot} \cdot e^{-\frac{t_\lambda}{\mu_0'}} \cdot (1-k_\lambda) \cdot \frac{p_\lambda'}{4\pi}(-\mu,\varphi;-\mu_0',\varphi_0') \cdot e^{-\frac{\tau_\lambda - t_\lambda}{\mu}} \cdot \frac{dt_\lambda}{\mu}$

Term 2 $\quad + \int_0^{\tau_\lambda} (1-k_\lambda) \int_0^{2\pi}\int_{-1}^{1} \frac{p_\lambda'}{4\pi}(-\mu,\varphi;\mu',\varphi') \cdot L_\lambda(t_\lambda,\mu',\varphi') \cdot d\mu' \cdot d\varphi' \cdot e^{-\frac{\tau_\lambda - t_\lambda}{\mu}} \cdot \frac{dt_\lambda}{\mu}$

Term 3 $\quad + \int_0^{\tau_\lambda} k_\lambda \cdot B_\lambda(T(t_\lambda)) \cdot e^{-\frac{\tau_\lambda - t_\lambda}{\mu}} \cdot \frac{dt_\lambda}{\mu}$ $\quad$ (29)

Term 7 $\quad + L_\lambda^{\odot}(-\mu_0',\varphi_0') \cdot \delta(-\mu,-\mu_0') \cdot \delta(\varphi,\varphi_0') \cdot e^{-\frac{\tau_\lambda}{\mu}}$

with $0 < \mu \leq 1$

These forms of the RTE makes it easy to see which processes contribute to the radiation field at level τ_λ in direction $(\pm\mu,\varphi)$ (Terms 0).

The terms 1 describe the scattered direct solar radiation, and terms 2 the contribution of the multiply scattered photons.

Terms 3 describe the emission of the atmosphere, that of the ground appears as term 4, but only in the upwelling radiances.

The terms 5 and 6 are the contribution of the direct and diffuse solar radiation reflected at the ground. They only appear explicity in the formula for upwelling radiances, while for the downwelling radiances they contribute to $L_\lambda(t_\lambda,\mu',\varphi')$ in term 2.

Term 7 only occurs in the formula for the downwelling radiances. It describes the attenuated direct solar beam.

The radiative transfer equation in integral form also has no analytic solution because it is still an integral equation. There are several methods that give the numerical solution which are not treated here.

Since the RTE requires that the optical effects can in fact be expressed by the optical depth (see Section 8.2), the transmission functions must be expanded as e-series no matter how small the wavelength intervals are, next the RTE must be solved separately for each term in the e-series and finally the results must be weighted and added.

9 THE RADIATIVE TRANSFER EQUATION AS THE BASIC EQUATION FOR REMOTE SENSING

The radiative transfer equation has been presented so explicitly for two reasons: firstly, to show how all the interactions between radiation and matter can be given a joint quantitative description so that the total radiative transfer in the atmosphere can be described by just one equation.

Consequently, it must be possible to derive all the radiation transport equations needed in remote sensing applications, from the RTE by adapting it to the particular problem (see the next article).

This enables us to put the different remote sensing techniques into a wider context and is the second reason for having presented the RTE so explicitly.

REFERENCES

1 S. Chandrasekhar, Radiative Transfer, Clarendon Press, Oxford, 1950
2 D. Deirmendjian, Electromagnetic Scattering on Spherical Polydispersions, Elsevier, New York, 1969
3 R.M. Goody, Atmospheric Radiation I: Theoretical Basis, Clarendon Press, Oxford, 1964
4 K.T. Kriebel, Reflection Properties of Vegetated Surfaces: Tables of Measured Spectral Biconical Reflectance Factors, Münchener Univ. Schriften, Meteorol. Institut, Wissenschaftliche Mitteilung Nr. 29, 1977
5 H. Quenzel, Computation of Luminance and Color Distribution in the Sky, in M. Nagel, (Ed.), Daylight Illumination-Color-Contrast Tables for Full-Form Objects, Academic Press, New York, 1978

Optical Remote Sensing of Air Pollution,
Lectures of a course held at the Joint Research Centre, Ispra, Italy, 12—15 April 1983,
P. Camagni and S. Sandroni (Eds). 27—43

PRINCIPLES OF REMOTE SENSING TECHNIQUES

H. QUENZEL

1 DEFINITION OF REMOTE SENSING

In general, remote sensing can be defined as follows:

From the state of a field at one location
conclusions are drawn
about quantities that influenced the field at a different location.

So far, nothing has been said about the nature of the field. Here we will deliberately restrict ourselves to electromagnetic radiation as the information carrier and will exclude acoustic and radioactive radiation.

Remote sensing, therefore, requires an interaction between the electromagnetic field and the matter whose properties are to be determined. As we saw in the previous lecture there are, strictly speaking, only two interactions between electromagnetic radiation and matter, namely absorption and emission. They are, however, usefully regrouped into four classes of interaction processes: scattering, reflection, absorption and emission. Here absorption means the permanent loss of a photon from the radiation field, while the temporary, usually very brief, removal of a photon from the radiation field by absorption and spontaneous re-emission is called either scattering or reflection.
The interaction can be understood formally as a functional relationship

measured radiation quantity = function (property of matter)

Concluding a property of matter (the wanted quantity) from the measured radiation quantity (the measured value, the optical quantity) is an 'inversion problem'. This can be written symbolically as

property of matter = inverse function (measured radiation quantity).

2 REMOTE SENSING AS AN INVERSION PROBLEM

In the simple case considered above, where the measured quantity depends on only one wanted quantity, the inversion problem is the solution of just one equation with one unknown. The functional relationship must, of course, be known whether from theory, ground or laboratory measurements.

In general however, even in the monochromatic case, the radiation field is influenced by not just one, but several substances. For instance, in the solar spectral region, the radiation which is reflected at the earth's surface is thus imprinted with a signature and this signature is influenced by the gases and particles in the atmosphere. Another example is the absorption by different gases at the same wavelength. The radiation field then, contains information about the wanted quantity but it is also influenced by 'perturbing quantities'.

This task of inversion becomes the problem of solving just one equation with several unknows, that is, there is no unique mathematical solution.

In the following we present the ways of solving the inversion problem using additional physical information (physical constraints).

The first way is to choose the measuring conditions so that the influence of the perturbing parameters is minimized. This can be achieved by selecting suitable wavelengths and measuring geometries. An example of this 'principle of small perturbations' is the determination of the O_3 concentration from the radiation emitted in the 9.6 μm band. The wavelengths are selected so that the emission of O_3 is much stronger than the emission and absorption of all the other substances in the stratosphere.

Another example is the determination of the atmospheric turbidity using the shortwave channels of geostationary satellites. This succeeds with reasonable accuracy when only cloudfree areas over water outside the sunglint are evaluated, because in this case the signal is determined predominantly by the aerosol particles in the atmosphere. For what we call 'principle of small perturbations' we sometimes say instead 'method of favorable conditions'.

A second way of solving the inversion problem does not require the perturbing quantities to be small, but their values must be known at the time of measurement, so that they can be taken into account during the inversion. This is the case, for instance, for the contribution of Rayleigh scattering in various remote sensing measurements, because it is determined by the number of air molecules, (the barometric pressure), which can vary only slightly. In general, therefore, the long term average is used. If the remaining uncertainty in the remote sensing result is too large, it can be reduced by measuring, in addition, the barometric pressure.

A third way of solving the inversion problem, which is mostly used with current radiometers, consists in obtaining additional information from additional measurements. These can be measurements at several suitable wavelengths or from several viewing angles. Often, the effect of the perturbing parameters is not just eliminated using this approach, it is also used as additional information to determine these perturbing parameters. Now these quantities can no longer be considered to be perturbing, rather they are further quantities which can be determined simultaneously. For this, an example is the determination of the concentration of several trace gases in the stratosphere from measurements at several wavelengths. Another example is the measurement of the chlorophyll content of sea-water, also from measurements at different wavelenghts. In this case, the perturbing influence of the atmosphere is measured at a wavelength where no radiation emerges from the water.

An example of a measurement where the wanted quantity is itself a perturbing quantity is the determination of vertical profiles of trace gas concentrations. Here, the information coming from lower layers is masked by the upper layers. To get a solution of the inversion problem despite this diffuculty, limb sounding measurements are made at different angles, each of whose signals is controlled by information from a different height. For the determination of the concentration in lower layers, the concentration determined in the layer above is allowed for.

(Normally one measures at different nadir angles, not to get additional information, but to sample further locations, that is, to do remote sensing over a large area).

Simulating the radiation field with the aid of a computer allows all the parameters that influence the radiation field to be varied independently, so that their effects can be studied separately. Such sensitivity studies do not just allow one to decide under which conditions (wavelength, geometry) the wanted quantity most strongly outweighs the perturbing quantities, they also allow one to decide how accurately the inversion problem can be solved when one has to accept measuring errors and a certain inaccuracy in the amount of the perturbing substances.

The inversion problem is also usually solved numerically with the aid of a computer because there are no analytical solutions.

The extraction of information from a satellite image by a skilled observer should also be regarded as remote sensing. This means a brain is solving the inversion problem. For qualitative analyses this approach is very wise because the brain is able to provide a great deal of a-priori knowledge, and it is also able to replace a certain amount of the information that is lost through perturbation. The visual interpretation of satellite images should be thought of as doing pattern recognition with the brain.

3 COMPLEXITY OF REMOTE SENSING TASKS

The remote sensing tasks can be arranged according to the degree of indirectness with which the measured radiation field and the wanted quantity are linked. This approach is more a structure for thinking with than pure physics, but it gives some insight into the remote sensing problem. Also it enables one to imagine how inaccurate are the quantities derived from the remote sensing measurements, because their inaccuracy increases as the link becomes more complex.

Every remote sensing operation starts with the measurement of radiation. If the radiation itself is already the wanted quantity, there are still two cases to be distinguished, namely whether the intensity of the radiation is wanted at the satellite, or elsewhere.

The first case has nothing to do with remote sensing at all, it is an in-situ measurement. As an example consider the radiation impact on a satellite, which must be known in order to design its radiative cooling system.

The second case we will call a remote sensing task of the first order. For example, take the determination of the radiation budget at the top of the atmosphere from satellite measurements. Here the radiation flux into space from the top of the atmosphere is arrived at from the radiances measured at the satellite. Additional information is needed, namely on the anisotropy of the radiation emerging from the atmosphere. In this example the additional information can be taken from measurements (here, in several different directions) or from model calculations.

Remote sensing tasks of the second order are those in which the measured radiation quantity is used to derive the strength of one of the interaction processes described in section 1. and this is already the required result. An example of this is the optical depth of the atmosphere caused by scattering at aerosol particles.

Remote sensing tasks of the third order are those then, in which the measured radiation quantity leads to the strength of the interaction process which in turn leads to another property of the substance. An example of this is the mass of the aerosol particles which can only be extracted from the optical depth with further additional information. In this example the aerosol size distribution and the refractive index of the particles must be known. Another example starts with the microwaves emitted from the sea-surface and moves to the amount of whitecaps and goes on to determine the wind-speed at sea-level.

A remote sensing task of the fourth order would be to go a stage further and derive the energy exchange between ocean and atmosphere from this wind-speed.

It is obvious that the errors in the result increase according to the complexity of the remote sensing task. If for the determination of the optical depth of the aerosol particles, the radiance has an error of 5%, then the optical depth will have an error of 20% and the mass of the aerosol will, in fact, have an error of 100%. An increase of this size in the error is typical but it does depend on the accuracy of the additional information that was used.

4 BASIC MEASURING GEOMETRIES AND MATTER-RADIATION INTERACTIONS

There is, of course, remote sensing from space but radiometers can also be used at ground stations and within the atmosphere (airplanes, balloons, rockets).

One of the advantages of remote sensing, namely that no alteration is made to the state of the atmosphere, is equally valid for remote sensing from space or from anywhere else. Nevertheless when one thinks of remote sensing, one thinks of remote sensing from space. This comes about, because the second great advantage of remote sensing, namely the monitoring of large areas of the earth with high spatial and temporal resolution, really only comes into its own with remote sensing from space.

This is also the reason why this article talks more of remote sensing from space than from the ground although it is introducing a course on remote sensing techniques from the ground.

Remote sensing techniques are called passive when they use radiation from natural sources (sun, earth, atmosphere). In contrast, one talks of active methods when artificial radiation sources are used (e.g. laser, microwave transmitter).

Fig. 4.1 shows the measuring geometry for passive remote sensing techniques from space and the interactions that can be used for remote sensing, namely emission, extinction, scattering (absorption plus scattering is called extinction) and reflection.

Satellite-borne radiometers can observe the earth-atmosphere system either towards the horizon (limb sounding, limb-viewing mode) or downwards (nadir sounding, down-viewing mode).

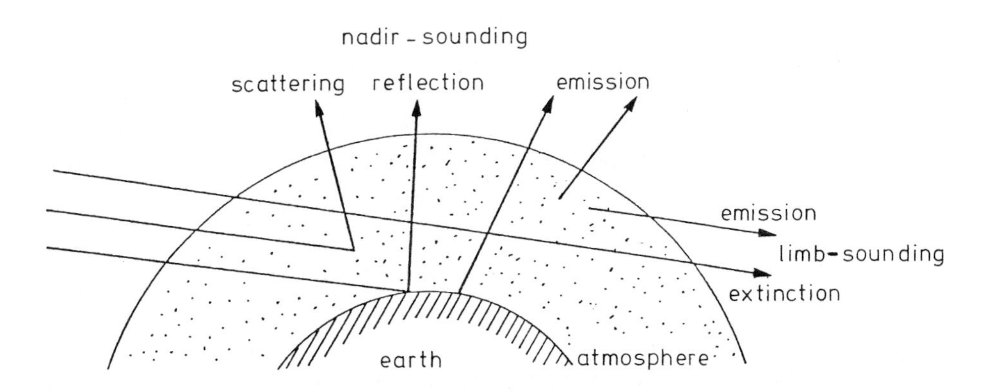

Fig. 4.1. Basic measuring geometries and matter-radiation interactions for passive remote sensing from space

Limb sounding is most suitable for remote sensing of the stratosphere because there are no clouds to cause perturbations. Since the effect of the atmospheric constituents in the stratosphere is weak because of their low concentrations, long optical paths are needed, which are only available when limb sounding. The resulting horizontal resolution is low, but this is not a disadvantage because the horizontal gradients in the stratosphere are also small. Nadir sounding is particularly suitable for the remote sensing of the troposphere, because there the interactions between matter and the radiation field are intense enough, and it gives good horizontal resolution.

Limb sounding can measure either emitted radiation (e.g. temperatures and trace gas concentrations e.g. with gas correlation radiometers) or, in the so-called occultation mode, it can measure the attenuated solar radiation (e.g. aerosol and again trace gas concentrations). The nadir sounding mode can measure the radiation emitted from the ground (e.g. surface temperature) or the atmosphere (e.g. temperatur profiles) according to the wavelengths, or it can measure reflected solar radiation. All according to the relative strength of the influence of the atmosphere compared to the reflection, scattering will predominate (over weakly reflecting surfaces, e.g. turbidity measurements over the ocean) or reflection will predominate (with atmospheres that cause little extinction, e.g. determination of the reflection function of the ground or the chlorophyll concentration in sea-water).

The basic measurement techniques from the ground are equivalent to the downward viewing mode from space but in upward viewing mode. Equally, the solar occultation technique from space does not, in principle, differ from that from the ground.

For the active methods there are only upward and downward viewing modes so that a separate picture is not necessary.

As examples of shortwave, and also longwave active methods, scattering is used to measure aerosols using lidar techniques, reflection in conjunction with extinction is used for measuring aerosols using the LPA (long path absorption) technique, and extinction is used for measuring trace gases using the DIAL (differential absorption lidar) method.

5 ADAPTATION OF THE RADIATIVE TRANSFER EQUATION FOR REMOTE SENSING TECHNIQUES

5.1 RADIATIVE TRANSFER EQUATION

The radiative transfer equation derived in the previous article allows the complete description of the radiation field in the atmosphere. Because it includes all the matter-radiation interaction processes, all the radiation transport equations which are needed for the different remote sensing methods can be derived from the radiative transfer equation. Depending on the wavelength, the geometry and the optical properties of the interacting substances, the different terms of the radiative transfer equation make different-sized contributions to the radiation field. So, for practical applications some terms can usually be omitted.

But never forget that in principle one must always consider all the terms of the RTE, and that when measuring under certain circumstances, remarkable errors will arise as a result of ignoring all the terms which are usually ignored for a particular remote sensing task.

5.2 USE OF THE RTE FOR PASSIVE REMOTE SENSING IN THE SOLAR SPECTRAL REGION

In the solar spectral region, there are so few photons that are emitted by the atmosphere and ground, compared to solar photons, that these few can be ignored. Thus terms 3 and 4 drop out of the radiative transfer equation (RTE) in its integral form (previous article).

In downviewing mode, of course, one makes use of the radiation field emerging from the top of the atmosphere ($\tau_\lambda = 0$). The appropriate radiative transfer equation (needed for shortwave passive methods that use upwelling scattered sunlight) is therefore the RTE for upwelling radiances where $\tau_\lambda = 0$:

Term 0 $\quad L_\lambda(0,\mu,\varphi) =$

Term 1 $$\int_0^{a_{E\lambda}} (1-k_\lambda)\cdot E_\lambda^{\odot}\cdot e^{-\frac{t_\lambda}{\mu_0'}}\cdot \frac{p_\lambda'}{4\pi}(\mu,\varphi;-\mu_0',\varphi_0')\cdot e^{-\frac{t_\lambda}{\mu}}\cdot\frac{dt_\lambda}{\mu}$$

Term 2 $$+\int_0^{a_{E\lambda}} (1-k_\lambda)\int_0^{2\pi}\int_{-1}^{+1} \frac{p_\lambda'}{4\pi}(\mu,\varphi;\mu',\varphi')\cdot L_\lambda(\tau_\lambda,\mu',\varphi')\cdot e^{-\frac{t_\lambda}{\mu}}\cdot\frac{dt_\lambda}{\mu} \qquad (1)$$

Term 5 $$+\gamma_\lambda(\mu,\varphi;\mu_0',\varphi_0')\cdot E_\lambda^{\odot}\cdot \mu_0'\cdot e^{-\frac{a_{E\lambda}}{\mu_0'}}\cdot e^{-\frac{a_{E\lambda}}{\mu}}$$

(formula continued next page)

Term 6 $$+\int_0^{2\pi}\int_0^1 \gamma_\lambda(\mu,\varphi;\mu',\varphi')\cdot L_\lambda(a_{E\lambda},-\mu',\varphi')\cdot\mu'\cdot d\mu'\cdot d\varphi'\cdot e^{-\frac{a_{E\lambda}}{\mu}}$$

where terms 1 and 2 apply to solar photons singly and multiply scattered, while terms 5 and 6 apply to the reflection from the ground, when illuminated with direct and diffuse solar radiation.

Because of the spectral dependence of the optical properties of the substances that influence the radiation field, at different wavelengths the different terms of the equation make different-sized contributions to the radiance to be measured. The contributions also depend on the geometry and, of course, on the type of the ground-surface. Obviously one attempts to select measuring conditions so that the signal is mainly controlled by the wanted quantities, but even so, in practice, one always has to consider all the terms.

A first example of nadir-sounding in the solar spectral region is the determination of the turbidity of the atmosphere. This is best carried out over water surfaces outside the sunglint, since then terms 5 and 6 are small. Naturally, this remote sensing method uses wavelengths at which there is least gaseous absorption.

In the example of the determination of the reflection function from a satellite (not yet done) the relationships are reversed. Here the optical depth of the atmosphere should be as small as possible so that terms 1 and 2 are small.

Nadir-sounding requires the use of cloud-free pixels since the optical depth of clouds is so great that the contribution of all other substances to the signal would be unrecognizable.

Limb-sounding in the solar spectral region involves, as shown in Fig. 4.1, the analysis of radiation that came from the sun and travelled through the atmosphere (solar occultation). To describe this situation, our radiative transfer equation (presently in a form valid for a plane atmosphere) will have to be rewritten for a spherical atmosphere. This would just be a transformation of coordinates and would not bring any new physical insight so it has been omitted.

Usually for sun photometer techniques only the extinction term of the RTE makes a contribution and so we take term 7 from the solution for the radiance exiting at the bottom of the atmosphere ($\tau_\lambda = a_{E\lambda}$). So the RTE for the sun photometer reads

Term 0 $\quad L_\lambda(a_{E\lambda}, -\mu, \varphi) =$

$$\text{Term 7} \quad L_\lambda^{\odot}(-\mu_0, \varphi_0) \cdot \delta(-\mu, \mu_0') \cdot \delta(\varphi, \varphi_0') \cdot e^{-\frac{a_{E\lambda}}{\mu}} \tag{2}$$

if all contributions of scattered photons to term 0 can be ignored.

Since also for occultations, only the extinction term of the RTE makes a contribution, the RTE for occultation is the same as for the sun photometer but with $a_{E\lambda}/\mu$ now understood as the optical depth in the path from sun to satellite. The direction coordinates will have to be redefined to correspond to the geometry.

An example of the application of limb-sounding is the determination of trace gas and aerosol concentrations in the stratosphere, where measurements are only taken at wavelengths where the substances cause sufficient extinction. The sun photometer is suitable for the equivalent measurements in the troposphere.

5.3 USE OF THE RTE FOR PASSIVE REMOTE SENSING IN THE TERRESTRIAL SPECTRAL REGION

It is typical of the terrestrial spectral region that one can ignore the contribution of the solar photons compared to that of those from the earth and atmosphere. Hence terms 1 and 5 drop away. Term 7 can also be ignored except the sun is directly viewed with an angle of observation as small as the solid angle of the sun. Term 2 contains also the scattering of the emitted photons. Nevertheless it makes no contribution, partly because at wavelengths where the atmosphere absorbs strongly, the scattering of the emitted photons is negligible in comparison to the absorption, and partly because in window-regions the photons nearly all origin from the ground and their scattering in the atmosphere can be ignored here on account of its small optical depth.

The RTE for nadir sounding now reads

Term 0 $\quad L_\lambda(0, \mu, \varphi) =$

$$\text{Term 3} \quad + \int_0^{a_{E\lambda}} k_\lambda \cdot B_\lambda(T(t_\lambda)) \cdot e^{-\frac{t_\lambda}{\mu}} \cdot \frac{d\tau_\lambda}{\mu} \tag{3}$$

(formula continued next page)

Term 4 $$+\ \varepsilon_\lambda \cdot B_\lambda(T_B) \cdot e^{-\frac{a_{E\lambda}}{\mu}}$$

Term 6 $$+\int_0^{2\pi}\int_0^1 \gamma_\lambda(\mu,\varphi;\mu',\varphi') \cdot L_\lambda(a_{E\lambda},-\mu',\varphi') \cdot \mu' \cdot d\mu' \cdot d\varphi' \cdot e^{-\frac{a_{E\lambda}}{\mu}}$$

The terms 3 and 4 account for the photons emitted by the atmosphere and by the ground. Term 6 accounts for the emitted photons reflected at the ground.

The exponential function in term 3 which describes the change in transmission with height, weights the contribution made to term 0 by the Planck radiation at each height. This term is therefore called 'weighting function'.

In the terrestrial spectral region, the transmission cannot, in general, be described by an exponential term. Instead, one uses the appropriate average transmission function for the wavelength interval under investigation. The weighting function is the change of this transmission function with height.

Choosing different wavelengths at which the atmosphere has different optical depths, gives rise to weighting functions each of which has a maximum at a different height. This enables one to determine the vertical temperature

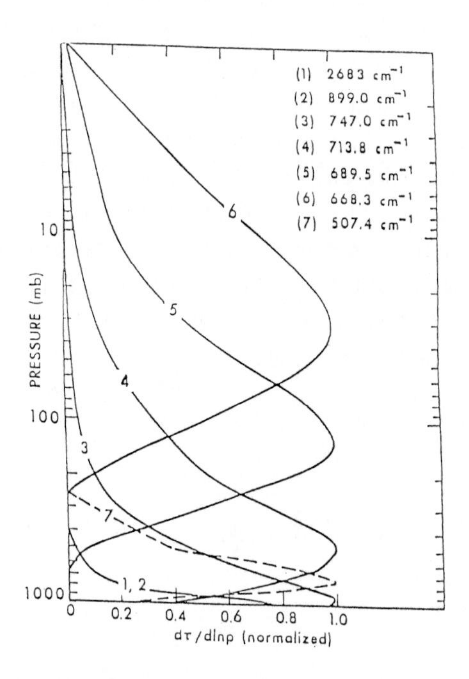

Fig. 4.2. Weighting functions of NIMBUS-5 ITPR

profile of the atmosphere. The determination of this temperature profile, an example of a passive multifrequency method using nadir sounding in the terrestrial region, is carried out at wavelenghts in the region of the CO_2 band. Carbon dioxide is chosen for temperature profiling because the mixing ratio of CO_2 hardly changes with height and so it does not need to be measured separately.

In this case the atmosphere is optically thick, so that terms 4 and 6 can be dropped. What remains is a Fredholm integral equation of the first kind because the quantities $L_\lambda(0,\mu,\varphi)$ are measured, the weighting functions are known and we want to derive B(T(z)) and hence T(z). The mathematical methods for the solution of these equations are not simple, but they cannot be discussed here. In a further step, when the vertical temperature profile has been determined, this method can also be used to get the vertical concentration profile of any atmospheric gas which has suitable absorption bands.

A further example of passive nadir-sounding in the terrestrial spectral region is the determination of the sea-surface temperature (SST). This is carried out using atmospheric windows, that is at wavelengths where the atmosphere is optically thin. Under these conditions term 3 makes only a very small contribution, and term 6 is also small because of the low reflectivity of water.

When limb sounding in occultation mode (against the sun), the same RTE holds as was used in the solar spectral region (section 5.3). Limb sounding with emitted radiation is described by term 3 of the RTE, once it has been recast to correspond to the geometry. Limb sounding is used in the terrestrial spectral region, again to determine stratospheric trace gases, aerosols and temperatures.

Microwaves which are a part of the terrestrial spectral region are subsumed above.

5.4 USE OF THE RTE FOR ACTIVE REMOTE SENSING

Remote sensing with artifical radiation sources requires the selection of measuring conditions so that the natural radiation is always negligible compared to the artificial radiation.

The radiation transport equations for all shortwave and longwave active methods do not differ so they can be treated together.

Since the irradiation of the atmosphere with microwave transmitters or lasers is carried out only in upward viewing mode or downward viewing mode (analogously to the irradiation by the sun in the solar spectral region) only terms 1 and 5 of the RTE need be used, with the solar radiation replaced, as appropriate, by the radiation of the artificial source.

Terms 3 and 4 vanish because they describe radiation that arises within the atmosphere. Since the transmitter irradiates only a small part of the atmosphere (unlike the illumination of the atmosphere by the sun) terms 2 and 6 also vanish, since the amplification of the signal by photons scattered into the measuring direction is negligible in most cases, but one must be aware that this simplification of the RTE involves ignoring multiple scattering processes.

In the following, the lidar equations for the LPA (long path absorption) method both with and without a reflecting target will be derived from the RTE.

In the LPA method with a reflecting target, radiation is received that has been emitted by a laser or a microwave transmitter (in the following these two sources are abbreviated to 'laser'), has traveled through the atmosphere to a target (e.g. retro-reflector, earth's surface), has been reflected there and has returned along the same path through the atmosphere to the detector which is usually directly adjacent to the laser. An example of this is the use of radar for the determination of surface structures. Here, one selects wavelengths where the atmosphere is optically thin, so that this configuration is describable by term 5 of the RTE, while even term 1 can be ignored, because it describes the radiation back-scattered by the atmosphere and not by the target.

$$\begin{aligned} &\text{Term 0} \quad L_\lambda(0,\mu,\varphi) = \\ &\text{Term 5} \quad \gamma_\lambda(\mu,\varphi;\mu_0',\varphi_0') \cdot E_\lambda^{\odot} \cdot \mu_0' \cdot e^{-\frac{a_{E\lambda}}{\mu_0'}} \cdot e^{-\frac{a_{E\lambda}}{\mu}} \end{aligned} \tag{4}$$

The appropriate lidar equation is derived from this RTE as follows.

If we imagine the laser and the detector are on board a satellite, that is outside the atmosphere, and replace the solar radiation $E_\lambda^{\odot}$ by the laser radiation E_λ^L and call the direction of the laser beam $(-\mu'_0, \varphi_0')$, and the reflection direction (μ_0,φ_0), then we get

$$L_\lambda(0,\mu_0,\varphi_0) = \gamma_\lambda(\mu_0,\varphi_0;\mu_0',\varphi_0') \cdot E_\lambda^L \cdot \mu_0' \cdot e^{-2\frac{a_{E\lambda}}{\mu_0}} \tag{5}$$

This is already the lidar equation, but not in its usual from. To get that, we multiply both sides by $A\cdot\Omega$, where A is the receiver surface and Ω is the solid angle subtended by the laser beam, that is $\Omega = F/R^2$ where F is the illuminated area perpendicular to the direction of propagation and R is the distance from the laser/detector to the target. This gives rise to

$$L_\lambda(0,\mu_0,\varphi_0)\cdot A\cdot\Omega = \gamma_\lambda(\mu_0,\varphi_0;\mu_0',\varphi_0')\cdot E_\lambda^L\cdot\mu_0\cdot e^{-2\frac{a_{E\lambda}}{\mu_0}}\cdot\frac{A}{R^2}\cdot F \tag{6}$$

Abbreviating $L_\lambda(0,\mu_0,\varphi_0)\cdot A\cdot\Omega = P_\lambda(R)$ which is the reflected power and $E_\lambda^L\cdot F\cdot\mu_0 = P_{0\lambda}$ which is what the laser emits, the so-called transmitted power, and with $\mu_0 = 1$ because the illumination is perpendicular to the target F, we now get

$$P_\lambda(R) = \gamma_\lambda(\mu_0,\varphi_0;\mu_0',\varphi_0')\cdot P_{o\lambda}\cdot\frac{A}{R^2}\cdot e^{-2a_{E\lambda}} \tag{7}$$

Just replacing $P_\lambda(R)$ by the received power $P_{r\lambda}(R)$, where $P_\lambda(R) = K^{-1}\cdot P_{r\lambda}(R)$ and K = optical systems efficiency, and replacing the reflection function by the reflection factor $\rho_\lambda = \pi\cdot\gamma_\lambda$, gives

$$P_{r\lambda}(R) = \frac{\rho_\lambda}{\pi}\cdot k\cdot P_{o\lambda}\cdot\frac{A}{R^2}\cdot e^{-2a_{E\lambda}} \tag{8}$$

or with

$$a_{E\lambda} = \int_0^R \alpha_\lambda(r)\cdot dr \tag{9}$$

we get

$$P_{r\lambda}(R) = \frac{\rho_\lambda}{\pi}\cdot k\cdot P_{o\lambda}\cdot\frac{A}{R^2}\cdot e^{-2\int_0^R \alpha_\lambda(r)\cdot dr} \tag{10}$$

This form of the lidar equation appears in every handbook of lidar techniques for the LPA method with a reflecting target, that is for non-range-resolved methods.

In the LPA method without a reflecting target, radiation is received that has been emitted by a laser, has traveled through the atmosphere to a particular range cell, has been reflected there by the air-aerosol mixture, and has returned to the detector. In contrast to the LPA method with a reflecting target, where a cw-laser (continous wave-laser) is used, here, in the LPA method without a reflecting target, a pulsed laser is used. Measuring the time between the emission and the reception of the pulse gives the distance from the detector to the range cell where the pulse is backscattered, and the pulse length defines the length of the backscattering cell ΔR. So the LPA method without a reflecting target is a 'range-resolved method'. An example of this is the remote detection of the intensity of precipitation using microwaves. In rain-showers, the atmosphere is optically thick to these microwaves, so term 1 of the RTE overwhelms term 5.

The RTE with just term 1 is

Term 0 $\quad L_\lambda(0,\mu,\varphi) =$ (11)

(formula continued next page)

Term 1 $$\int_0^{a_{E\lambda}} E_\lambda^{\odot} \cdot e^{-\frac{t_\lambda}{\mu_o'}} \cdot (1-k_\lambda) \cdot \frac{p_\lambda'}{4\pi}(\mu,\varphi;-\mu_o',\varphi_o') \cdot e^{-\frac{t_\lambda}{\mu}} \cdot \frac{dt_\lambda}{\mu}$$

Again we replace $E_\lambda^{\odot}$ by E_λ^L, consider the direction (μ_0,φ_0) and get

$$L_\lambda(0,\mu_0,\varphi_0) = \int_0^{a_{E\lambda}} E_\lambda^L \cdot e^{-\frac{t_\lambda}{\mu_o'}} \cdot (1-k_\lambda) \cdot \frac{p_\lambda'}{4\pi}(\mu_o,\varphi_o;-\mu_o',\varphi_o') \cdot e^{-\frac{t_\lambda}{\mu_o}} \cdot \frac{dt_\lambda}{\mu_o} \tag{12}$$

Since we want range-resolved information, we ignore the integral from $a_{E\lambda}$ to 0 and instead consider the path ΔR along the range cell, which makes us write dt_λ as $\Delta\tau_\lambda$. The symbol $\tau_\lambda(R)$ represents the optical depth of the atmosphere from the top to the relevant range cell. This gives rise to

$$L_\lambda(0,\mu_0,\varphi_0) = E_\lambda^L \cdot e^{-\frac{\tau_\lambda(R)}{\mu_o'}} \cdot (1-k_\lambda) \cdot \frac{p_\lambda'}{4\pi}(\mu_o,\varphi_o;-\mu_o',\varphi_o') \cdot e^{-\frac{\tau_\lambda(R)}{\mu_o}} \cdot \frac{\Delta\tau_\lambda}{\mu_o} \tag{13}$$

or since $\mu_0' = \mu_0$

$$L_\lambda(0,\mu_0,\varphi_0) = (1-k_\lambda) \cdot \frac{p_\lambda'}{4\pi}(\mu_o,\varphi_o;-\mu_o,\varphi_o) \cdot \frac{\Delta\tau_\lambda}{\mu_o} \cdot \Delta R \cdot E_\lambda^L \cdot e^{-2\tau_\lambda(R)/\mu_o} \tag{14}$$

At the scattering volume $1-k_\lambda = \alpha_{s\lambda}/\alpha_\lambda$ with $\Delta\tau_\lambda = \alpha_\lambda \cdot \Delta R$.
Since ΔR lies in the direction of propagation, that is the volume is illuminated perpendicularly, $\mu_0 = 1$. This gives

$$L_\lambda(0,\mu_0,\varphi_0) = \frac{\alpha_{s\lambda}}{\alpha_\lambda} \cdot \frac{p_\lambda'}{4\pi}(\mu_o,\varphi_o;-\mu_o,\varphi_o) \cdot \alpha_\lambda \cdot \Delta R \cdot E_\lambda^L \cdot e^{-2\tau_\lambda(R)} \tag{15}$$

Now we abbreviate

$$\alpha_{s\lambda} \cdot p_\lambda'(\mu_0,\varphi_0,-\mu_0,\varphi_0) = \beta_\lambda \tag{16}$$

so that β_λ is the value of the absolute scattering function at a scattering angle of 180°, the so-called backscatter coefficient. Now we get

$$L_\lambda(0,\mu_0,\varphi_0) = \frac{\beta_\lambda}{4\pi} \cdot \Delta R \cdot E_\lambda^L \cdot e^{-2\tau_\lambda(R)} \tag{17}$$

This equation is converted to the usual form, in the same way as above for the LPA method with a reflecting target except that R now stands for the distance from the laser/detector to the range cell, and not as with the LPA method with a reflecting target, the distance to the ground.

$$P_{r\lambda}(R) = \frac{\beta_\lambda}{4\pi} \cdot \Delta R \cdot k \cdot P_{o\lambda} \cdot \frac{A}{R^2} \cdot e^{-2\int_0^R \alpha_\lambda(r)\,dr} \tag{18}$$

This form of the lidar equation is that for the LPA without a reflecting target, that is for range-resolved methods.

These two lidar equations completely describe the reflected or back-scattered laser signal. But when working with short-wave radiation one must be aware that during the day the received radiation may not solely consist of the laser signal as it may also contain a contribution from scattered solar radiation. In this case the correct radiation transport equation is the RTE consisting of terms 1, 2, 5 and 6. This measuring scheme should be avoided as already stated. The relative contribution of the laser signal and the scattered solar radiation depends, of course, on the wavelength and the power of the laser.

The lidar equations are difficult to solve since from one equation, more than one essential unknown has to be derived, the backscatter coefficient and the vertical profile of the extinction coefficient. So, one can only obtain information that combines the backscatter coefficient and the extinction coefficients, but they can be separated if one additionally makes plausible assumptions about the state of the atmosphere, makes use of the lidar signals from different distances, or if one has additional independent information to hand.

6 ACTIVE MULTIFREQUENCY METHODS

Multifrequency methods are the method of choice for determining gas concentrations. The two-wavelength method using lasers relies on the principle of measuring at two adjacent wavelengths, one with strong gas absorption, the other with as little gas absorption as possible. This principle can be applied to both the LPA method with reflecting target and the LPA method without a reflecting target which is called the DIAL (differential absorption lidar) method. Since both methods are extensively treated in this book there is no need to elaborate on them here, nor yet to give the principal formulae. But obviously the basic measuring geometries and matter-radiation interactions fit into the schema outlined above.

7 SPECTRAL FILTERING METHODS

Normal spectrometers achieve a wavelength dispersion by means of filters, gratings or prisms, and have detectors that measure at one place in the spectrum. In addition there are methods that can simultaneously account for all parts of the spectrum where the gas to be measured has absorption lines. This approach, called spectral multiplexing, is intended to improve the signal-to-noise ratio of the measuring instrument and is particularly practical for measuring the often low concentrations of the gases of interest.

The most prominent spectrometers of this type are the correlation spectrometers and the gas correlation radiometers.

The essence of the correlation spectrometer is that instead of one slit, a mask containing several slits which let pass all those parts of the spectrum where there are absorption (or emission) lines associated with the gas to be measured. This requires a special mask for each gas. By measuring the throughput of all the slits simultaneously, the spectrum is filtered for just one specific gas in order to obtain the highest signal-to-noise ratio.

The gas correlation radiometer in contrast to the correlation spectrometer above, has no dispersing element. Instead, the incoming radiation is split into two beams one of which passes through an evacuated cell, the other of which passes through a cell filled with an appropriate mass of the gas that is to be measured. The filled gas cell changes the radiation at exactly those parts of the spectrum where the absorption lines of the gas lie. Here too the filtering is specific to the particular gas, and simultaneous use is made of the influence on the radiation of all the absorption lines.

Both these sorts of instruments do not represent a special method of remote sensing which was not introduced above, instead they are sensors with a particular method for filtering the radiation field. Since these are passive methods, the RTE is directly applicable, exactly how, depends on the specific type of remote sensing scheme.

With active methods, optical filtering is usually unnecessary since lasers emit monochromatically in any case.

8 FLUORESCENCE AND RAMAN SCATTERING

Fluorescence and Raman scattering are two more possibilities for determining the concentrations of gases or particles consisting of specific substances. Both types of scattering are inalastic scattering, by which two or more photons with less energy (at larger wavelengths) are released, with a time delay of 10^{-9}s for Raman scattering and up to 10^{-4}s for fluorescence scattering (Section 3 of the first article).

The Raman wavelength depends on the specific scatterer and so carries information just on this substance. This positive feature is counteracted by the negative feature of the very small Raman scattering cross-section, resulting in a low signal-to-noise-ratio. So the method is used only with active techniques and predominantly at night time. For filtering the radiation at the Raman wavelength out of the background radiation, devices with very high spectral resolution have the be applied. Even with sophisticated instruments, at present, measurements are feasable only over short distances and at rather high concentrations (a typical example is a chimney plume). Nevertheless Raman scattering is a promising technique because the wavelength of the illuminating laser may lie anywhere below the Raman wavelength.

Raman techniques are extensively treated in this book, not so fluorescence Raman. The latter has the additional disadvantage that because of its long lifetime the minimum range that can be resolved is only 15 km.

Again Raman techniques are no special basic remote sensing method not introduced above, but a basic matter-radiation interaction, which can be understood as a special kind of filtering.

9 FINAL REMARKS

In this paper a classification of the different remote sensing techniques has been presented. Further it has been shown how all the mathematical formulae used in conjunction with remote sensing tasks can be derived from the radiative transfer equation.

This provides a framework into which all remote sensing methods may be embedded.

Optical Remote Sensing of Air Pollution,
Lectures of a course held at the Joint Research Centre, Ispra, Italy, 12—15 April 1983,
P. Camagni and S. Sandroni (Eds). 45—67

ABSORPTION CORRELATION SPECTROMETRY

Millán M. MILLÁN

1. INTRODUCTION

Correlation spectrometry is now a well established remote sensing technique for the measurement of some atmospheric pollutants. The basic instrumental concept is to make use of knowledge about the target gas spectrum to "minimize the effect of perturbing parameters" (ref. 1) in the radiation transfer problems involved in the remote sensing process. The a priori information about the target gas, possible interferents and available backgrounds is also used to gain a detection advantage and to guide the instrumental design.

This work is intended to be a critical review of the principles and developments which led to the establishment of this technique and to the present line of instruments. Some operational procedures hitherto unreported and some recent applications will also be presented. However, the scope of the work will be limited to dispersive correlation systems only.

The implementation of the correlation spectrometry concept into a workable instrument is but one example of how knowledge from various fields was combined to fulfill a newly identified research area, i. e. to characterize the behaviour of pollutants and other gases in the atmosphere. It also means that several disciplines have played, at one time or another, leading and conflicting roles in its development. Some of these are: spectroscopic multiplexing, its application to astronomy and its further extension to geophysics; the concept of cross-correlation as a signal processing technique and its extension to dispersive spectroscopy; the optimization of the signal to noise ratios as an optical design task, etc.

A very important role in shaping technical developments has also been played by feedback information from field use, experimental results and disgruntled customers. These include the effects of changing backgrounds as well as atmospheric radiation transfer and meteorological interpretation problems, etc., some of which were only nebulously envisaged, or simply not even foreseen "ab initio". All of this has resulted in an overall development by leaps and bounds, of which some users are aware, but which will not be discussed any further here.

In this presentation the approach to dispersive spectrometry has been structured as if all the required knowhow had been available at the start. Some of the basic problems involved in attempting to make a spectroscopic passive remote measurement are addressed first, together with some initial concepts in signal to noise ratios in spectroscopic systems. This is followed by a presentation of the concept of spatial cross-correlation techniques as applied to spectroscopic systems and how they can be used to enhance the signal and average out undesirable effects. The instrumental embodiment of these concepts is then introduced as the logical merging of those areas, via a historical review. Finally, the generalized response function and some operational characteristics of these instruments are reviewed. The atmospheric interactions and meteorological interpretation procedures are addressed in two other chapters in this book.

2. SOME BASIC CONCEPTS IN SPECTROSCOPIC REMOTE SENSING

The use of absorption spectroscopy for the measurement of gas concentrations in the atmosphere is (initially) based on the optically thin approximation to the equation of radiative transfer. This approximation results in the well known Bouguer-Beer-Lambert law of absorption, viz:

$$N_\lambda(L) = N'_\lambda(0)\exp[-\tau_\lambda] \; \exp[-k(\lambda)cL] \tag{1}$$

where $N'_\lambda(0)$, $N_\lambda(L)$ are, respectively, the spectral radiances at wavelength λ before and after traversing the path L ($Wm^{-2}sr^{-1}nm^{-1}$).

$k(\lambda)$ is the absorption coefficient of the target gas at wavelength λ (m^2).

L is the pathlength along the line of propagation (m).

c is the average concentration of the target gas along the line of propagation of the radiation (m^{-3}).

τ_λ is an optical depth at wavelength λ including the absorption by all other species.

Both $N'_\lambda(0)$ and the extinction due to absorption by other species can be effectively grouped into a background radiance

$$N_\lambda(0) = N'_\lambda(0)\exp[-\tau_\lambda] \tag{2}$$

One of the basic problems in passive remote sensing is that $N_\lambda(0)$ is not usually available or known. A first approach to obtain cL is to measure (1) at two neighbouring wavelengths, viz:

$$N_2(L) = N_2(0) \exp[-k_2 cL]$$
$$N_1(L) = N_1(0) \exp[-k_1 cL] \quad (3)$$

from where one can obtain:

$$cL = \frac{-1}{k_2-k_1} \left[\ln \frac{N_2(L)}{N_1(L)} - \ln \frac{N_2(0)}{N_1(0)} \right] \quad (4)$$

In Eq. (4), the problem of finding the unknown background radiance has been traded for one of determining the ratio between two equally unknown radiances. However, this means that the burden (cL) of the target gas can be measured provided that two wavelengths can be found for which (a) the absorption coefficients of the target gas are different, and (b) the effective background radiances are very nearly the same, or the ratio is well known, or it is expected to be well behaved around an approximate value. Both assumptions imply a foreknowledge of the spectrum of the target gas and of the available background. The very last assumption has been quite widely used on the basis that backgrounds do not change too significantly over spectral intervals of one or two nm or so.

Natural backgrounds, however, do have a significant amount of structure in them on account of Fraunhofer lines in the sun spectrum, and their spectral distribution can also change with time of day, atmospheric conditions, etc. These render the second approximation rather suspect regardless of the spectral resolution used for the bands selected. The accuracy in the measurement of cL is directly dependent on the ratio being one (or any other known value). Possible errors in the measurement become increasingly important, the smaller the cL to be measured.

A second approach is to use the ratio between the radiances in (3) directly to obtain:

$$\left\{1 - \frac{N_2(L)}{N_1(L)}\right\} = \left\{1 - \frac{N_2(0)}{N_1(0)}\right\} + \frac{N_2(0)}{N_1(0)} \left\{1 - \exp[-(k_2-k_1)]\, cL\right\} \quad (5)$$

In this equation, one minus the ratio becomes the physical observable chosen for the measurement. This expression indicates even more explicitly that an accurate measurement of <u>cL</u> is highly dependent on the ratio being very close to unity, or known to remove the bias (or offset).

There are some other aspects of Eq. (5) worth examining. In the first place, for small values of cL, the right hand term becomes:

$$\left\{1 - \frac{N_2(0)}{N_1(0)}\right\} + \frac{N_2(0)}{N_1(0)} (k_2 - k_1)\, cL \qquad (6)$$

This is the equation of a straight line with a zero intercept equal to one minus the ratio of the background radiances, i. e. the direct background reading, which determines the basic uncertainty in the measurement. The target gas term, the second, has a slope given by the product of the ratio between the unknown radiances and the value $(k_2 - k_1)$. The overall slope can be determined by calibration with known values of cL, and equation (6) makes it quite explicit that the measurement sensitivity is a function of $(k_2 - k_1)$.

The second aspect of equations (5) and (6) is that it opens the possibility of using several wavelengths simultaneously so that some type of averaging can be used. This concept forms the basis of the correlation spectroscopic approach and will be explored more closely in the next section.

Among other points to consider, the first one is that in any dispersive instrument the spectral radiant power, or dispersed spectrum is a multi-wavelength image of the entrance slit. With no aberrations considered, it becomes

$$P_\lambda = \Omega N_\Lambda * hE(\lambda - \Lambda) \qquad (7)$$

where

*	denotes convolution
Ω	is the acceptance solid angle of the optical system (sr)
h	is the entrance slit height (m)
N_Λ	is the incident spectral radiance ($Wm^{-2}sr^{-1}nm^{-1}$)
E	is the entrance slit width defined as:

$$E(\Lambda) \begin{cases} 1 \text{ for } |\Lambda| < \Delta_e \\ 0 \text{ for } |\Lambda| > \Delta_e \end{cases}$$

Δe is the entrance slit half width (m)

The spectral radiant power becomes

$$P_\lambda = \Omega h \int_{-\Delta e}^{\Delta e} N(\Lambda + \lambda)\, d\Lambda = 2\Delta_e h\Omega N_\lambda \quad (W\, nm^{-1}) \qquad (8)$$

where

$2\Delta_e h\Omega$ is nominal etendue or throughput of the optical system, defined by the product of an acceptance solid angle Ω(sr) and an entrance aperture $2\Delta_e h$ = Area of entrance slit (m^2)

N_λ is an "average" spectral radiance within the spectral interval defined by the entrance slit width, and should be considered as such throughout this presentation.

The actual sampling or selection of the wavelengths (or wavebands) chosen is performed by means of exit slits - in the monochromator section of a dispersive system - or narrow bandpass filters in a non-dispersive system. Either of these will have a spectral bandpass Δ_λ(nm), so that the slits sample, or transmit, a radiant power given by

$$P = A\Omega N_\lambda \Delta_\lambda \quad (W) \qquad (9)$$

In as much as both entrance and exit slits have finite width, one can only speak of spectral radiances or of absorption coefficients "averaged" by an instrumental transfer function defined jointly by the image of the entrance slit and width of the exit slit.

The conversion of the sampled radiant power into an electronic signal is performed via a suitable photodetector, and the signal is further processed to yield an output as given by either of equations (4) or (5).

In general terms, the signal current and the photon noise associated with it are given by

$$I_s = PS_{kr} \quad \text{(signal current)} \qquad (10)$$

$$(i_{rms})^2 = 2eI_s\Delta v = 2ePS_{kr}\,\Delta\nu \quad \text{(rms noise current)}$$

where

P is the sampled radiant power (W)

S_{kr} is the photo cathode radiant sensitivity ($A\ W^{-1}$)

e is the electron charge (C)

$\Delta\nu$ is the electronic bandwidth of the processing system (s^{-1}).

When two wavelengths are used to derive a response like the one in Eq. (5), it can be shown (ref. 2) that the Signal to Noise Ratio (SNR) on the basis of photon noise limitations is:

$$\mathrm{SNR} = \sqrt{\frac{S_{kr}d}{2e\Delta\nu}} \cdot \frac{|P_1 - P_2|}{\sqrt{P_1 + P_2}} \qquad (11)$$

which is a particular case of the more general expression given by

$$\mathrm{SNR} = \sqrt{\frac{S_{kr}\cdot d}{2e\Delta\nu}} \cdot \frac{|P_1 - P_2|}{\sqrt{(P_1 + P_2) + \frac{N^2}{(2eS_{kr}d)}}} \qquad (12)$$

where d is the duty cycle of the electronic synchronous detector waveform, and N^2 is an rms noise current per unit electronic bandwidth (or an rms sum of such noise currents) generated by other (uncorrelated) noise sources.

The requirements for both: large sensitivity (or signal to noise ratio) and an accurate (specific) measurement, can thus be summarized as follows:

(1) The difference between the sampled radiant powers should be large, i.e. the $(k_1 - k_2)$ value must be large.

(2) The overall sampled power should be as large as possible within the constraints required to maximize the applicable SNR equation, either (11) or (12).

(3) The ratio of the background radiances should be very nearly unity.

It should also be mentioned that in any dispersive system the high spectral resolution required for large values of the difference $(k_2 - k_1)$ is usually associated with narrow slits, while large radiant powers require wide slits. In the next two sections we will examine how, and to what extent, these (somewhat conflicting) requirements can be met simultaneously with the concept of correlation spectrometry.

3. CROSS-CORRELATION FUNCTIONS IN SPECTROMETRY

The space correlation function (ref. 3) is the integral of the product of two (generating) functions as one of them is displaced with respect to the other the correlation variable, or coordinate, being the displacement between them. If the two functions are identical, there will be a value of the correlation variable for which they exactly overlap with the result that every ordinate, positive or negative, yields a positive product. The total sum or integral of all products is therefore large. If, however, the functions are dissimilar, or displaced, some products would be positive and some negative, some cancellations take place and the final sum is smaller. The cross-correlation function is, then, a measure of their similarity or coincidence. Specifically, the autocorrelation function measures the integral of the product of a function with itself, and some of its characteristics are as follows:

(a) It has a positive maximum at the value of the correlation variable for which the functions exactly overlap. This value of the correlation variable is usually taken as the zero.

(b) It is an even symmetrical function with respect to this value of the correlation coordinate.

(c) If the generating function is periodic, the autocorrelation is also periodic and has the same period.

(d) Its shape also depends on the frequency content of the generating function. The broader its spectrum, or the more complicated the function is, the more sharply defined is the maximum and the quicker the autocorrelation falls off from that value.

In the case of the dispersed spectrum of the incoming radiation, the cross-correlation between the spectral radiant power and a sampling function can be written as:

$$R(\xi) = \int_0^{\infty} P_\lambda M(\lambda + \xi)\, d\lambda \qquad \text{(W)} \qquad (13)$$

where ξ is the cross-correlation coordinate and $M(\lambda)$ is the sampling function. For our own purposes, we will consider it to be an array of exit slits, or a mask, and its value can be defined as:

$$M(\lambda) \begin{cases} 1 \text{ for } \lambda_i - \Delta_i \leqslant \lambda \leqslant \lambda_i + \Delta_i \\ 0 \text{ for any other wavelength} \end{cases} \qquad (14)$$

where λ_i, Δ_i are, respectively, the centreline position and half width of the slit $\underline{i}$ of the mask (nm).

Defined in this way, the cross-correlation function between the dispersed spectrum or spectral radiant power at the incident radiation and the exit slit(s), is exactly the same as the radiant power transmitted through the slit(s) of the mask and, correspondingly, has the dimensions of Watts.

The cross-correlation function can, thus, be detected or measured by locating a suitable photodetector system behind the mask. Scanning the spectrum across the mask, or viceversa, yields a detector output as a function of the scan, or cross-correlation, coordinate which can be plotted to give a correlogram.

This concept can be explored further by examining in a step by step fashion how the cross-correlation masks can be obtained. The slits widths, centre positions and number are the key parameters to be determined.

Substituting (14) into (13) and re-arranging, the cross-correlation function becomes:

$$R(\xi + \Delta_i, \Delta_i) = \sum_i \int_{-\Delta_i}^{\Delta_i} P(\xi + \lambda_i + \lambda)\, d\lambda \tag{15}$$

It can be easily shown that each term in this sum reaches a maximum or minimum when:

$$P(\xi + \lambda_i + \Delta_i) \equiv P(\xi + \Delta_i - \Delta_i) \tag{16}$$

for each slit _i_

i.e., for each value of the cross-correlation coordinate that places the centreline position of slit _i_ in a position $-\xi+\lambda_i$ such that both ends of the slit intercept equal values of the radiant power P_λ. This can be used as a first criterion (a) for the selection of the slits in a mask.

Figure 1(a) shows a transmission spectrum of SO_2 in the near UV and three masks of eighteen slits each. Figure 2(a) shows the transmission spectrum of NO_2 in the blue region of the spectrum and two masks with six slits each. In both cases the slits shown have been made to coincide with an arbitrarily selected number of absorption minima (troughs). The only other limiting criterion imposed on the slits has been with respect to their widths, which have been made narrower than the distance between the two encompassing peaks. This is criterion (b) which will be discussed later. It should be noted that once a slit width has been selected, fulfilling criterion (b), application of criterion (a) automatically determines the slit centreline position.

Figure 1(b) shows the three SO_2 correlograms for the masks shown. Figure 1(c) shows the correlograms for three masks with the same number of slits having approximately the same individual slit widths as in the trough masks but designed for the consecutive, longer wavelength peaks. The reasons for the peak masks will become apparent later in this section. Similar examples for NO_2 are shown in Figures 2(b) and 2(c).

In as far as all of these masks have been designed using some of the features in the spectrum of the target gas, these graphs are (quasi) autocorrelograms of both spectra. One set emphasizes the effect of the transmission minima (the troughs) and the other, the transmission maxima (the peaks).

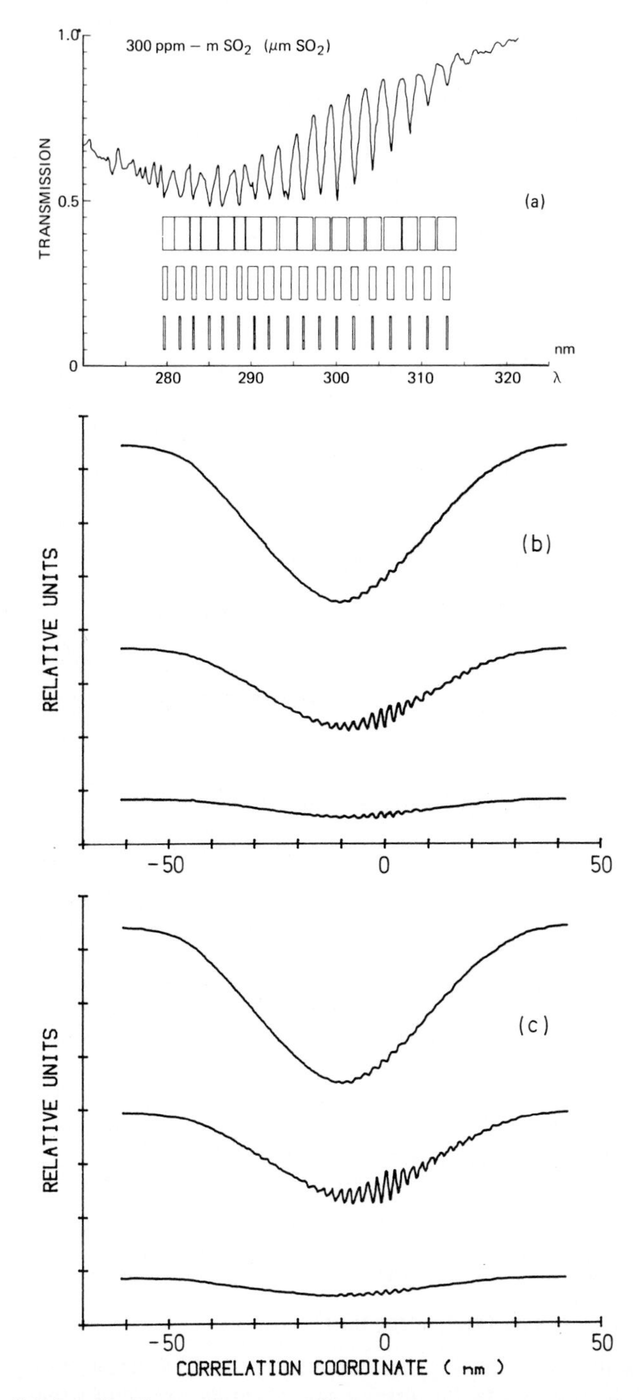

Fig. 1 - Sulphur dioxide correlograms. a) Transmission spectrum of 300 ppm-m of SO_2 and three masks of 18 slits used for the correlograms. b) Trough mask correlograms obtained from masks with 318, 15.6 and 3.6 nm total width, respectively. c) Peak mask correlograms obtained from masks of the same total width.

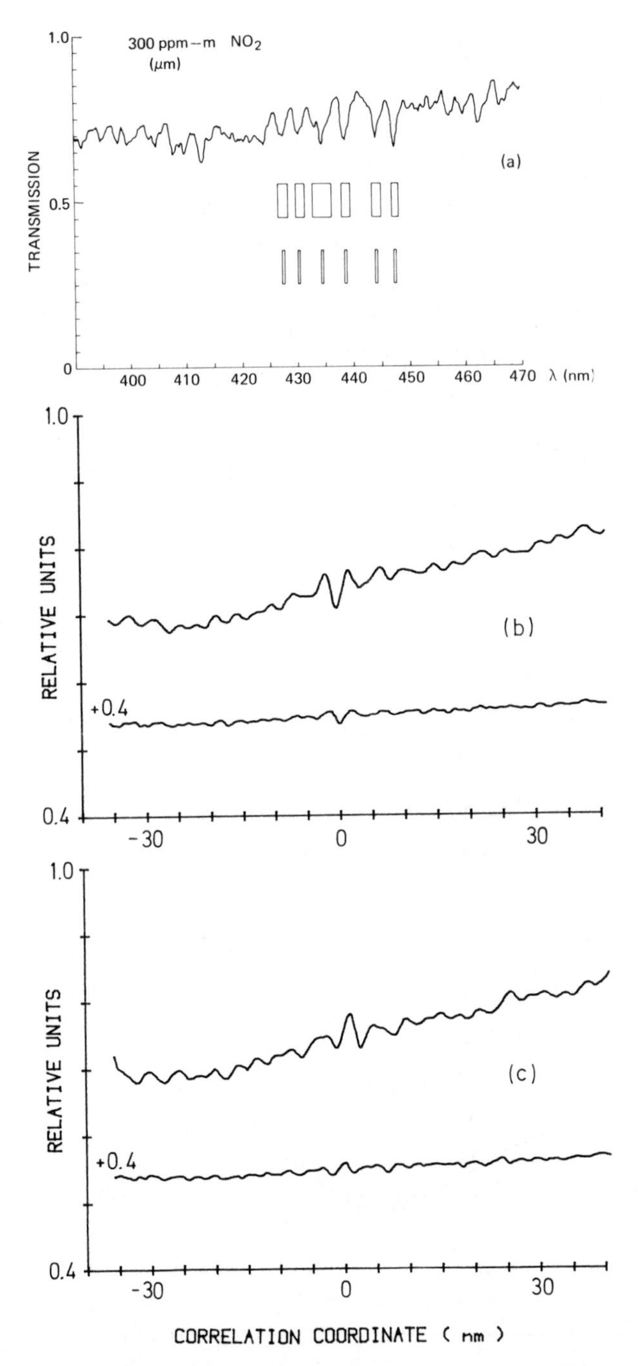

Fig. 2 - Nitrogen dioxide correlograms. a) Transmission spectrum of 300 ppm-m of NO_2 and two masks of six slits each used for the correlograms. b) Trough mask correlograms obtained from masks with 1'.8 and 2.4 nm total width, respectively. c) Peak mask correlograms obtained from masks of the same total width.

With respect to the properties of the autocorrelation function, some even symmetry can be observed with respect to, and in the neighbourhood of, the designed position (correlation extrema) which brings all the slits into register with their respective peaks, or troughs. This symmetry is much more apparent in the case of NO_2 which has a more irregular spectrum and thus a more distinctively defined autocorrelogram. Conversely, the SO_2 presents a clear case of a quasi-periodic spectrum (the vibration bands) which yields a corresponding periodicity in its autocorrelogram.

Finally, in the case of the SO_2 correlograms, there are other important features to observe. The mask(s) with the wider slits is (almost) a wide slit by itself. Its correlogram enhances the large scale structure of the spectrum, such as the overall shape of the electronic band, and smears out the finer structure. The masks with slit widths nearly half the interpeak distances enhance their corresponding feature, i. e. the vibrational band structure which has similar spatial frequency components, etc. This can be summarised by indicating that in the correlation function, each spatial component in one of the generating functions will enhance, or bring out, the corresponding component in the other function.

At this point, we have introduced the concept of peak and trough masks, and we can re-consider and generalize the concept of using two neighbouring wavelengths to measure the target gas. For this, we examine the idea of subtracting directly the output from a peak and trough masks having a common scan coordinate. For example, this can be achieved by engraving the two sets of slits on each half of a disc which is rotated in front of the photo detector and process its output accordingly.

Two more complementary criteria are now required to choose the slits in the masks. The third one, criterion (c) is that each peak slit and following (or previous) trough slit should have equal widths, i. e. paired slits. This condition achieves zero difference from a spectrally flat background when no gas absorption is present. A fourth criterion (d) follows directly from maximizing the difference between the cross-correlation extrema for the two masks.

The difference function becomes:

$$G(\xi+\lambda_i,\delta_i,\Delta_i) = \sum_i \int_{-\Delta_i}^{\Delta_i} [P(\xi+\lambda_i+\lambda) - P(\xi+\lambda_i+\delta_i+\lambda)]\, d\lambda \qquad (17)$$

where δ_i are the distances between centres of the paired slits. The conditions that yield a maxima or minima for each slit pair are:

$$P(\xi + \lambda_i + \Delta_i) = P(\xi + \lambda_i - \Delta_i)$$

$$P(\xi + \lambda_i + \delta_i + \Delta_i) \equiv P(\xi + \lambda_i + \delta_i - \Delta_i) \qquad (18)$$

i. e. criterion (a), requiring equal intercept at both ends of the slit, is returned for both slits of the pair, and

$$P(\xi + \lambda_i + \Delta_i) \equiv P(\xi + \lambda_i + \delta_i - \Delta_i) \qquad (19)$$

which requires that the values of P_λ must be equal at all ends of the paired slits. This becomes criterion (d) for maximum difference (positive or negative). Eq. (19) also yields $\delta_i = 2\Delta_i$, i. e. the distance between centres is equal to the slit width. As soon as a peak and trough of a band are selected, application of the four criteria yields an unique value for the width of the paired slits and their centre positions. This is illustrated in Figure 3.

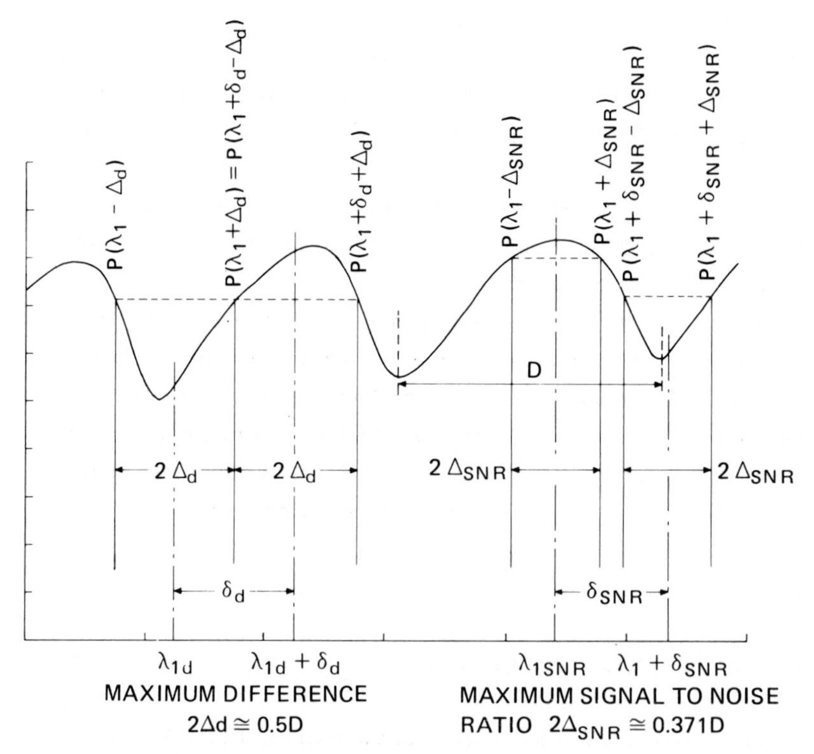

Fig. 3 - Optimum slit parameters for maximising either the difference signal or the signal-to-noise ratio.

Figure 4 shows the SO_2 difference correlogram for two paired masks of eighteen slits each. Basically, the same as the intermediate masks in Fig. 1. Each slit pair fulfills the four criteria with respect to its trough and successive (longer wavelength) peak. The corresponding difference correlogram for NO_2 is shown in Figure 5.

Up to this point the constraints on the slit width (criterion (b)) have been applied with respect to the vibrational bands. Relaxation of criterion (b) means that a maximization of the difference by applying the remaining criteria can take place at three main scales, each with a solution and each having certain advantages in some types of gas detection.

The first solution is with respect to the overall transmission spectrum, i. e. the large scale structure including the overall effect of the continuum absorption, bands and lines. The solution consists of two slits, each as wide as the total spectrum of the target gas: one over it, and the other just outside (in any direction). This pair also gives the maximum attainable difference for any amount of absorption by the target gas. However, it is the least specific, since any other non-uniform absorption will produce a difference between the energy sampled by the two macro slits.

This scale solution would provide a gas sensor most suited to in-situ monitoring when interferences from competing absorbers are known to be negligible. Because of its simplicity, in fact, the two macro slits can be substituted by suitable band pass filters; a sensor using this approach appears most attractive for process control applications.

The second solution, emphasized in this presentation, is with respect to the vibrational bands. Figure 6 compares the results for two scales and shows the difference correlograms in the case of SO_2. The larger scale solution is for two slits 31.8 nm wide each (much less than optimum) and positioned, respectively, one centered, the other just outside the electronic band. For the second graph (as in Fig. 4), the masks were optimized to get the maximum difference from the eighteen strongest vibrational bands. Figure 6 illustrates the magnitude of the maxima for a first scale solution vs the second. The widths for the first scale were selected to encompass a spectral span comparable to that for the slits in the second solution.

As could be expected from the increased frequency (or structure) content of the masks, an increased number of slits produces a difference correlogram with more sharply defined extrema, which extend over a narrower span of the cross-correlation coordinate. This is clearly noticeable in Fig. 6, where the effect of the large structure, band and continuum, which dominates

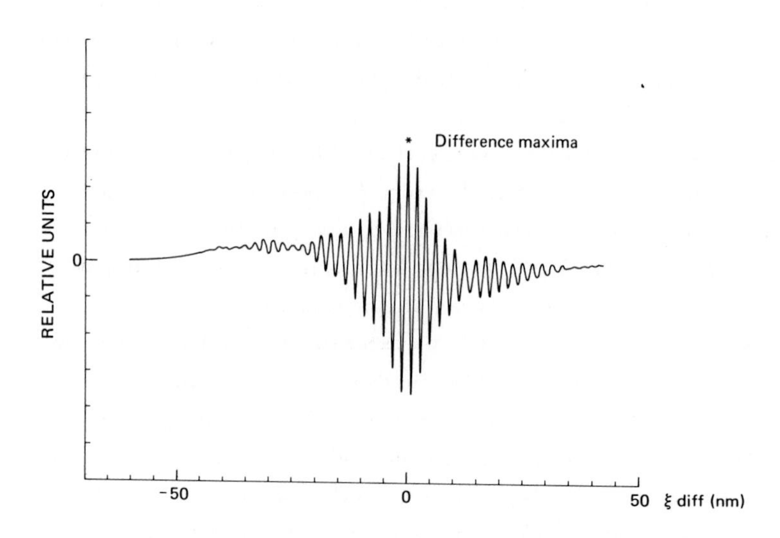

Fig. 4 - SO_2 difference correlogram obtained with two paired masks of eighteen slits each.

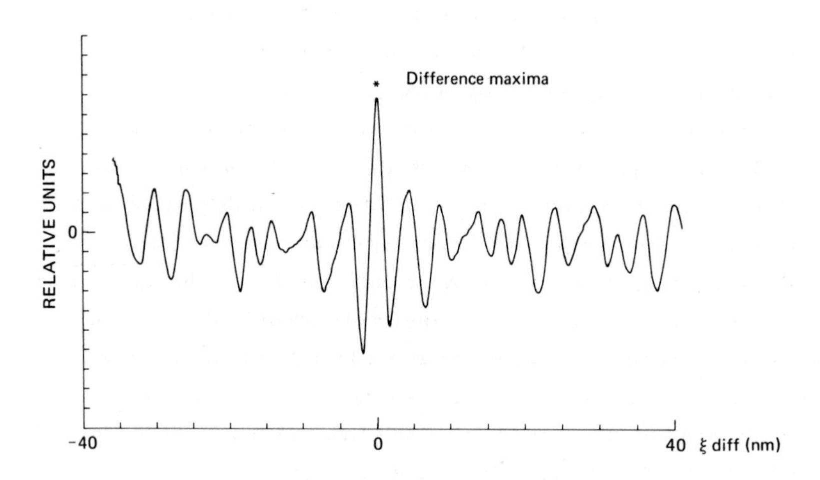

Fig. 5 - NO_2 difference correlogram from two paired masks of six slits each.

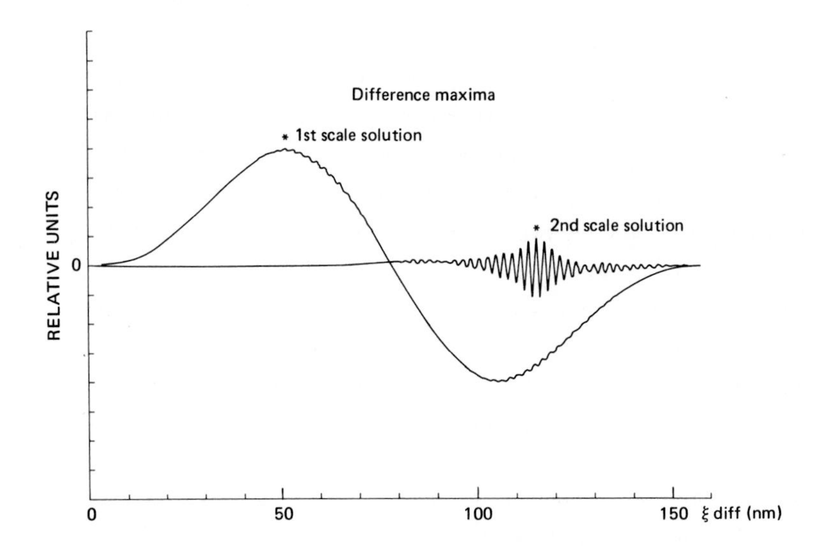

Fig. 6 - SO_2 difference correlogram showing the relative amplitudes of the first and second scale solutions, and the narrowing of the region where the peaks are reached.

the first scale solution, has almost disappeared in the difference correlogram for the second scale. This also illustrates that the more complicated the mask (solution) is, the more specific (unique) the cross correlation peak (extremum) becomes.

If the rotational lines are resolved, a third scale solution exists with a pair of masks having a large number of slits. Depending on their number, these could have the greatest specificity and provide the highest level of rejection from competing absorbers. On the other hand, the available signal, proportional to the peak amplitude of the difference correlogram, would be the smallest. Because the rotational structure must be resolved, the third scale solution would be easier to implement in the ir wavelengths.

We have just explored how the concept of cross-correlation can be applied within the context of sampling (or measuring) radiation from optical spectra by means of selected elements or masks. These are derived from the spectral structure of the target gas by following certain criteria and can be used, in the process of measuring the incoming radiation, to enhance the presence of that very same structure. This enhancement, in the form of the cross-correlogram peak(s), serves to detect the target gas.

4. THE COSPEC, A HISTORICAL PERSPECTIVE

The cross-correlation function between the spectrum and the mask(s) can be used in several ways to perform the detection of the target gas. In principle, we know that the cross-correlation peak (positive or negative) includes the information collected simultaneously by all the slits in the mask. In a first step the (quasi) autocorrelogram can be used directly to identify the presence of the target gas. This can be performed by using an oscillatory scan of limited amplitude centered over the expected position of the cross correlation peak to record its presence and amplitude. Because the cross-correlation function is photodetected, one can also expect a significant increase in the signal to noise ratio of the measurement by approximately $\sqrt{N}$, where N is the number of slits, as compared with a normal spectrum obtained by scanning the spectral radiant power with a single exit slit.

This type of approach was proposed by the spectroscopist W.S. Benedict to J. Strong around 1958 (refs. 4, 5), as a means of increasing the SNR of the measurement without decreasing the spectral resolution when he suggested the use of 21 exit slits aligned with absorption lines in the 1.13 μm band of water vapour, to detect its presence in the atmosphere of Venus (refs. 5, 6).

A similar concept was proposed by A. R. Barringer (ref. 7) and by P. B. Kay (ref. 8) for the detection of iodine vapour for mineral exploration and SO_2 in the atmosphere.

Since the most specific information about the target gas is centralized in the correlogram extrema, successive systems developed means of oscillating the relative position mask-spectrum over selected amplitudes and methods to process the photodetector signal. In spectra with regularly spaced bands, a large amplitude scan could be used, taking advantage of the periodicity of the correlogram, to analyse the harmonic content of the signal. In this case, the physical observable chosen for the detection was the presence, and relative amplitude, of the signal at certain harmonics of the mask-spectrum oscillation frequency (refs. 8, 10). For more irregular spectra oscillations of very small amplitude were used and the photodetector signal was processed to yield the second harmonic component (ref. 9). This approach was further generalized into a branch of "derivative spectroscopy" (refs. 14, 15).

A parallel approach consisted in moving the mask between two fixed points. One corresponding to the (quasi) autocorrelation peak (positive or negative) and the other to a neighbouring extrema of opposite sign.This is equivalent to sampling the spectrum at the number of selected troughs (or peaks) and then moving the mask to the neighbouring peaks (troughs). The displacement was initially achieved with a refractor plate located behind the entrance slit and oscillated about its vertical axis to produce the required spectral displacement. Alternatively, the mask was oscillated across the spectrum. In either case, the mask spends a considerable amount of time per detection cycle travelling between the optimum positions.

It was quickly discovered that an increased duty cycle and, thus, an improved SNR could be obtained with a jumpwise alternation between the two selected positions. This was achieved with two refractor plates joined together at a fixed angle, corresponding to the two end positions of the oscillating plate, and mounted on a tuning fork which oscillated them across the entrance slit of the monochromator box. This system became the basis of the Barringer Refractor Plate Correlation Spectrometer Remote Sensor (refs. 12, 13) which became operational in 1967.

Up to this time no specific criteria existed to select the mask parameters, i.e. number of slits, positions and widths, other than: there should be as many as possible, they should be "aligned" with the spectral features of the target gas (usually the troughs), and that they should have "high contrast" with the spectrum of the target gas. Those, and other information gaps plus some data interpretation problems forced a re-evaluation of the whole instrumental concept by mid 1968.

A systematic study of this technique was thus initiated late in 1968 under the joint sponsorship of the National Research Council of Canada and the NASA Manned Spacecraft Center (Houston). The results of this work (refs. 2, 11, 18) can be summarized as follows:

(a) A comprehensive theoretical study of the technique was performed.

(b) The instrumental transfer function and signal to noise ratio equation were derived and used to emphasize the interplay between the incoming radiation and the instrumental design parameters (optical and electronical).

(c) The SNR equation was used to derive the optimum mask parameters for a maximum of the signal to noise ratio.

(d) The atmospheric radiative transfer problem and variability at natural backgrounds were evaluated within the context of the instrumental transfer function to explain the observed response behaviour under actual operational conditions. Work in these same areas was continued during the early 70's in the Atmospheric Environment Service of Canada and by other scientists working in Europe. In this context, the work by P. Hamilton at CERL (UK), Onderdelinden at RIV (Holland) and the Vittori group in Italy, should be acknowledged.

It was also established early in 1969, that the mask-in-two-positions approach yielded less than optimum performance, particularly with irregular spectra. This occurred simply because it could only be optimized either as a peak or a trough mask for the one position, but it would be less than optimum for the other position. This guided the concept and introduction of paired mask, as mentioned in the previous section. It also forced the development of other mask-spectrum modulation techniques.

The diurnal variation of the available backgrounds and their effect on the instrument were also documented extensively during this period, and resulted in the introduction of the Multistable concept (ref. 2), i. e. the use of four masks in two pairs in order to minimize the effect of diurnally varying backgrounds (ref. 16). This is discussed further in the chapter on atmospheric effects.

The use of the SNR equation to obtain the mask design parameters (refs. 2, 17) showed that: there is an optimum number of slits per mask pair as a function of the target gas spectrum and of the available backgrounds. The slits width for maximum SNR are narrower than those for a maximum of the signal (or the difference), as illustrated in figure 3, and the entrance slit has also an optimum width comparable to that of the exit slit pairs, and of the order of 37% of the band period (inter peak distances).

Further work in this area indicates that the optimum parameters vary with the dominant noise source, whether photon noise alone or from other sources, with a trend for the optimum mask parameters to approach those which maximize the signal if the predominant noise source is other than photon noise (ref. 18).

A parallel development concerned the use of multiple entrance slits, i. e. the optical multiplexing can also be performed at the front end of the spectrometer. With multiple entrance slits, the spectra for each are displaced by distances corresponding to the (spectral) separation between the entrance slits and are partially overlapped in the exit plane. The sampling (or cross-correlation) is thus performed against an artificially produced spectral radiant power, which can also be selected (calculated) to fulfill certain criteria with respect to the (original) spectrum of the target gas in order to optimize the SNR (ref. 19), specificity, or other including, for example, the optimization of the use of areas in the entrance and exit planes to minimize aberrations and simplify the optical design, etc. It should be quite clear that the full potential of this technique has not yet been fully developed.

The embodiment of the Multistable concept was made in the present line of COSPEC instruments (II to V series) which became commercially available in mid 1970. The instrument is shown schematically in Figure 7. It consists of a telescope which focuses the incoming radiation onto the entrance slit(s). The radiation is further collimated, dispersed by a grating, and the selected spectral region is focused onto the exit plane where the masks are. These are photoetched on a thin aluminium layer deposited on a quartz disc. Four masks are engraved on the disc corresponding to a peak-trough-trough-peak sequence, or its alternate, depending on the spectrum of the target gas. A Fabry lens, to image the grating onto the photomultiplier, the calibration cells, optical filter and photomultiplier complete the optical system.

5. OPERATIONAL CHARACTERISTICS

The radiant powers passing through the masks strike the photocathode and generate four current pulses per revolution of the disc, viz:

$$I_i = P_i S_{kr} G + I_{oa} \tag{20}$$

where I_i is the anode current for the incident radiant power P_i transmitted by the mask $\underline{i}$, S_{kr} is the photocathode radiant sensitivity, G is the photomultiplier gain and I_{oa} is the anode dark current. In what follows, we will assume $I_{oa} = I_{ok}G$, where I_{ok} is the photocathode dark current.

The gain in the present COSPEC configuration is related to the photomultiplier applied voltage V_b by:

$$G = 1.5\left(c\frac{V_b}{11.25}\right)^{10} \tag{21}$$

where

c   is the secondary emission factor of the dynodes = $\frac{\text{secondary electron}}{\text{incident electron x Volt}}$

10   is the number of dynodes in the P. M. tube.

The four pulses are pre-amplified and converted to voltages. During the passage of the first mask, transmitting radiant power P_1, the gain is adjusted as follows:

$$V_i = I_1 R = \text{fixed and selectable} = (P_1 S_{kr} + I_{ok})G$$

which yields

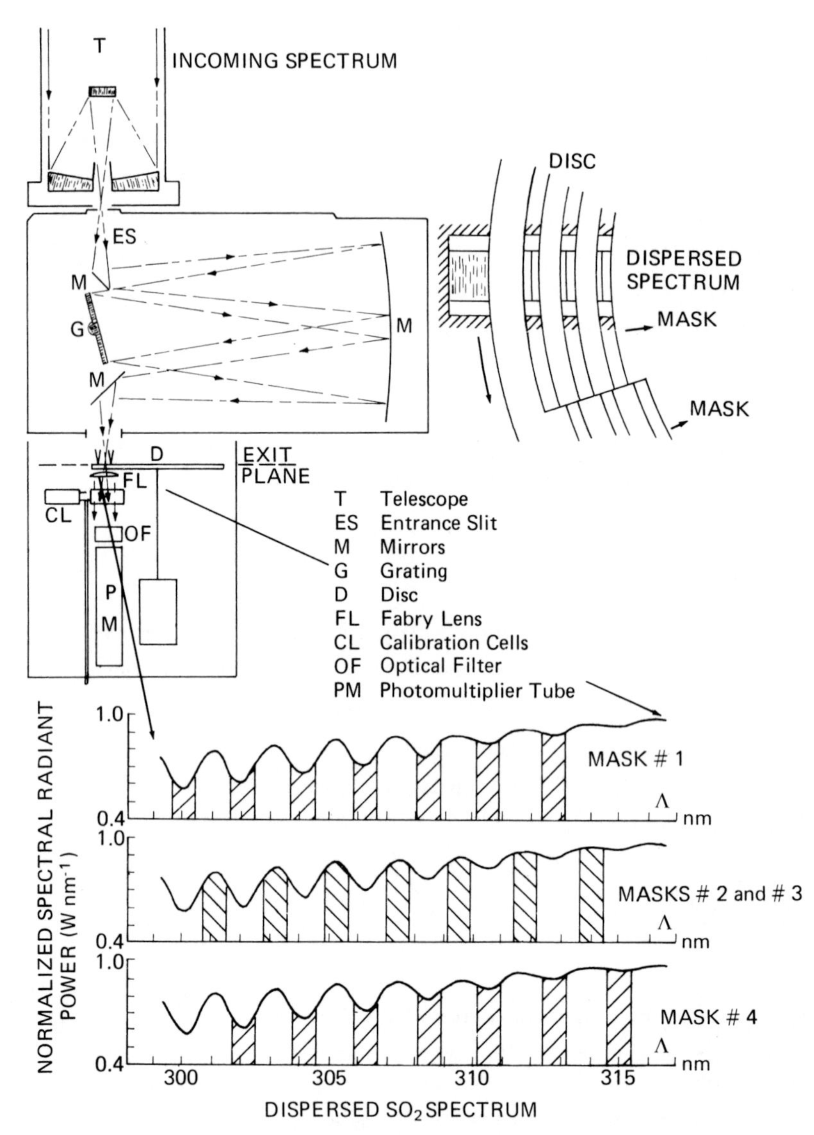

Fig. 7 - Schematic layout of the COSPEC optical system, slits on the disc end, slits on the SO_2 absorption spectrum.

$$G = \frac{I_1 (\text{fixed})}{P_1 S_{kr} + I_{ok}} \tag{22}$$

The COSPEC's Automatic Gain Control voltage output is proportional to V_b and is thus

$$V_{AGC} \; \alpha \; G^{1/10} \; \alpha \left(\frac{I_1}{P_1}\right)^{1/10} \tag{23}$$

i.e. proportional to the one tenth of the incoming (and available) radiant power.

The gain Eq. (22) is maintained during the detection cycle. This essentially normalizes all the pulses by radiant power seen through mask No. 1 and the pulses become:

$$V_1, \; V_1 \frac{P_2}{P_1}, \; V_1 \frac{P_3}{P_1}, \; V_1 \frac{P_4}{P_1}$$

where the assumption $P_i \gg S_{kr}/I_{ok}$ has been used.

Pulses 3 and 4 are re-routed through a second Automatic Gain Control circuit which makes:

$$V_1 \frac{P_3}{P_1} G_2 = V_3 \; (\text{fixed and selectable}) \tag{24}$$

This second gain is applied to the fourth pulse, and the final waveform consists of:

$$V_1, \; V_1 \frac{P_2}{P_1}, \; V_3, \; V_3 \frac{P_4}{P_3}$$

The waveform is then synchronously detected and amplified to yield an output of the form

$$V_o = \alpha \left[V_1 (1 - \frac{P_2}{P_1}) - V_3 \; (1 - \frac{P_4}{P_3}) \right] \qquad \text{or}$$

$$V_o = \alpha \; V_1 \left\{ (1 - \frac{P_2}{P_1}) - k(1 - \frac{P_4}{P_3}) \right\} \tag{25}$$

where α is an amplification factor and k is the ratio between the voltages V_3 (selected to control the 2nd AGC) and V_1 (selected to control the 1st AGC).

Eq. (25) consists of two terms, each of the form given by Eq. (5). The difference is that instead of spectral radiances (actually radiant powers) at one wavelength, we now use the radiant powers P_i transmitted simultaneously by all the slits of mask $\underline{i}$, selected to enhance the effects of the target gas.

Essentially, each term in Eq. (25) yields a normalized difference correlogram, with the slits optimized for a maximum of the SNR. Also, as in Eq. (5), each term can be split into a direct background signal and a target gas signal with some scaling effects. This will be discussed further in the chapter on Atmospheric effects.

To finalize this chapter, and to add a practical point of view we should examine what Eq. (22) means. Substituting the gain from Eq. (21) one has:

$$\frac{I_1 \text{ (fixed)}}{P_1 S_{kr} + I_{ok}} = F(V_b)^{10} = G \tag{26}$$

Because a photomultiplier can work properly only between two voltages (typical values 1200 V max and 900 V min), there is only a range at values of P_1 which will fulfill this equation for each I_1 (or V_1) selected. This means that: selecting a low value for the 1st AGC controlling voltage the COSPEC can work with low light levels, but could saturate at high levels (e.g. mid day in June at low latitudes). Selecting a higher value of V_1, the instrument will operate with higher light levels, but will not work with low light levels, e.g. low solar elevation angles at high latitudes. Selection of a proper V_1 for each desired operational range should be done experimentally in each case. One should also be aware that for a normal PMT, the range Maximum Gain to Minimum Gain is of the order of 300 (e.g. for the commercial COSPECs), and for a very good tube could be of the order of 1000, so that a better tube can operate with a much wider light range for the same value of V_1.

With this discussion we have illustrated how the abstract concept of cross-correlation applied to a spectroscopic measurement leads to an instrument with real interactions with the outside world and with real limitations imposed by the available hardware. In the two other complementing chapters we will discuss how it interacts with a real atmosphere and how to interpret the data.

REFERENCES

1 H. Quenzel, Principles of remote sensing techniques, in this text.
2 M. Millan, Ph. D. Thesis, Univ. of Toronto, Ont., Canada, 1972.
3 F. H. Lange, Correlation techniques, Van Nostrand-Reinhold, New York, 1967.
4 J. Strong, Multiplex spectrometry, in Essays in Physics, Vol. 5, Academic Press, London, 1973.
5 J. Strong, Balloon telescope optics, Applied Optics, 6, 179-189, 1967.
6 M. Bottema, W. Plummer and J. Strong, Water vapour in the atmosphere at Venus, Astrophysical J., 139, 1021-1022, 1964.
7 A. R. Barringer, Developments towards the remote sensing of vapours as an airborne and space exploration tool, Proc. 3rd Symp. in Remote Sensing. U of Michigan, 279-292, 1964.
8 R. B. Kay, Absorption spectra apparatus using optical correlation for the detection of trace amounts of SO_2. Applied Optics, 6, 776-778, 1967.
9 D. T. Williams and B. L. Kolitz, Molecular correlation spectrometry, Applied Optics, 7, 1968.
10 M. Millan, Analysis of the in-stack correlation spectrometer signal, Tech. Rep. TR70-143, Barringer Research Ltd., Toronto, Canada, 1970.
11 M. Millan and R. G. Quiney, final report "Absorption spectroscopy flight instrumentation study and SO_2 investigation", prepared for NASA Manned Spacecraft Center. Under contract NAS9-11478, Barringer Research Ltd., Toronto, Canada, report TR71-194, 1971.
12 M. M. Millan, S. J. Townsend and J. H. Davies, Study of the Barringer refractor plate correlation spectrometer as a remote sensing instrument. UTIAS rep. No. 146. Univ. of Toronto, Ont., Canada, 1970.
13 G. S. Newcomb and M. M. Millan, Theory, applications and results of the long-line correlation spectrometer. IEEE Trans. Geosc. Elec. GE-8, 149-157, 1970.
14 D. T. Williams and R. N. Hager, Jr., The derivative spectrometer. Applied Optics, 9, 1597-1605, 1970.
15 R. N. Hager, Jr. and R. C. Anderson, Theory of the derivative spectrometer. J. of the Optical Soc. of America, 60, 1444-1449, 1970.
16 M. M. Millan and R. M. Hoff, How to minimize the baseline drift in a COSPEC remote sensor, Atmospheric Environment, 11, 857-860, 1977.
17 M. M. Millan and R. M. Hoff, Dispersive correlation spectroscopy: A study of mask optimization procedures. Applied Optics, 16, 1609-1618, 1977.
18 M. M. Millan, Dispersive correlation spectroscopy for atmospheric monitoring. D. Reidel - Publishing Co., Dordrecht (Holland), in preparation.
19 L. A. Alonso, Doctoral Thesis: Industrial Engineering School of Bilbao, Bilbao, Spain (1982).

Optical Remote Sensing of Air Pollution,
Lectures of a course held at the Joint Research Centre, Ispra, Italy, 12–15 April 1983,
P. Camagni and S. Sandroni (Eds). 69–93

EFFECTS OF ATMOSPHERIC INTERACTIONS IN CORRELATION SPECTROMETRY

Millán M. MILLÁN

1. INTRODUCTION

The use of "freely available", and naturally occurring radiation sources for the remote sensing of atmospheric pollutants, or other species, is not necessarily as free or as simple as some have proclaimed. In fact, it exacts a heavy toll in at least two areas: it requires a comprehensive knowledge of the instrumental operation and settings, and it also requires a careful interpretation of the results. In general, the processes associated with the transfer of radiation through the (whole) atmosphere affect the ability of a correlation spectrometer to detect the presence of the target gas. Consequently, they also affect the data collection and interpretation procedures and ultimately, as we will examine in this chapter for the case of the COSPEC, the design of the instrument itself.

Just to put some things in their proper perspective, one can consider that the spectroscopic measurements of upper atmospheric ozone with the Dobson spectrometer are considered accurate to within 5% or so. This amounts to an uncertainty of some 100 - 200 ppm-m over total values of 2500 to 4000 ppm-m (0.250 - 0.400 Atm-cm) using integration time constants in the order of tens of seconds to minutes. In contrast, the tracking of a plume from a medium sized power plant for a few tens of kms, may require sensitivities of better than 5 ppm-m with integration time constants of the order of seconds, which represents a required sensitivity increase of approximately two orders of magnitude with respect to other spectroscopic techniques. This is not a reflection on the methods themselves but rather on what is required from them as new applications emerge.

In the case of the COSPEC, the effects caused by the diurnal variation of the sky-solar spectrum and the sky-to-cloud changes in backgrounds, had to be corrected with substantial modifications of the original instrument and led to the implementation of the multistable concept in the present line of commercial instruments.

The characteristics of the radiation transfer process involved in the instrumental observation also determine the boundary between a qualitative observation, or simple identification of the presence of the target gas, from its quantitative measurement. In the latter case, when the "inversion" of the data can be performed, they also determine to what extent the data must be manipulated to extract the measurement and the margin of error that can be expected. Another important aspect of the interaction between the radiation field and the instrument is that the results can be broadly classified in two categories: those effects which appear more readily as instrumental operational characteristics, i.e. baseline drift, calibration and signal to noise variations during the day, and those which directly affect the way in which the results must be processed and/or interpreted.

In this chapter, some basic aspects of the available backgrounds are introduced first. These will be presented in the context of how they affected the evolution and design of the present day COSPECs. What effects were compensated, to what extent, and which ones still remain will also be presented as part of the COSPEC's operational baggage. Its generalized response equation will be developed and subsequently used to study its operational characteristics and how the radiation transfer problem affects the measurement of real plumes in a real atmosphere.

The work presented here is very heavily biased (almost exclusively) towards the atmospheric and instrumental problems associated with SO_2 measurements. This is so because this author's research effort has been mostly devoted to this area, which was the most problematic during the initial period of COSPEC development and usage. The author is not aware of any extended work of similar characteristics for the NO_2 region, except for that of Van der Meulen and colleages (ref. 1), or for other regions and other gases. These are still open research areas and, in this spirit, other outstanding problems will be mentioned in the hope that this may stimulate further research. Foremost in the approach to this chapter is the quote from ref. 2 that "once we understand the limitations of each analytical technique we can get on with the problems of studying variations of atmospheric quantities".

2. ATMOSPHERIC BACKGROUNDS

Ground level radiation in the 300 to 320 nm spectral region has as its ultimate source the sun, but reaches the ground via two main paths: a directly transmitted solar component, and a diffuse, or sky component, which originates from the scattering of solar radiation in the upper atmosphere. An important aspect of the latter component is that the multiple scattered radiation which reaches the ground originates mostly in the troposphere from primary scattered radiation arriving from about, or below, the ozone layer (ref. 3).

This important effect results from a combination of the atmospheric density distribution and the selective absorption of uv radiation by stratospheric ozone in the Hartley bands and continuum. For the largest solar elevation angles the solar radiation path is the shortest through the O_3 layer. The transmitted radiation is then scattered mostly below the ozone layer and propagates downwards. As the solar elevation decreases, the slant path through the O_3 layer increases and so does its (selective) attenuation. Radiation at the longer wavelengths is less attenuated and is scattered downward from an ever increasing height. This process continues until the slant path through the ozone layer is so long that no useable uv remains to be scattered below the layer. At this point, the only available uv proceeds from scattering in the less dense upper atmosphere above the ozone layer. Those radiation components also propagate downward via a vertical (shortest) path through the O_3 layer. The end result of these processes is that the bulk of the primary scattered radiation occurs in a broad but reasonably well defined atmospheric layer whose shape and average height vary during the day as a function of the solar zenith angle θ (preferred to the elevation angle by atmospheric physicists), and of the wavelength of the radiation. The height of this "effective scattering layer" is higher for large θ (low solar elevations), and for the shorter wavelengths (ref. 4) which are depleted more rapidly. The vertical displacement of this scattering layer and the associated spectral effects are used in the Umkehr method to determine the burden and distribution of stratospheric ozone (ref. 5). In certain situations it may also affect the SO_2 data reduction (ref. 23).

From the point of view of using the resulting radiation field as the background to measure tropospheric SO_2, and in terms of the instrument response behaviour, the most important effects to consider are as follows:

- There is a pronounced diurnal variation in the spectral content of the downward radiation from the sky. The variation is nearly symmetrical with respect to solar noon. An example is shown in Figs. 1 and 2.
- The variation in the spectral distribution (slope) of the radiation is non-monotonic. The distribution is flatter for the (weak) uv radiations, scattered above the O_3 layer, which prevails for very low or zero solar elevation angles. Thereafter, the slope increases as the sun rises and the longer wavelengths are first to penetrate the O_3 layer. With even higher

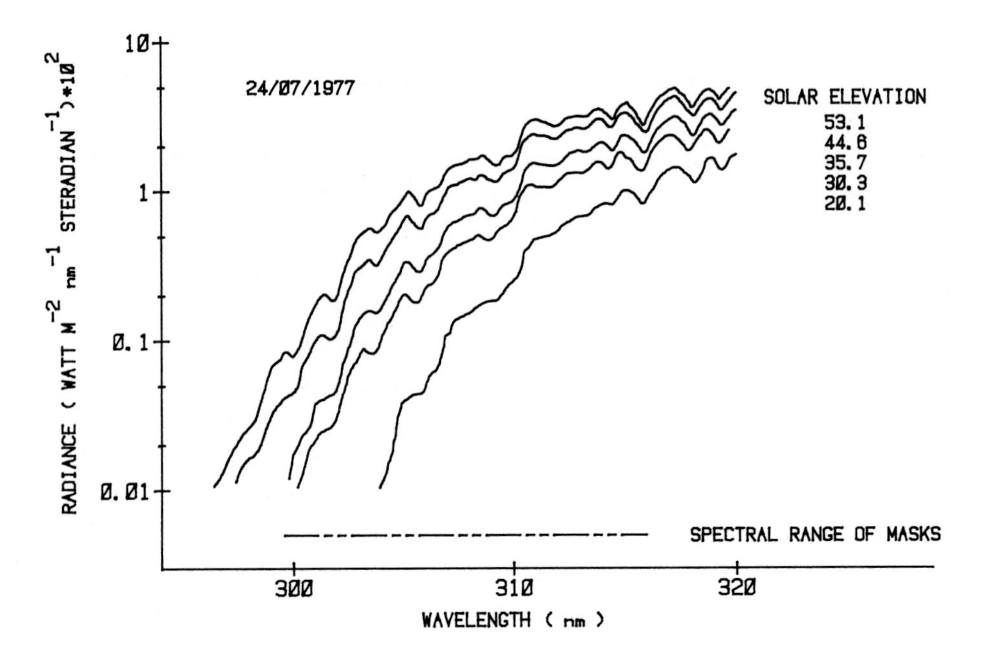

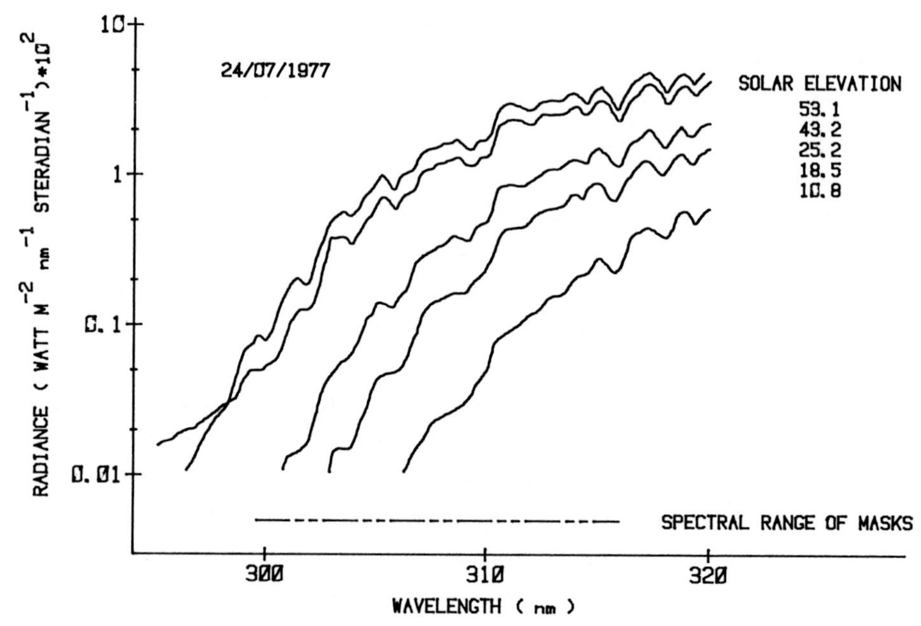

Fig. 1 and 2.

Zenith sky spectral radiances over Toronto on a cloudless day Jul 24, 1.977. Fig. 1, shows the variation for increasing solar elevations to noon. Fig. 2, shows them for decreasing solar elevations. Data courtesy of Drs. K. Anlauf and R. Hoff.

solar elevations the intensity increases proportionally more for the shorter wavelengths (as can be observed in Figs. 1 and 2) and the slope decreases again to reach a minimum near the solar noon. This mechanism is the same one responsible for the Umkehr, or inversion, effect (refs. 4, 5).

- There is an upper atmospheric ($\sim$ 50 km) ozone maximum during the night which disappears by photodissociation within an hour or two from sunrise (ref. 6). This higher O_3 burden during the early part of the day is responsible, in part, for a weak asymmetry in the diurnal variation described above, and affects both the direct solar and diffuse radiation components.
- Superimposed on the diurnal variation, there is also an annual variation caused by the seasonal changes in the atmospheric ozone burden which, for northern latitudes, reaches a maximum in early spring and a minimum in autumn.
- With clear skies, the spectral content and state of polarization of the sky radiance vary for each region of the sky relative to the solar position in a manner that is nearly symmetrical with respect to the solar azimuth plane (ref. 7).
- The state of polarization decreases with multiple orders of scattering and thus diminishes with atmospheric turbidity and some types of cloud cover.
- Most cloud types act as good diffusers of solar radiation, in particular cirrus types, altostratus, low level fog, marine haze and the edges of cumulus clouds. In particular, important and abrupt changes in the spectral distribution can take place in the transition from clear sky to the silver linning of cumulus.

Throughout this chapter we will refer to these effects time and again as they relate to the instrumental response behaviour.

A common denominator to all of the effects mentioned is the change in the spectral content of the background, i.e. intensity and its distribution as a function of wavelength.

Effects of similar nature do occur in <u>all</u> spectral regions as a result of the interaction between atmospheric phenomena and the available radiation field. However, what effects predominate, i.e. scattering vs thermal emission, changes in the spectral properties of sky to clouds, etc., do vary with the spectral region, as indicated by Quenzel in this text.

In the case of the COSPEC there is, among others, a very important difference between the atmospheric effects in the spectral regions selected for SO_2 (300 - 316 nm) and for NO_2 (420 - 455 nm). In all the effects mentioned for the SO_2 region, the intensity increases with wavelength under all conditions so that in spite of changes in the spectral distribution, its overall slope is always positive, i.e. intensity increases with wavelength. This is not the situation in the NO_2 spectral region. The spectrum of the atmospheric scattered radiation for clean and dry air peaks further into the blue (~ 400 nm) than the original solar distribution (at ≅ 480 nm). This, which is responsible for the blueness of the sky, also results in a flat, or slightly downward sloped spectral distribution in this region for very clear skies (refs. 8, 9). In this situation, diurnal changes move the intensity distribution up and down without much of a spectral (slope) change. Clouds and hazy skies, however, change this state of affairs drastically since they all act as good diffusers of the solar spectrum and reproduce its spectral distribution more faithfully, which has a positive slope in the same region.

Sky to cloud changes in the NO_2 region, therefore, imply both changes in intensity as well as in the slope (included sign). This can be further complicated when the amount of water vapour and/or atmospheric haze increases, because then large spectral variations can occur with simple changes in the orientation of the instrument field of view, e.g. from pointing upwards to the zenith sky to looking towards the horizon. The associated spectral changes can seriously affect both sensitivity and baseline stability as we will discuss now.

3. GENERALIZED RESPONSE FUNCTIONS AND DIRECT INSTRUMENTAL EFFECTS

To study how changes in the background spectral distribution affect instrumental performance and signal behaviour, we require a workable instrumental response function which we will develop now. At the same time, we will follow closely the COSPEC response characteristics, reviewing first how a bistable system is affected, to be followed by the behaviour of the present line of COSPEC remote sensors.

We recall first that the dispersed spectrum or spectral radiant power is the multiwavelength image of the entrance slits and, second, that the sensor field of view (FOV) consists of the image of the entrance slit(s) formed onto the background by the telescope in the front optics. We can expect, therefore:

1) Partial field of view effects caused by sources or backgrounds with smaller dimensions than the FOV;
2) Effects caused by the changes in the spectral distribution of the various background sources present within the FOV.

Partial FOV effects are most important when using the remote sensor in an aiming mode, either with controlled light sources or looking at small areas (e.g. stack tops). These effects have been discussed elsewhere (ref. 10) and will not be presented here.

To discuss the second, some simplifying assumptions must be made to keep the equations tractable. From the chapter in Correlation Spectrometry (Eqs. 13, 15) we recall that the radiant power sampled by mask-j- can be written as:

$$P_j = \int_{\text{mask-j-}} P_\lambda d\lambda = \sum_i \int_{-\Delta i}^{\Delta i} P(\xi + \lambda i + \lambda)\, d\lambda \tag{1}$$

where P_λ is the spectral radiant power, ξ and λ_i are, respectively the reference position of the mask and centreline coordinate of slit-i- within the mask, and Δi is the half width of slit-i-.

We will first assume that the mask is spectrally located in its design position (either band peaks or troughs) and thus we can omit the parameter ξ.

Secondly, Eq. (8) in lecture III of this Course shows that the spectral radiant power can be written as

$$P_\lambda = \Omega h \int_{-\Delta e}^{\Delta e} N(\Lambda+\lambda) d\Lambda = 2\Delta e h \Omega N_\lambda = A\Omega N_\lambda \tag{2}$$

where $A = 2\Delta_e h$ is the nominal entrance aperture of the system, h and Δe being, respectively, the entrance slit height and half width, and N_Λ is the background spectral radiance (in true wavelength units).

Equation (2) is a simplified expression for the imaging process within the monochromator system and represents, essentially, a running average of the incoming spectrum performed across the (spectral) image of the entrance slit. In a real system, aberrations, etc., must be included, and the equivalent of Eq. (2) are the results from a suitable raytrace program.

For our purpose, we will simplify Eq. (1) to

$$P_j = \sum_i p_{ji} \tag{3}$$

where p_{ji} is the average radiant power sampled by slit-i- of mask-j-, and P_j the radiant power sampled by the total of the mask.

Following a similar procedure, for the case that a cloud of the target gas fills the FOV, we can also write

$$P_j = \sum_i p_{ji} \exp[-a_{ji} cL] \tag{4}$$

where a_{ji} is the average absorption coefficient seen by slit-i- of mask-j- and cL is also the concentration times pathlength product (assumed uniform) within the FOV.

Finally, for the case that several radiation components are present and fill the field of view (e.g. direct background and atmospheric scattered components), with varying degrees of absorption by the target gas, we can write

$$P_j = \sum_{\ell} \sum_i p_{j\ell i} \exp[-a_{ji}(cL)_{\ell}] \tag{5}$$

where for the components with null target gas absorption, the values of $(cL)_{\ell}$ would be zero. A further assumption here is that the value of a_{ji} holds for a range of values of <u>cL</u>.

3.1 Bistable Systems

In these, one mask in two spectral positions, or two masks are used to obtain the signal (ref. 11). In terms of the radiant powers trasmitted by each mask P_j, the photomultiplier (PMT) current output is given by:

$$I_j = P_j S_{kr} G + I_{oa}(G,T) \tag{6}$$

where I_j is the anode current, S_{kr} and G are, respectively, the photocathode radiant sensitivity and photomultiplier gain, and I_{oa} is the anode dark current, a function of temperature and gain.

It will be further assumed that $I_{oa}(G, T) = I_{ok}(T) \cdot G$, i.e. the anode dark current equals the cathode dark current amplified by the same gain as the signal current. This is not exactly true for most photomultipliers (refs. 12, 13), but it is reasonable for our purposes.

The I_j pulses are converted to voltages in the pre-amplifier section. In the COSPEC type of instruments, the PMT gain is controlled via an automatic gain control circuit (AGC) which performs the process:

$V_1 = I_1 R$ = AGC controlling voltage, a selectable constant (see lecture III), where R is the resistive load equivalent of the pre-amp. section. The two pulses thus become

$$V_1 = (P_1 S_{kr} + I_{ok}) RG \quad \text{which yields} \quad G = \frac{V_1(\text{fixed})}{P_1 S_{kr} + I_{ok}} \tag{7}$$

and

$$V_2 = (P_2 S_{kr} + I_{ok}) RG = V_1 \frac{P_2 S_{kr} + I_{ok}}{P_1 S_{kr} + I_{ok}} \tag{8}$$

These are synchronously detected and amplified to yield the final output as:

$$V_o = \alpha d(V_1 - V_2) = \alpha d V_1 \{1 - \frac{P_2 + DCEP}{P_1 + DCEP}\} \tag{9}$$

where α is the final (inverting)*) amplifier gain, d is the synchronous detector duty cycle (ref. 14) and DCEP = I_{ok}/S_{kr} is the Dark Current Equivalent Power. The latter is a figure of merit in the selection of a photomultiplier and should be as small as possible. The optical design of the sensor, on the other hand, should also be aimed at making $P_j \gg$ DCEP for all operating conditions. Alternatively, the term DCEP can be completely eliminated with a suitable electro-optical design at the cost, however, of a reduced duty cycle of the synchronous detector. This is not available in the commerical COSPECs. For most COSPEC systems with V_1 between 1.5 to 2 Volts and with the optics property aligned, the condition $P_1 \gg$ DCEP holds for all operational ranges, and we will neglect this term henceforth.

Finally, in all COSPEC systems the limiting electronic bandwidth is that of the final amplifier section which has a selectable time constant. In terms of the time response, the output can be written as:

$$V_0(t) = c_1 \{1 - \frac{P_2}{P_1}\} (1 - \exp[-t/\tau]) \tag{10}$$

where $c_1 = \alpha d V_1$ and τ is the electronic time constant. In the text that follows the time dependent factor will not be used.

3.2 Spectral Effects in Bistable Systems

Taking into account (3) and (4), and the simplifying assumptions made, the response function (10) can be expressed as

---

*) COSPEC users interested in modelling its output and comparing theoretical vs experimental results, should be aware of the change of sign thus implied.

$$V_0 = c_1 \{1 - \frac{P_2}{P_1}\} = c_1 \{1 - \frac{\Sigma p_{2i}}{\Sigma p_{1i}}\} ; \quad \text{with no gas} \tag{11}$$

$$V_0(cL) = c_1 \{1 - \frac{\Sigma p_{2i} \exp[-a_{2i}cL]}{\Sigma p_{1i} \exp[-a_{1i}cL]}\}; \quad \text{with target gas} \tag{12}$$

Equations (11) and (12) can be written in a variety of manners in order to emphasize certain aspects of the response as well as the spectral effects. Initially we will rewrite Eq. (12) as follows:

Add and substract P_2/P_1 to obtain

$$V_0(cL) = \underbrace{c_1 \{1 - \frac{P_2}{P_1}\}}_{\substack{\text{no-gas term} \\ \text{(direct background signal)}}} + \underbrace{c_1 \frac{P_2}{P_1} \left\{ 1 - \frac{\Sigma \frac{p_{2i}}{P_2} \exp[-a_{2i}cL]}{\Sigma \frac{p_{1i}}{P_1} \exp[-a_{1i}cL]} \right\}}_{\substack{\text{gas-signal term} \\ \text{(background scaling effects)}}} \tag{13}$$

An expression is obtained in the same form as Eq. (5) in lecture III and with the same implications, i. e. the sensor yields an output as soon as there is a difference between the sampled radiant powers. The progress from that equation to this point is: on the averaging imposed by the multiplicity of the slits, specificity and the increase in signal to noise ratio.

The first term in the right hand side is the no gas term, or direct background response present as soon as $P_2 \neq P_1$. It corresponds to the zero gas offset of earlier reports (refs. 11, 15). The second, which includes the scaling factor P_2/P_1 is the direct gas-signal term. For reasons that will become obvious in the next section, it is convenient to divide the output into these two terms.

We examine first the no-gas term for the case of a two mask (bistable) system. For that we rewrite Eq. (11) in the form:

$$V_0 = c_1 \{1 - \frac{\Sigma p_{2i}}{\Sigma p_{1i}}\} = \Sigma \left\{1 - \frac{p_{2i}}{p_{1i}}\right\} \frac{p_{1i}}{(\Sigma p_{1i})} \tag{14}$$

Equation (14) shows that the output is a sum of the contributions by each slit pair and each is weighted by the ratio of the radiant power sampled by the first slit of the pair to the total radiant power sampled by the whole first mask of the pair.

The SO_2 masks in the 300 to 316 nm region are those shown in Fig. 7 of lecture III. Mask no. 1, which controls the AGC, has the slits placed at shorter wavelengths than those for mask no. 2.

For very low, or zero, solar elevation angles, the spectral distribution of the available radiation is that imposed by Rayleigh scattering of the solar radiation in the upper atmosphere above the O_3 layer. In the processes:

- Rayleigh scattering of the solar radiation followed by:
- Transmission through the O_3 layer to
- Multiple scattering in the troposphere and
- Downward flux of radiation to ground level,

The overall slope of the spectral distribution is that imposed by the wavelength dependence of the molecular scattering followed by the ozone absorption. The latter occurs via the shortest path possible (i. e. the vertical). Other things considered equal, this means that the spectral distribution is the least sloped of those available in the mid latitudes, albeit with very low intensities.

For the masks considered, $P_2 \gg P_1$ in an upward slope*), and the no-gas term starts the day, and always remains, negative. As the solar elevation increases, the longer wavelengths with weaker O_3 absorption penetrate first through the O_3 layer to be scattered below, more strongly, in a denser atmosphere. The initial result is that this gives rise to an ever increasing slope which affects mostly the longer wavelength slit pairs of the masks, and the no-gas term grows more negative.

With further increases in the solar elevation a point is reached when the shorter wavelength end of the available radiation starts filling in proportionally faster than for the longer wavelength end, as shown in Figs. 1 and 2, so that the overall slope decreases again. The no-gas term reverses its negative growth towards positive values and reaches a maximum (still with negative values) near midday. After the solar noon, a nearly symmetrical variation takes place. This process is illustrated in curves c and d in Fig. 3. The depth of the "bow" in the variation is of the order of 400 - 600 ppm-m.

Traditionally, the time variation of the no-gas term has been denoted as the instrument baseline in the present COSPEC systems. In the old bistable system it was denoted as the zero-offset and attributed to either drifty electronics or optical misalignments before its diurnal variation was discovered in early 1969 during an extended series of tests which took place once the spectral effects became suspect (ref. 15). In what follows baseline will be used indistinctly for both: bistable and multistable systems.

---

*) For individual slit pairs this may not hold on account of the presence of deep Fraunhoffer lines in the solar spectrum.

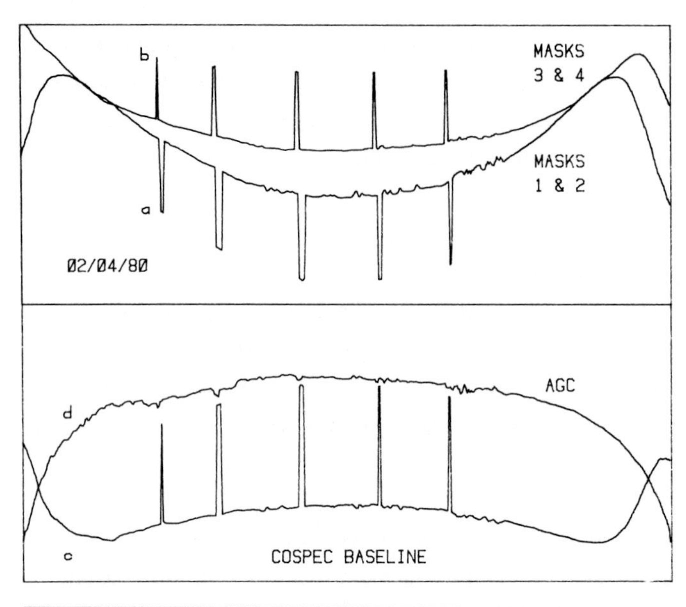

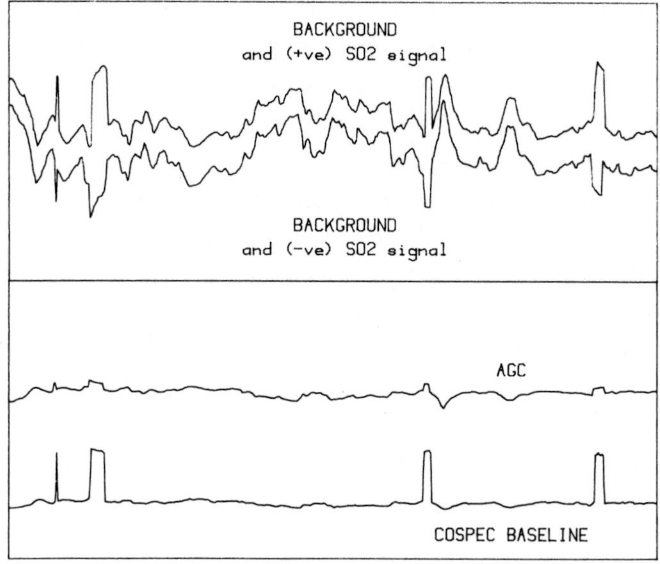

Fig. 3 and 4.

Multistable concept. The signals for the first mask pair and for the second have similar baselines while their target gas signals have opposite polarities. Their (weighted) difference, the COSPEC output, undergoes a smaller diurnal variation, which can be minimized (see Figs. 6 and 7). Fig. 4, with a proper scaling of the signal from the second pair, cloud effects can be cancelled.

Its smooth variation for a clear sky (curves c or d in Fig. 3) changes significantly under broken cloud conditions, as shown in curves a and b in Fig. 4. As can be seen, a two mask (or two wavelengths) bistable system is unable to distinguish between the signals imposed by sharp spectral changes, i. e. sky to cloud effects, from those of genuine target gas signal. The sky to cloud changes in spectral composition are equivalent to sudden displacements within points in the baseline. As we will discuss next, they are also accompanied by changes in the slope of the calibration signal, and are more accentuated for low solar elevation angles. A bistable passive spectroscopic system (in the SO_2 region) is burdened with a diurnal baseline variation of 400 to 600 ppm-m and the possibility of a sky-to-cloud "noise" envelope of the order of 100 to 200 ppm-m.

The target gas effects can be explained with the second term in Eq. (13). It shows that the gas-signal term is generated from the ratio between two weighted sums. For each slit-i- in positions j = 1, 2, their contributions to the signal -exp $[-a_{ji}cL]$- are weighted by the ratios p_{ji}/P_j between the radiant power available to the slit to that available to the whole mask. This is more clearly emphasized by rewriting Eq. (12) as:

$$V_0(cL) = c_1 \Sigma \left\{ \underbrace{(1-\frac{p_{2i}}{p_{1i}})}_{\text{no-gas term}} + \underbrace{\frac{p_{2i}}{p_{1i}}}_{\text{background scaling}} \underbrace{(1-\exp[-(a_{2i}-a_{1i})cL])}_{\text{intrinsic signal term}} \right\} \underbrace{\frac{p_{1i}\exp[-a_{1i}cL]}{(\Sigma p_{1i}\exp[-a_{1i}cL])}}_{\text{weighting factor}} \quad (15)$$

As previously, the two terms inside the outermost bracket in Eq. (15) represent the sum of a no-gas term and a gas-signal term. However, Eq. (15) shows very explicitly that the output is a weighted sum of the contributions by <u>each</u> slit pair. The intrinsic signal term represents the target gas signal that would be obtained with each slit pair from an ideally flat spectral background. The weighting factors:

$$\frac{p_{1i}\exp[-a_{1i}cL]}{\Sigma p_{1i}\exp[-a_{1i}cL]}$$

represent, as before, the ratio between the radiant power sampled by slit -1i- of the mask to the total radiant power sampled by all slits in mask no. 1.

Explanation of the spectral effects follows naturally: In the case of the SO_2, and because of its band structure, the slit pairs at the shorter wavelengths have greater values of $(a_{2i} - a_{1i})$ and, therefore, they also possess the larger intrinsic signals. On the other hand, their contribution to the overall signal depends on the radiant power available to them.

As the solar elevation increases so does the radiation available to the slit pairs at the shorter wavelengths. This results in an increase of the calibration and target gas signals. A (relative) maximum of the weighting occurs near midday and, correspondingly, another one for the target gas or calibration signals. Subsequently, these decrease with the solar elevation during the afternoon. This situation continues until the scattering above the ozone layer starts to dominate the available uv and the steepness of the slope decreases. The weighting of the shorter wavelengths becomes favoured once more, and a reversal of the decreasing trend takes place so that the calibration and target gas signal increase again. The reversal of the calibration is shown in Fig. 5. For most COSPEC set-ups the instrument could run out of light before this reversal can be detected just about or before sunset. This effect should be considered if the operational range is extended to operate with low solar elevations (ref. 16) as may be required to work in high latitudes*). In any case, it should also be indicated that in these "marginal" effects the diurnal asymmetry of the O_3 burden starts playing an important role. In fact, the calibration reversal may not be noticeable in the early morning on account of the local O_3 maximum which occurs during the night at high altitudes.

3.3 Multistable Systems: the COSPEC

In its basic form, the multistable concept is illustrated in Figs. 3 and 4. Its initial conception was to combine two bistable outputs for which the background signals had equal values and the target gas signal had opposite polarities. The cancellation of the direct background term was the driving concept behind this development.

On the basis of the signal processing discussed in the previous section (Eq. 15), target gas signals of opposite polarity with respect to the baseline require a change in the sign of the factors $(a_{2i} - a_{1i})$, i. e. if the absorption coefficients for the slits of mask no. 1 are smaller than those for mask no. 2, i. e. a peak-trough mask pair, the second mask pair requires the opposite for masks no. 3 and 4, i. e. to be a trough-peak pair. The first pair yields a gas signal positive with respect to baseline, and the second pair yields a

---

*) For values of V_1 less than approximately 1-1.5 Volts, depending on the quality of the photomultiplier, the approximation $P_j \gg I_{ok}/S_{kr}$ may no longer hold. In this case, the reversal may not be detected and some other effects can take place, i. e. the diurnal variation will be less pronounced and the calibration signal will decrease very strongly towards the end of the day.

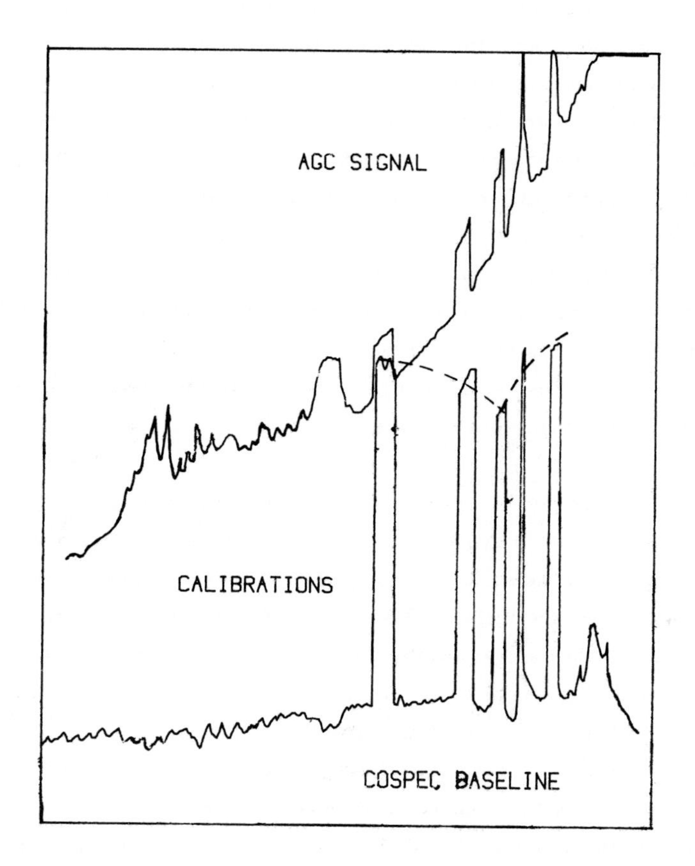

Fig. 5.- Calibration reversal. It occurs when the COSPEC is set to operate at very low solar elevations. The spectral distribution of the radiation scattered above the Ozone layer favours the contribution of the shorter wavelengths and produces this effect.

gas signal negative with respect to its baseline. The other key concept is that the two mask pairs should sample the spectrum in the same sense. In the case of the SO_2, this is from the shorter to the longer wavelengths.

If the masks are suitably selected, i. e. mask no. 3 is nearly the same as mask no. 2 and mask no. 4 is similar to mask no. 1 displaced by one band head, one can expect a similar behaviour for the no-gas term during the day. They cannot be exactly equal on account of the fact that the one pair is displaced one band-head with respect to the other and it sees a somewhat different spectrum due to: the presence of Fraunhoffer lines, wavelength dependent properties of the optical system imaging and components, etc.

On the basis of the spectral effects just described one can also expect:

- a generally less pronounced diurnal baseline variation for the mask pair at the longer wavelengths;
- a weaker target gas signal and calibration, since the second pair looses a band with strong absorption at the short λ end, and gains a weaker band and the long λ end with increased weighting;
- the point of reversal of the no-gas term occurs at lower solar elevations for the mask pair at the longer wavelengths.

All of these effects can be seen by comparing the curves c and d in Fig. 3.

After initial attempts to design mask pairs with exactly the same diurnal variation of the no-gas term, and failing for the reasons stated above, the next approach was to take full advantage of whatever could be of use. In this case, we used the near symmetry of the observed diurnal behaviour. This allowed the development of a compensation technique (refs. 16, 17) to minimize the variation of the residual baseline during most of the day, and to optimize the design of the mask pairs for best SNR (refs. 14, 18).

Although the multistable concept was known as far back as 1969 when it was implemented in the present COSPEC series of instruments, a means of providing the COSPEC user with the two individual bistable signals that make up its output has not been generally available until the COSPEC V model of 1980. These two signals are the outputs of the Supplementary Signal Processor, SSP.

In these terms, the COSPEC signal output is then:

$$V_0 = c_1 \left\{ (1 - \frac{P_2}{P_1}) - k(1 - \frac{P_4}{P_3}) \right\} \tag{16}$$

and as a function of the voltage pulses after the second AGC (Eqs. 22 and 24 in lecture III.

$$V_0 = \alpha dV_1 \left\{ (1 - \frac{P_2}{P_1}) - \frac{V_3}{V_1} (1 - \frac{P_4}{P_3}) \right\} \qquad (17)$$

where $\alpha dV_1 = C_1$ and $V_3/V_1 = k$ is the baseline compensation factor. Either term in Eq. (17) can be expressed as in Eqs. (13), (14) or (15).

Compensation of the baseline is illustrated graphically in Figs. 3, 6 and 7. Curves c and d in Fig. 6 show the two bistable outputs, their difference (with some scale change), and primary AGC output (Eq. 23 in previous lecture) which reaches a minimum at solar noon. A suitable compensation value k is calculated from those curves to minimize the variation during a part of the day (ref. 17) and a $V_3 = kV_1$ is set in the electronics. This is shown in Fig. 6 (time progresses from left to right), where a straight baseline is obtained after electronic setting. Finally, Fig. 7 shows the new baseline after compensation. With respect to these figures it should be mentioned that the present COSPEC SO_2 mask design, Mask No. 1 in troughs, results in a negative signal and that the graphs have been arranged accordingly. See first bottom note of previous section.

If the use of four masks, in two pairs, can minimize the diurnal baseline variation, it does not correct for some of the other effects. Changes in sensitivity including the reversal of the trend in the calibration signal occur just the same as for the bistable signals that compose the COSPEC output. An added, and somewhat serindipitous, benefit of the four masks is the compensation of polarization and, in particular, sky-to-cloud effects on the baseline, as shown in Fig. 4 for a COSPEC V.

In the case of the SO_2, the compensation factor(s) that minimize the baseline variation during a great part of the day is (are) nearly the same one that cancels the sky-to-cloud baseline effects[*)] which vary during the day. Sky-to-cloud effects can be compensated entirely during a limited part of the day. If performed at noon, residual effects will appear towards the end of the day (low solar elevations) and viceversa. Similarly, a compromise compensation intended to minimize this effect may leave a bowed baseline which requires, as discussed below, other approaches to compensate the measurements.

Finally, little information of this type appears to be available for the case of NO_2 measurements. The lack of the supplementary (bistable) signals in

---

*) The author is aware that this may change from instrument to instrument on account of possible spectral dependent differences in the optical systems, alignments, etc.

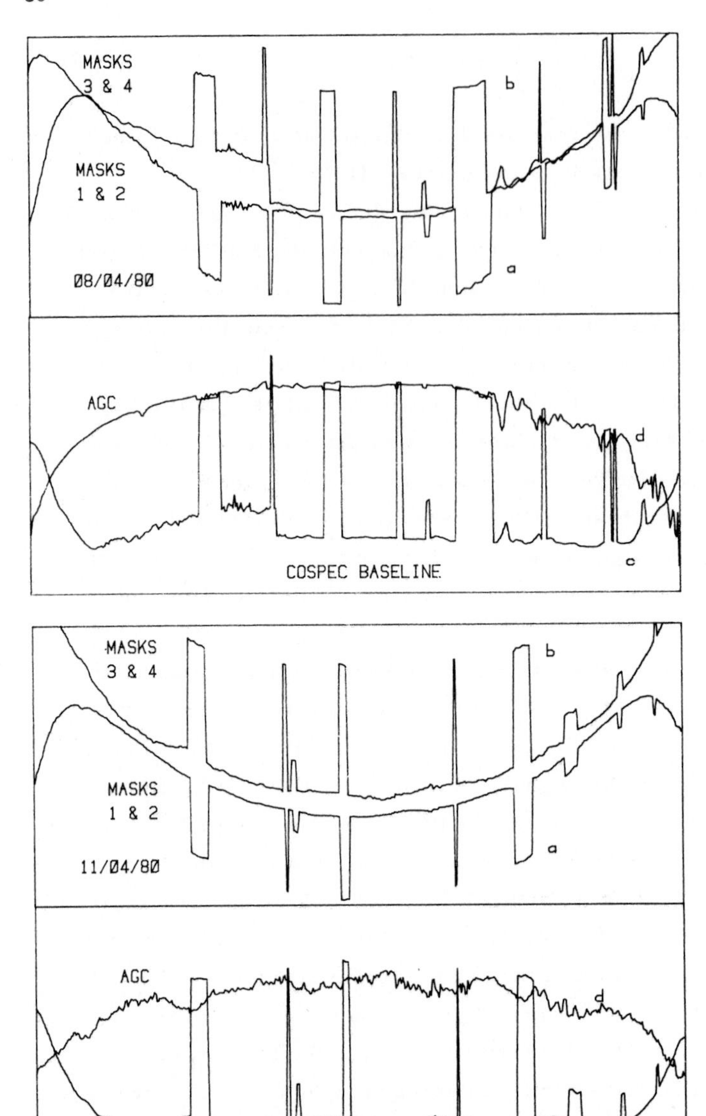

Fig. 6 and 7.

COSPEC baseline compensation technique (ref. 17). With the proper factor calculated from the graphs in Fig. 4, the signal for the second mask pair can be adjusted so that its baseline tracks that of the frist pair. In Fig. 6, the correction is applied before midday 08/04/80. Fig. 7, shows the bistable baselines and the COSPEC's on the 11/04/80.

COSPEC models previous to the Mk V, have obviously played an important part. Research in this area is still required.

3.4 Some practical hints

Compensation of the COSPEC baseline requires the determination of the two baselines for the outputs of each mask pair. These were not generally available until 1980, and compensation had to be made after a careful recording of the video signals (Vi pulses) in an oscilloscope (ref. 17).

In any case, with the present SO_2 masks in their designed positions, i.e. first slit of mask No. 1 centered at ≅ 300 nm, a baseline stability to within 15 ppm-m can only be achieved during the central part of the day, i.e. for solar elevations greater than approximately 15°. Thereafter the baseline variation can be quite pronounced until the unit runs out of light, as shown in Figs. 6 and 7.

Several methods can be used to minimize these problems. In the first place, and because of the periodicity of the SO_2 bands, several mask positions can be selected towards longer wavelengths. These yield reduced target gas-signals due to the increased weighting of the weaker absorption bands past 315 nm. This procedure, however, displaces the points of (bistable) baseline reversal further towards the ends of the day. In most COSPECs, one or two positions towards the visible can push the point of reversal to before sunrise and after sunset, leaving the central part (with similar curvatures) correspondingly wider. Baseline compensation methods (ref. 17) can then be used to obtain a diurnal variation within 40 ppm-m for the whole day.

A compensated baseline, however, still has a number of ripples within it, i.e. an even number of inflexion points and, in general, an overall tilt towards the later part of the day. Plume traverses at short distances from the source can be completed within a short enough time interval so that no significant variations in the baseline take place. This may not suffice for perimeter surveys or traverses at larger distances which require a much longer time period during which the baseline may change.

For plume traverses at large distances, i.e. 30 to hundreds of km (refs. 20, 21), a modified data gathering procedure can be used. This consists in recording the data in two modes: time based and distance based. The concept is shown schematically in Fig. 8. The distance proportional record of a weak plume is not sufficient to determine when the edges of the plume (baseline) have been reached. The time based record, however, allows the instrument operator to see that the end of his traverse coincides with the "expected" baseline. This procedure also requires a statistical knowledge of the baseline behaviour for whatever electronic setting previous to the field work. In general terms, and for the work that this author has been more familiar with, a suggestion is to compensate sky-to-cloud effects in the best possible manner and correct for any residual baseline variation in this way.

4. REMOTE SENSING IMPLICATIONS

Up to this point, atmospheric interactions with the radiation field have been presented within the context of the so-called direct instrumental effects, i.e. baseline, calibration and sensitivity effects, etc. A completely different and yet simultaneous set of problems arise when a correlation spectrophotometer is used in a real atmosphere.

The use of passive spectroscopic remote sensors in air pollution studies is based on the (initial) assumption that the target gas signal is proportional to the line-integrated concentration of the target gas along the radiation path*). As we have indicated, there are basically two radiation sources: a direct solar component, and a diffuse (solar-atmospheric scattered) sky component.

For any geometry of observation one can thus expect:

- transmission of the background(field of view) radiation to the sensor;
- scattering of radiation from the sun and other areas of the sky into the acceptance solid angle of the sensor;
- the sensor accepts all radiation components available to it and processes them as described previously.

In general, the radiative transfer problems involved can be quite complicated and not necessarily solvable. On the other hand, there may be certain situations for which the atmospheric effects are acceptable and for which a calibration of the signal is still acceptable. The study of even the most common situations is beyond the scope of this presentation and we will refer the reader to some available analyses of the problems involved (refs. 1, 22).

The simplest, and most widely employed viewing geometry is that of upward (vertical) looking, and the prevailing problems will be reviewed. A typical situation can be as the one shown in Fig. 9. For the situation shown there are three main sources of radiation, viz:

---

*) In this situation a calibration of the signal is all that is required to retrieve the signal or "invert" the data.

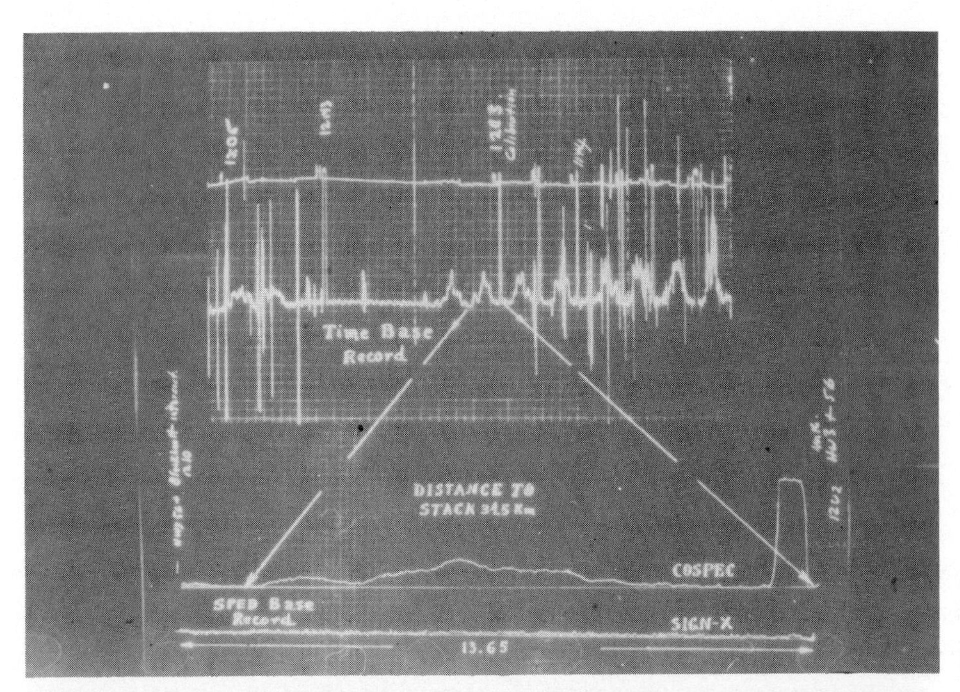

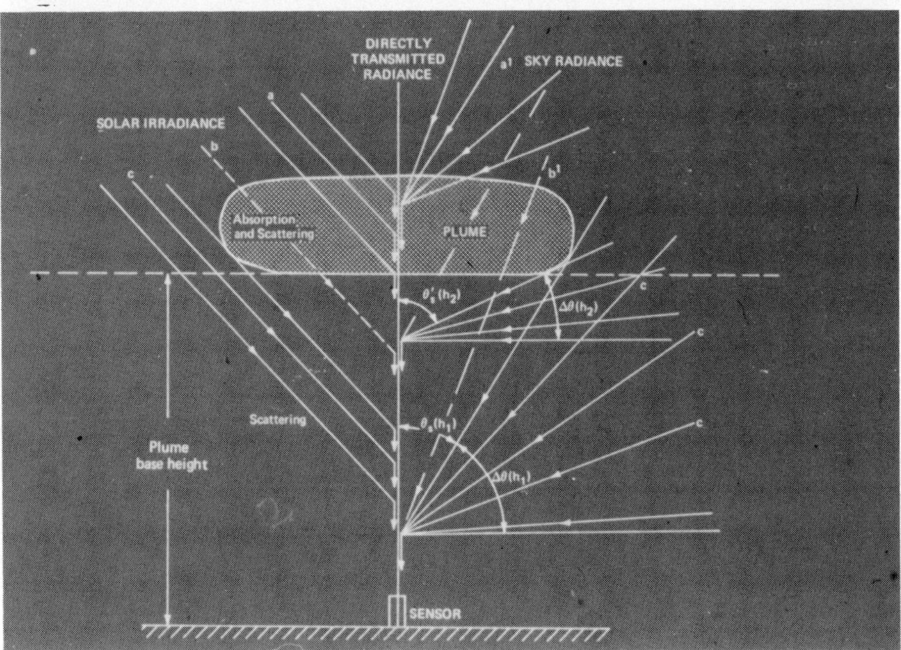

Fig. 8.- Baseline drift correction for plume measurements at large distances from the source. The simultaneous recording of a slow time-based signal can be used to establish when the edges of the plume have been reached, as shown.

Fig. 9.- Geometry for calculating atmospheric scattering effects under a plume (ref. 22).

- a directly transmitted component from the FOV over the zenith sky. This is the ideal source;
- scattered components proceeding from the whole of the sky, considered now as a primary source, and scattered within the acceptance solid angle of the sensor, i. e. sky-scattered components;
- components proceeding from scattering of the direct solar component, i. e. solar-scattered components.

These components can be further subclassified as: primary radiations components, scattered within the target gas layer and acquiring information (absorption) about it, and side radiation, scattered between the target gas and the sensor and with no target gas information.

The key to understand the effects of these components is in Eqs. (14), (15) and (17). Although the first two were developed to emphasize the contributions from each slit pair, its generalization for use with each radiation component can easily be made. Initially the Multistable (COSPEC) response equation (17), can be written as:

$$V = c_1 \sum_{\ell} \left\{ (1 - \frac{P_{2\ell}}{P_{1\ell}}) \frac{P_{1\ell}}{\pi_1} - k(1 - \frac{P_{4\ell}}{P_{3\ell}}) \frac{P_{3\ell}}{\pi_3} \right\} \tag{18}$$

where

$P_{j\ell}$ is the radiant power sampled by mask-j- from the radiation component-ℓ-

π_1, π_3 are the total radiant powers sampled from all radiation components by mask 1 and 3, respectively.

In turn, each term between parenthesis can be expressed as in Eqs. (14) or (15). It must be noted that the "component" weighting factors $P_{1\ell}/\pi_1$ and $P_{3\ell}/\pi_3$ now affect all terms inside the parenthesis including the individual slit weighting factors. It is not really necessary to expand these equations any further here to note that the overall response is dominated by the component that contributes the largest amount of radiant power and not necessarily by that with the largest intrinsic signal. This fact is tremendously important for several reasons:

- Multiple scattered components collect a large cL value and acquire a large intrinsic signal, however, their weighting, also decreases as the factors $p_{j\ell i} \exp[-a_{ji}(cL)_{\ell}]$, $j = 1, 3$ are attenuated accordingly (Eq. 15);
- The available radiant power also diminishes with increased scattering orders.

The overall result is that the resulting signal is dominated by those components which are attenuated the least and this means, in turn, that from the same original component, first order scattering is very heavily favoured against second order, and so on. It also means that the favoured components are those with the shortest path through the target gas layer.

The importance of this resides in the fact that for upward looking below a plume, the effective contributors tend to be those which acquire an intrinsic signal closest to the "ideal", i. e. a direct vertical path through the layer. This being clearly one of those few cases when mother nature is somewhat helpful.

Several calculations using first and second order scattering models (refs. 1, 22) show that an increased signal of the order of 10 to 15% can be expected under most conditions.

So far, this discussion is applicable to target gas layers much larger than the height sensor-to-cloud. Under normal conditions other effects can take place, viz:

- Plume edge effects, particularly with narrow and elevated plumes, where dilution of the signal can take place due to the scattering of side radiation between the plume and the sensor (ref. 22);
- Calibration changes. These can become very important with intermittently condensed plumes. The effects arise because the calibration off to the side of the plume uses spectral components different from those under the plume. It is a sky-to-cloud change that can be compensated in the baseline but which leaves the calibration effects. It is important to indicate that in a condensed plume its upper part can act as a good diffuser of the direct solar component and increase the weighting of the short wavelength bands in the masks and, correspondingly, the target gas signal;
- Signal pre-loading. Can occur when a layer of target gas lies above low level fog. In this situation, the direct solar component acquires a $\underline{cL\sec\theta}$ signal (θ being the sun zenith angle) before being diffused down within the fog layer. The radiation that reaches the sensor has already in it the $\underline{\sec\theta}$ increase, which can lead to anomalously high observed values (ref. 23).

Finally, it should be emphasized that this is an open research area and that there are many problems to be solved. Knowledge of the instrumental behaviour is the key to a comprehensive effort in modelling and experimental work still required to advance this technique.

REFERENCES

1 A. van der Meulen and D. Onderdelinden. Optical pathlength of zenith skylight in passive remote sensing of air pollution. Atmospheric Environment, 17, 417-428 (1983).
2 S. S. Butcher and R. J. Charlson. An Introduction to Air Chemistry, Academic Press (1972).
3 Z. Sekcra and J. V. Dave, Diffuse transmission of solar ultraviolet radiation in the presence of ozone. Astrophys. Journal, 133, 210-227 (1961).
4 C. L. Mateer. A study of the information content of Umkehr observations. Tech. Rep. No. 2, Dept. of Meteorology, Univ. of Michigan, USA (1964).
5 C. L. Mateer. On the information content of Umkehr observations. J. of Atmospheric Sciences, 22, 370-381 (1965).
6 A. E. S. Green (Editor). The Middle Ultraviolet - its Science and Technology, John Wiley and Sons Inc., N. Y. (1966).
7 K. Bullrich. Scattered radiation in the atmosphere and the natural aerosol. Advances in Geophysics, Vol. 10, Academic Press, N. Y. (1964).
8 K. YA. Kondratyev. Radiation in the Atmosphere. Academic Press, N. Y. (1969).
9 S. T. Henderson. Daylight. American Elsevier Publishing Co., N. Y. (1970).
10 M. M. Millan and R. M. Hoff, Remote sensing of air pollutants by correlation spectroscopy - Instrumental response characteristics. Atmospheric. Environment, 12, 853-864 (1978).
11 G. S. Newcomb and M. M. Millan. Theory, applications and results of the long-line correlation spectrometer, IEEE Transactions on Geoscience Electronics, G. E. -8, 149-157 (1970).
12 J. M. Schonkeren. Photomultipliers. Publications Department, N. V. Philips Gloeilampenfabrieken, Eindhoven, the Netherlands (1970).
13 Electron Tubes (Photomultiplier tubes) Data Handbook N. V. Philips Eindhoven, the Netherlands (1976).
14 M. M. Millan. A study of the operational characteristics and optimization procedures of dispersive correlation spectrometers for the detection of trace gases in the atmosphere. Ph. D. Thesis, Univ. of Toronto, Toronto, Ontario, Canada (1972).
15 M. M. Millan, S. J. Townsend and J. H. Davies. A study of the Barringer refractor plate correlation spectrometer as a remote sensing instrument. M. A. Sc. Thesis, UTIAS Report 146, Univ. of Toronto, Ont. Canada (1970).
16 M. M. Millan and R. M. Hoff. The COSPEC remote sensor, II Electronic set-up procedures. Rpt. ARQT-6-76. Environment Canada, Atmospheric Environment Service, Ontario, Canada (1976).
17 M. M. Millan and R. M. Hoff. How to minimize the baseline drift in a COSPEC remote sensor. Atmospheric Environment, II, 857-860 (1977).
18 M. M. Millan and R. M. Hoff. Dispersive correlation spectroscopy: A study of mask optimization procedures. Applied Optics, 16, 1609-1618 (1977).
19 Barringer Research Ltd.. COSPEC IV and V Users' Manual. Toronto, Ontario, Canada.
20 M. M. Millan, A. J. Gallant and J. Markes. Nanticoke environmental study: COSPEC/SIGN-X project . Report ARQT-6-78, Atmospheric Environment Service, Downsview, Ontario, Canada (1978).
21 R. M. Hoff, N. B. A. Trivett, M. M. Millan, P. Fellin, K. G. Anlauf and H. A. Wiebe. The Nanticoke shoreline diffusion experiment June 1978-III. Ground-based air quality measurements. Atmospheric Environment, 16, 439-454 (1982).
22 M. M. Millan. Remote sensing of air pollutants - a study of some atmospheric scattering effects. Atmospheric Environment, 14, 1241-1253 (1980).

23 M. M. Millan, A. J. Gallant, J. S. Chung and F. Fanaki. Analysis of the Mount St. Helens SO_2 cloud observation over southern Ontario, Atmospheric Environment (in press) (1984).

Optical Remote Sensing of Air Pollution,
Lectures of a course held at the Joint Research Centre, Ispra, Italy, 12—15 April 1983,
P. Camagni and S. Sandroni (Eds). 95—122

SOME APPLICATIONS OF OPTICAL REMOTE SENSING TO METEOROLOGY

R. E. W. PETTIFER B.Sc., PH.D., F.R.Met.S

1. INTRODUCTION

The fundamental basis of experimental study is the making of measurements and one of the inherent difficulties associated with measurement is the degree to which the act of making the measurement influences the state of the measured quantity. This is a particular problem in atmospheric physics since almost all conventional techniques for sampling the state of the atmosphere rely upon immersion methods: that is, the instrument used to make the measurement is immersed in the atmosphere at the point at which the measurement is required. The effect of this upon the result of the measurement can be large and is often only poorly known. It is a major attraction, therefore, of remote optical sensing that the influence of the interaction between the measurement device (an optical beam or, perhaps locally emitted optical energy) and the atmosphere being measured is on a scale that is normally orders of magnitude below that of the interest which required the measurement to be made.

This has led to attempts to apply optical remote sensing methods to most of the common measurements of meteorology both on the surface (i.e., within the boundary layer) and at heights up to 40 km. Most of these methods rely upon the effects of scattering (both elastic and inelastic), absorption or more rarely, the detection of naturally emitted radiation.

In this discussion, mainly for reasons of space, we shall concentrate upon active optical techniques which have been applied to the measurement of cloud characteristics, visibility, precipitation, humidity, wind and temperature. The basic theory of both lidar techniques and the various scattering processes will be assumed; they are thoroughly discussed in the standard literature (ref. 1, ref. 2, ref. 3) and elsewhere within these lectures.

2. MEASUREMENTS OF CLOUDS

2.1 Cloud height

The simplest application of optical remote sensing to measurement of the atmosphere is probably the use of lidar to determine the height of cloud. The basic principle of this technique is shown in Fig. 1(a). A laser pulse, emitted at zero time, is scattered back by the cloud and received at a time Δt so that, for the simple case of a specularly reflecting target, the range R, (i.e., the cloud height) is simply $\frac{1}{2}\, c\, \Delta t$ where c is the velocity of light.

Optically, clouds provide a sufficiently dense target, even when they are visually very tenuous, to allow easy detection of the energy back-scattered from a powerful visible pulse. The earliest system of this

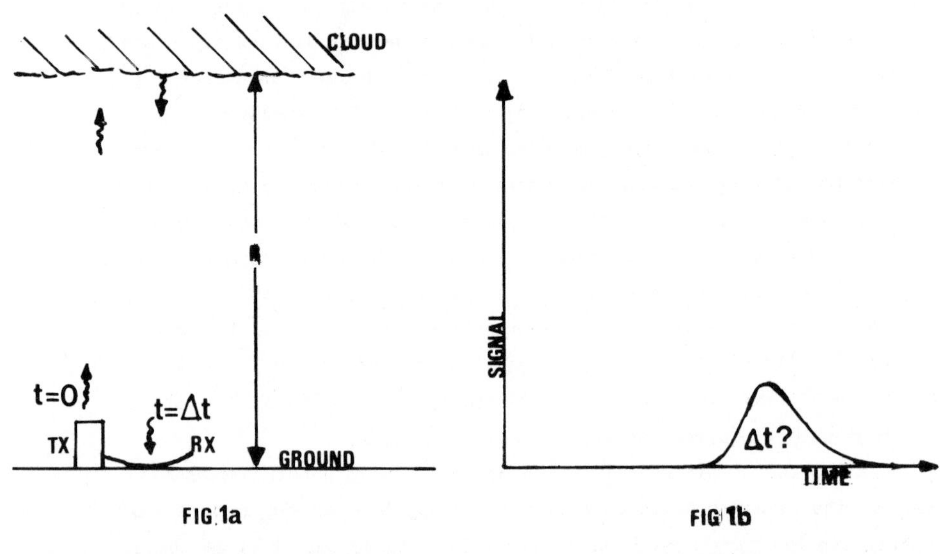

FIG 1a

THE LASER RADAR CLOUD BASE DETECTOR

FIG 1b

IDEALIZED SIGNAL FROM CLOUD

type (ref. 4) operated in an analogue detection mode with a receiver aperture a few inches across and a Q-switched laser output of 400 mJ at a rate of 2 pulses per minute. The device was capable, in daylight, of observing clouds from about 100 ft up to 10,000 ft with a range uncertainty of about ± 20 ft. Unfortunately the instrument was not considered eye-safe and since its principal operational rôle would have been on airfields it could not be accepted for operational use. Present devices are based upon low power, infra-red junction diode lasers which are eye-safe at reasonable viewing distances and which rely upon a very high pulse rate, low pulse energy and integration over many pulses to yield cloud height data.

There are, however, still some problems associated with this method. Among the most troublesome are the definition of "cloud base" and the spurious returns which arise from precipitation and fog. The laser pulse will, in general, penetrate into the cloud so that the back-scatter signal builds up and then decays (Fig. 1(b)). It is tempting to assign the term "cloud base" to the level corresponding to the maximum in the return signal but unfortunately this may sometimes be sufficiently far into the cloud that it does not accord with a conventional, visual assessment of the cloud base. For example a pilot flying at this level may not, in his terms, be "out of the cloud". The problem of spurious return from fog or precipitation is also very difficult to handle, especially if, as is increasingly the case, the equipment is required to operate automatically on unmanned sites so that the data user has no prior information about the existence of fog or precipitation. In spite of these problems, laser cloud base detection is now becoming a widely-used operational tool and it is so far the only laser sounding technique to attain this status in meteorology.

2.2 Cloud structure and properties

Laser radar methods have also been used to study the structure and properties of high-level ice crystal clouds. The technique, pioneered by C. M. Platt (ref. 5) is normally combined with ground-based measurements of the infra-red radiative properties of the clouds. The measurements are important because these clouds can be a significant contributory factor in the radiation schemes adopted in numerical models of atmospheric behaviour.

A typical back-scattered return from cirrus cloud is shown in Fig. 2. To obtain the true back-scattered signal, this return must be corrected for the two-way attenuation through the cloud but given this correction Platt has shown that the visible extinction optical depth of the cloud, $\bar{\delta}(h)$, can be found from

$$\bar{\delta}(h) = \frac{1}{k}\int_{Z_0}^{Z_T} NB(\lambda)dZ$$

where k is the back-scatter to extinction ratio, assumed invariant with height, and Z_0 and Z_T are the heights of the cloud base and cloud top.

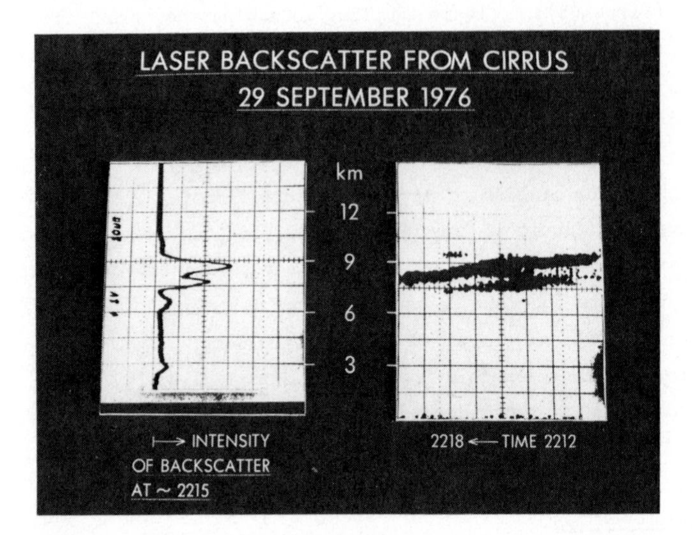

FIG 2

We may write the laser radar equation in the form:

$$C(\lambda,Z) = \frac{P_o(\lambda)T^2(\lambda)AQ(\lambda)\delta Z \quad NB(\lambda)}{Z^2} \tag{1}$$

where $C(\lambda Z)$ is the signal (number of photons) detected at the laser wavelength, λ, from a height interval, δZ, at height Z, $P_o(\lambda)$ is the number of photons per laser output pulse, $T(\lambda)$ is the one-way transmission through the atmospheric path at wavelength λ, A is the receiver collection area $Q(\lambda)$ is an instrument calibration factor, N is the number density of scatterers and $B(\lambda)$ the appropriate elastic back-scattering cross-section.

From this we have

$$T^2(\bar{\lambda})NB(\bar{\lambda}) = \frac{C(\bar{\lambda},Z)Z^2}{P_o(\bar{\lambda})AQ(\bar{\lambda}) \ Z} \tag{2}$$

For this work, it cannot be assumed that $T^2(\bar{\lambda})$ is a known constant because of the significant pulse attenuation through the cloud and Platt (ref. 5) has developed a technique for estimating this effect.

A knowledge of the visible extinction optical depth, $\bar{\delta}(h)$, and the infra-red emissivity allows a reasonable numerical model of the radiative effects of the cloud to be built up and so far, ground-based lidar is the only economic way of obtaining a large body of values of $\bar{\delta}(h)$. It should be noted, however, that the accuracy obtained in $\bar{\delta}(h)$ is dependent upon the accuracy with which the laser radar can be "calibrated" (i.e., the error in $Q(\lambda)$) and this depends upon the validity of the assumptions made about the absence of aerosol at the chosen normalization level. Pettifer et al have pointed out (ref. 6) that these assumptions need to be treated with some circumspection.

3. MEASUREMENT OF VISIBILITY

At first sight, one might suppose that of all the meteorological measurements for which one could use laser sources, that of visibility would be the most obvious candidate. After all, the manual method of measuring visibility relies on the existence of a powerful light source – the sun – how readily then would this measurement lend itself to the use of lasers? In fact, the measurement of horizontal visibility benefits hardly at all from laser properties. Indeed the small spatial size of the source makes integration over a large field of view difficult and the coherence of the source gives rise to effects such as speckle and scintillation which complicate the whole problem of measuring the extinction of a beam directly. It is difficult though by no means impossible to use conventional incoherent light sources to make visibility measurements but it is more difficult to use lasers in the same way.

However, there is one case in which a conventional, two-ended extinction measurement of visibility is simply not possible and that is the assessment of the slant visual range on an aircraft glide-path. The problem is illustrated in Fig. 3. The single-ended, pulsed lidar system has been used to attempt to relate the slant visual range to the back-scattered intensity.

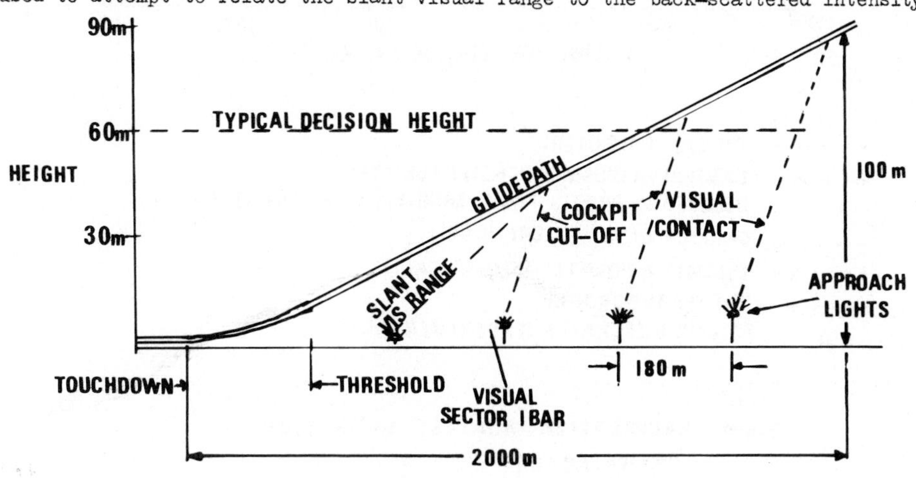

FIG 3 THE SLANT VISUAL RANGE PROBLEM

To do this one must establish a relationship between the back-scattering and the extinction of the beam as it traverses the slant path.

In water fog, it can reasonably be supposed that at visible wavelengths the extinction is almost entirely due to scattering and we may write an

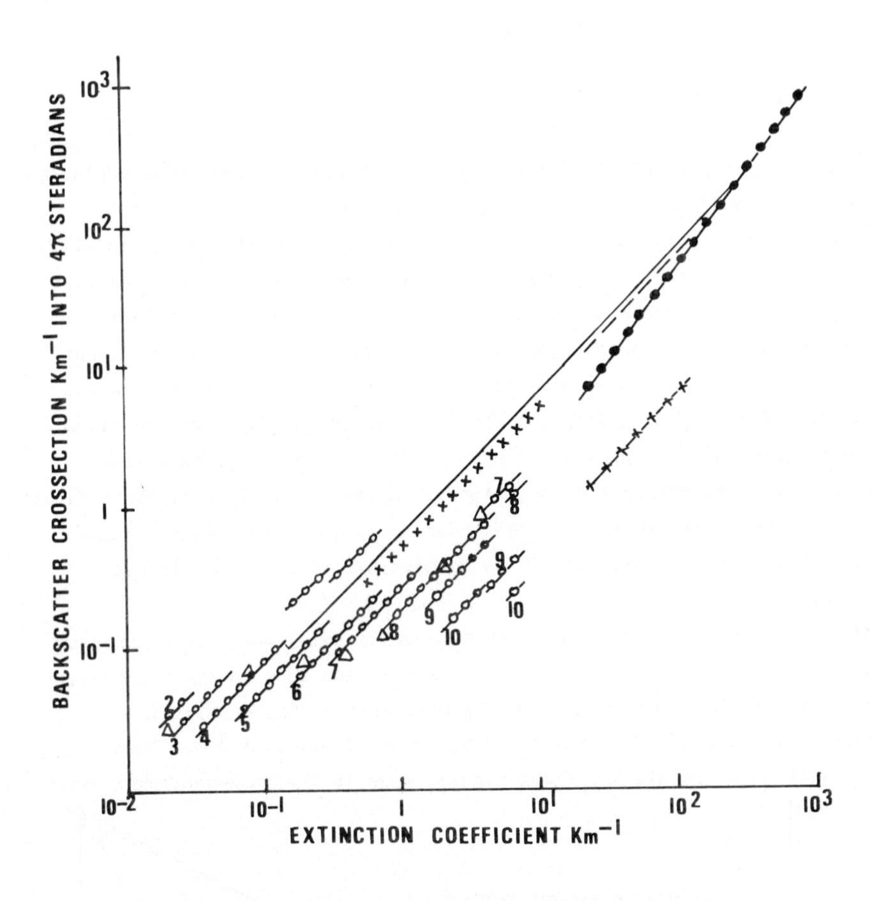

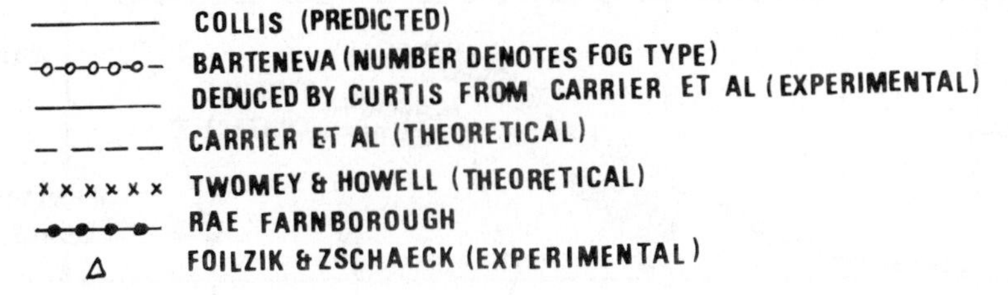

FIG 4 BACKSCATTER AGAINST EXTINCTION FOR WATER FOG

empirical relationship of the form

$$B(\lambda) = K\,\sigma(\lambda)^n$$

where $B(\lambda)$ is the back-scatter cross-section, $\sigma(\lambda)$ the extinction cross-section and K and n are to be determined. A great deal of work has gone into trying to establish agreed values for K and n and some typical results can be seen in Fig. 4. It is clear from this set of data that there is considerable variability in these results. This variability arises principally because of the wide variations of droplet size distribution both within one fog and from fog to fog. These distributions are governed by a number of interacting factors such as the moisture content of the air, the type of nucleii available for condensation, the amount of turbulence within the fog and so on and are essentially unpredictable in detail. As the result mainly of this degeneracy in the relationship between $B(\lambda)$ and σ this method for determining slant visual range has so far proved unsuccessful.

Although not principally aimed at measuring visibility per se, the work of Collis et al aimed at mapping aerosol distributions in the boundary layer has obvious connections with it. This group have used elastically back-scattered light from ruby laser pulses to provide signatures from aerosol layers trapped beneath the low level inversion in stable conditions. This technique has been applied to such problems as the dispersal of gaseous and particulate effluent from chimneys, the dispersal of clouds of insecticide and the dispersal of motor vehicle exhaust products. An introductory review of these techniques has been given by Collis and Russell (ref. 7).

4. THE MEASUREMENT OF WIND

Velocimetry has become one of the widest application areas for laser-based remote sensing devices in the past ten years or so. Among the many techniques which have been developed, Doppler anemometry (ref. 8) is probably the one of greatest interest to us although we will look at some other, less widely-used, methods as well.

The principles behind the laser Doppler anemometer are quite different from those used for other atmospheric optical measurements. Back-scattered radiation is, as usual, the information carrier but the information is extracted from the Doppler shift in the back-scattered frequency which arises from the radial component of motion of the aerosol in the atmosphere. On the (reasonable) assumption that this aerosol moves freely with the wind, the radial component of the wind velocity may be deduced.

This radial component, V_R, is given from the Doppler frequency shift, $\delta\nu$, by $\delta\nu = \frac{2V_R}{\lambda}$

where λ is the transmitted wavelength. For a transmitted wavelength of 500 nm, a 1 $msec^{-1}$ radial wind component will yield a shift of 4 MHz which is easily detectable. For reasons of signal to noise, the power needed in the output of a visible laser for this work makes their use unacceptable for atmospheric use on safety grounds alone. For this reason and for reasons of frequency stability it is customary to use as the laser source CO_2 or CO lasers operating in the infra-red. At an output wavelength of 10.6 μm, the Doppler shift for a 1 $msec^{-1}$ radial component reduces to about 200 KHz and to detect such shifts reliably it is usually necessary to use heterodyne techniques (ref. 9).

Both pulsed and CW lasers have been used for this work. In the CW case, range information for a single-ended system can be obtained from the FM-CW method familiar in microwave radar in which the transmitted frequency is modulated as a function of time so that a particular transmitted frequency is uniquely associated with a time of transmission. The returned frequency is mixed with a stable local oscillator and the resulting difference frequency is a measure of range. The Doppler shift which arises from the moving target is of constant sign and will alternately add to and subtract from the range-proportional beat frequency.

To obtain the transverse wind component from Doppler systems it is usual either to employ two lasers (and detectors) illuminating a common volume orthogonally, or to traverse a single transmitted beam so that it illuminates a second atmospheric volume at the same height as the first but in an orthogonal direction. This latter arrangement assumes a measure of horizontal homogeneity in the atmospheric wind over distances which increase with the range of the illuminated volume; an assumption which needs to be viewed critically for any particular experiment.

A weakness of the normal Doppler technique is that the transmitted and scattered beams will be subjected to different and variable atmospheric path effects. For example, varying refractive index along the path will cause differential distortion between the beams (ref. 10). A new development of the Doppler method which will overcome these problems has been reported by Eberhard and Schotland (ref. 11). They transmitted two beams either coaxially or at a small angle of intersection ($< 10^{-5}$rad) and measured the differential Doppler shift between them, $\Delta\nu = \delta\nu_1 - \delta\nu_2$, where ν_1 and ν_2 are the transmitted frequencies and $\delta\nu_1$, $\delta\nu_2$ the associated Doppler shifts. From this signal a component of the wind velocity in the scattering region may be obtained. The direction of the component depends upon whether the two beams are exactly superimposed or intersect at a small angle.

For the case of scattering from a single particle from two exactly superimposed beams it can be shown that

$$\Delta\nu = -2K \sin(\theta/2) V_L \qquad (3)$$

where θ is the scattering angle, V_L is the component of the wind velocity parallel to the bisector of the transmitted and scattered beams and

$K = \frac{\lambda_f}{2\pi}$ where λ_f is the spacing of sinusoidal interferences fringes generated by the temporal interference of the transmitted beams. Notice that $\Delta\nu$ is independent of the transmitted frequencies and depends only on the difference between them.

The choice of transmitted frequencies such that their difference $\nu_1 - \nu_2$ lies in the microwave region, allows the resulting signal to be processed by the standard techniques of microwave radar as if the transmitted frequency were $(\nu_1 - \nu_2)$ and this were shifted to $(\nu_1 - \nu_2) + \Delta\nu$ before reception.

It is necessary to generalize the relation, (3), to account for the finite beam shapes, scattering from many particles within the scattering volume and for their random motions. These effects lead to a spectral broadening which requires statistical treatment to extract the wind component. The techniques for this are tedious but relatively well-established and can be found discussed in Eberhart and Schotland's original paper (ref. 11). The efficacy of this method for detecting winds over short ranges ($\sim$20 m) is shown in Fig. 5 in which the results from the dual frequency Doppler lidar are compared with those from a propeller anemometer positioned $\sim$20 cm from the scattering volume. The agreement shown in this figure is striking and is probably among the best obtained so far between remote sensing and direct sampling of wind.

There are several other, less well-established, optical remote sensing methods for measuring wind components. Sroga and Eloranta (ref. 12) have measured a time series of back-scattered profiles from three closely-spaced azimuth angles scanned rapidly in succession. By using a correlation technique the time taken for recognizable aerosol inhomogeneties to drift from one azimuth to the next can be found and this is related to the wind component transverse to the lidar beam. This technique is a variation of an earlier reported method in which the beam from a low power CW laser such as a He Ne, is expanded and transmitted into the far field through a Fabry-Perot interferometer. Correlation of the scintillations of the resulting fringe pattern can then be used to derive the traverse wind component.

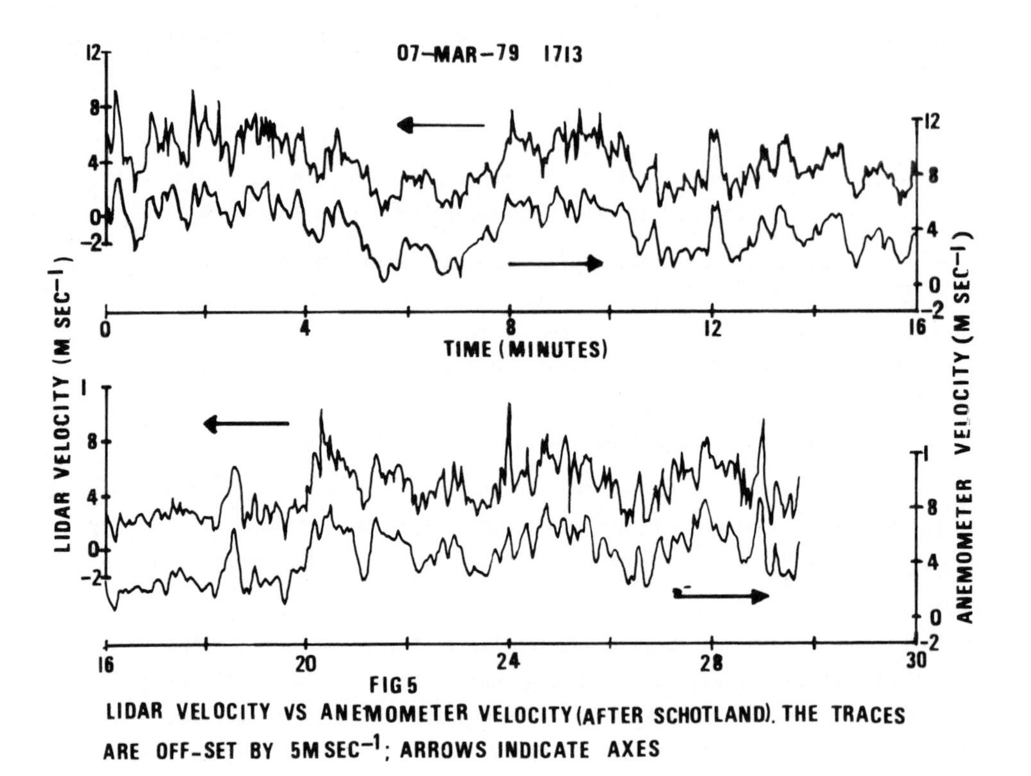

FIG 5
LIDAR VELOCITY VS ANEMOMETER VELOCITY (AFTER SCHOTLAND). THE TRACES ARE OFF-SET BY 5M SEC^{-1}; ARROWS INDICATE AXES

Path averaged winds can be obtained over distances of tens of metres from a technique reported by Lawrence (ref. 13) in which variations in beam irradiance due to refractive index variations along the path can be correlated at the output of a pair of closely-spaced parallel slits. The time lag between the correlated output fluctuations from detectors placed behind the slits can be used to derive the wind averaged over the transmission path. This is an earlier version of the method discussed below for detecting precipitation.

An alternative correlation method has been reported by Zuev et al (ref. 14). Two lasers and two detectors are used in the arrangement shown in Fig. 6 and the signals in the detectors are correlated. The time lag between well correlated features of the intensity of the forward scattered or transmitted light is determined and on the assumption that the aerosol density distribution in space retains its shape over the short time intervals needed to advect the aerosols across the two optical paths, a knowledge of the separation of the scattering volumes allows the transverse wind speed to be determined. For this work, the much greater efficiency of forward scattering or transmission compared to back-scattering allows low power, visible lasers such as He Ne to be used, with significant advantages in both cost and safety.

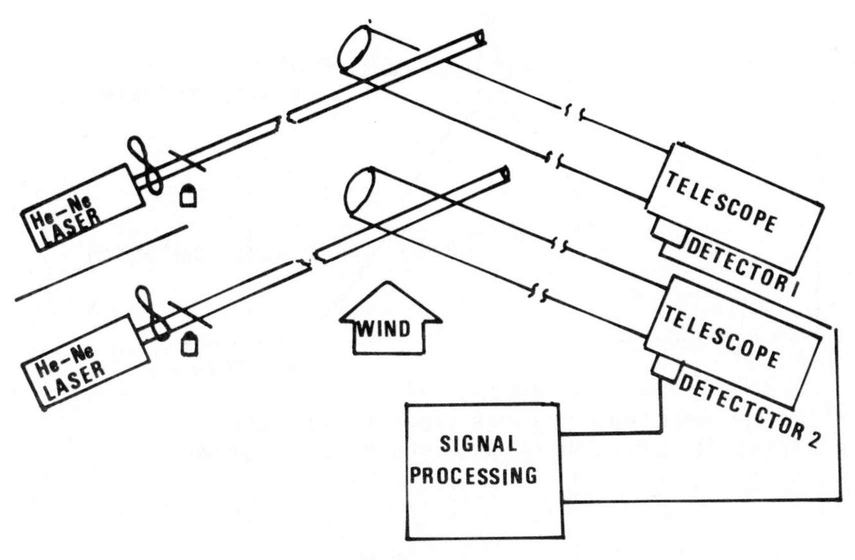

FIG 6
BLOCK DIAGRAM OF FORWARD SCATTER-CORRELATION METHOD DUE TO ZUEV

A most difficult experimental spectroscopic technique developed by Benedetti Michelangeli et al (ref. 15) involves the use of a confocal spherical Fabry-Perot interferometer to make a direct measurement of the spectral distribution of the back-scattered laser light. The emitted frequency must be highly stable ($\pm$50 MHz). The spectral distribution in the returned signal is shown schematically in Fig. 7. λ_0 is the transmitted wavelength; the broad band return arises from molecular (thermal) broadening and, at a temperature of 25°C and for visible radiation, will be about 0.02Å, while the much more massive aerosols produce the narrow band signal superimposed upon it with a width of about 0.005Å For a radial wind of 10 m sec^{-1} the Doppler shift at the Ar^+ wavelength will be about 7.1×10^{2}Å Although a most difficult method, results which claim an accuracy of 0.27 m sec^{-1} at a range of 500 m have been published (ref. 15).

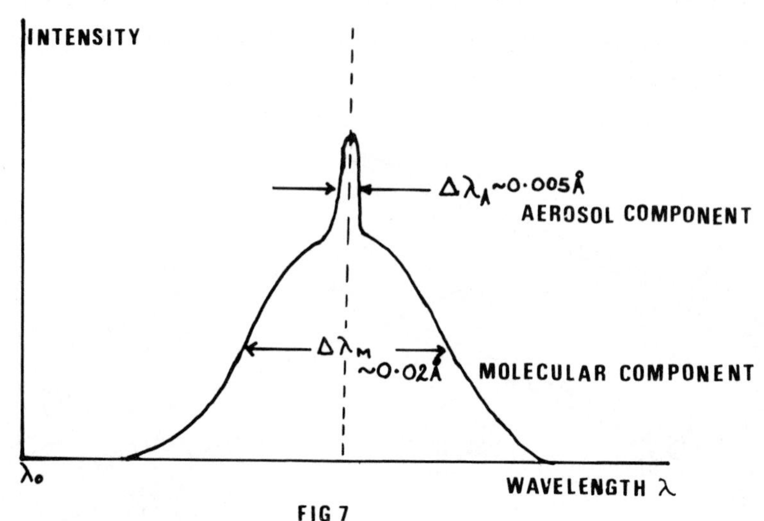

FIG 7
DOPPLER SPECTRUM OF LASER LIGHT ELASTICALLY BACKSCATTERED FROM THE ATMOSPHERE; $\lambda_0 \sim 4600$ Å

5. THE MEASUREMENT OF HUMIDITY

Of all the physical measurements of interest to atmospheric science, that of the water vapour content of the air is among the most difficult. We are concerned to measure molecular water vapour concentrations in the parts per million range with an accuracy of the order of a few per cent, a task which most experimentalists would regard as formidable. Not surprisingly, devices for making these measurements successfully are not plentiful and most of the usual ones, such as wet and dry bulb thermometers, hair hygrometers and chemical hygristors suffer either from their essentially manual operation, or from the need for extensive and clumsy processing for their output, or from fundamental instability, or indeed from a combination of any or all of these. Therefore a device which could make these measurements reliably and perhaps automatically is much to be desired.

An additional difficulty in the measurement of humidity is that the immersion instruments used, if contaminated with water, can seriously affect the measurement they make and routine exposure of such instruments in conditions of fog, cloud, rain or other precipitation can, unless very careful precautions are taken, lead to just this problem.

When considering how optical remote sensing can be applied to these measurements, it is important to remember that in the case of humidity, we are really only concerned with a relative measurement. We wish to know the ratio of the partial pressures of water vapour and dry air.

This potentially mitigates the calibration problems which are a source of difficulty with many lidar measurements.

Two principal techniques have been used to measure relative humidity by remote optical methods. The most common is the same as that used for temperature (see Section 7) and is simply the use of the vibration/rotation Raman effect to make relative measurements of the number density of nitrogen and water vapour molecules. We may write the laser radar equation for the case of a Raman-shifted signal wavelength in the form

$$C(\lambda',Z) = \frac{P_o(\lambda)T(\lambda)T(\lambda')AQ(\lambda')\,\delta Z}{Z^2}\;N\;\frac{d\sigma(\lambda)}{d\Omega} \qquad (4)$$

where now λ is the incident (laser) wavelength, λ' is the scattered (Raman-shifted) wavelength, N is the number density of the scattering species and $\frac{d\sigma(\lambda)}{d\Omega}$ is the differential Raman cross-section for the species at wavelength, λ. Notice that the transmission terms are now different and that we can deal with only one scattering species - there is no aerosol term. By writing (4) for a second scattering species and dividing the two equations we obtain

$$\frac{C(\lambda'Z)}{C(\lambda''Z)} \propto \frac{T(\lambda')}{T(\lambda'')}\;\frac{N_1(Z)}{N_2(Z)}\;\frac{d\sigma_1(\lambda)}{d\Omega}\Big/\frac{d\sigma_2(\lambda)}{d\Omega} \qquad (5)$$

where λ'' is the Raman-shifted wavelength for specis 2.
In writing this equation we have assumed that $\frac{Q(\lambda')}{Q(\lambda'')}$ is a constant and that $P_o(\lambda)$ is a constant for all firings of the laser which are integrated to obtain the total number of counts used in the analysis. It is necessary to examine these two assumptions critically.

A method of ensuring the first condition is met is to establish two receiver channels, one tuned to λ', the other to λ'' and ratio their outputs for each laser pulse. The problem is that this reduces the signal to noise available in the detector which will be an important limitation on the vertical range of the device. Although only the ratio $Q(\lambda')/Q(\lambda'')$ is now required, and a knowledge of the ratio of the cross-sections, it is necessary, unless $Q(\lambda)$ can be directly measured, to measure at least the wavelength dependence of $Q(\lambda)$. Indeed, if two separate receiver channels are used, a direct experimental determination of the Q ratio may still be required so that the problems of this method are by no means trivial.

The output energy of even the best regulated lasers is seldom constant from shot to shot to within 10% and it is often much worse than this for the large, powerful lasers. Such a variation makes the experimental measurement of $P_0(\lambda)$ necessary for every shot - a difficult measurement to make with certainty.

In addition to these instrumental problems, the constancy of $T(\lambda)$ with time is not often a marked feature of the lower atmosphere and there is little that can be done about this except to use a coaxial system so that the incident and scattered light travel together a common path.

In spite of all these difficulties, some results of this technique have been published (ref. 16) and Fig. 8 shows the water vapour mixing ratio obtained in 30 m height increments over the first 1500 m of the atmosphere compared to a concurrent profile obtained from a conventional radiosonde ascent.

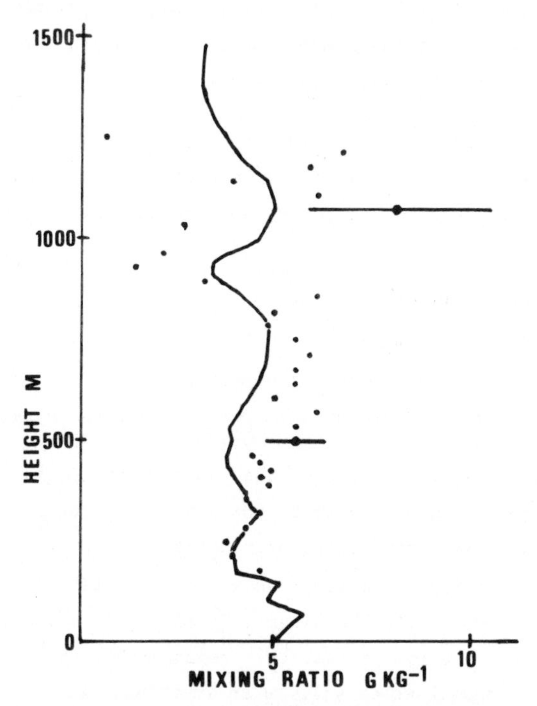

FIG 8
WATER VAPOUR PROFILES FROM POURNY ET AL
LIDAR • VS RADIO SONDE ———

Although the agreement shown is quite good below 500 m, the large uncertainty of the data at higher levels makes them of questionable value in atmospheric physics. Because of the rapidly diminishing mixing ratio with height and the R^{-2} effect in the transmitted beam, very large increases in system efficiency or transmitted power (or both) are needed to improve this type of result substantially.

6. THE MEASUREMENT OF PRECIPITATION

It has long been an aspiration of meteorological instrumentalists to devise a remote, automatic method for the measurement of the occurrence and nature of precipitation and this aim was an early application of optical remote sensing experiments.

The first work attempted was really no more than the use of laser sources in the now well-established microwave radar method for measuring rainfall (or other precipitation). The technique is aimed at obtaining a measure of the total liquid water (or water equivalent) which has been deposited over an area during the observed precipitation event. The use of microwave radar to make these measurements is now such a well-established procedure that there seems little point in prosecuting the use of lasers as alternative sources, particularly since in general they are only capable of considerably smaller ranges than is radar. However, the small size and potential cheapness of laser sources relative to microwave radars have tempted some to try the method.

The rate of rainfall, R, is related to the rainfall optical extinction coefficient, β_r, by an empirical relationship of the form

$$\beta_r = GR^{\gamma} \tag{6}$$

β_r is then related to the back-scatter cross-section via the back-scatter phase function $P_r(\pi)$

$$P_r(\pi) = \frac{C}{\beta_r}\int_0^{\infty} N_r(\alpha)\, i\pi(m,\alpha)\, d\alpha \tag{7}$$

where $\alpha = \frac{2\pi x}{\lambda}$ and x is the raindrop diameter. $i\pi(m,\alpha)$ is an intensity function dependent on α and the refractive index, m. This formulation of the back-scatter theory can be found in Kerker's book (ref. 2).

Some results of this type of measurement have been published (ref. 17) and Fig. 9 is an example of them.

A more interesting approach than this is that adopted by Wang et al. They have attempted to interpret forward scattered laser light from low power helium neon or CO_2 lasers so as to identify the type of precipitation particle in the laser beam and to determine its size and fall speed. The method is based upon the different diffraction patterns created by forward scatter from different types of precipitation particle and the correlation between the signals one gets from two parallel horizontal slit-shaped detectors which view the diffraction pattern sequentially as the precipitation particle falls past them.

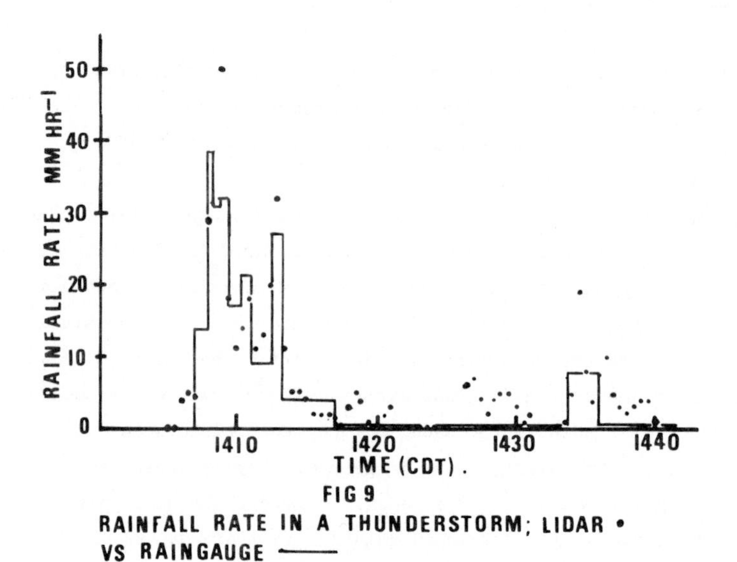

FIG 9

RAINFALL RATE IN A THUNDERSTORM; LIDAR •
VS RAINGAUGE ——

The typical amplitude-scintillation pattern caused by the refraction of a plane parallel wave at a spherical drop is given by:

$$I(xyz) = -Io \, \frac{a}{\rho} \, J_1\left(\frac{2\pi a \rho}{\lambda x}\right) \sin\left(\frac{\pi \rho^2}{\lambda x}\right) \tag{8}$$

Where $\rho = \sqrt{y^2 + z^2}$, a is the drop radius, λ the wavelength of the incident light, I_o the intensity of the incident light and $J_1(\ )$ is the first order Bessel function. At large distances from the drop, this pattern is a series of alternate bright and dark rings which will sweep downwards past the two parallel slit detectors at the fall velocity of the drop given by

$$V = 200 \sqrt{x\,a} \tag{9}$$

The time varying signal from a 2 mm drop at a distance of 50 m is shown in Fig. 10.

When a large number of drops fall through the laser beam then the individual signals add in a random way. If all the drops were the same size, the signals in the two channels would be identical in shape but time delayed by the traverse time of the drops across the detector slits. There is of course a size distribution in raindrops which results in the signals from the two channels having no obvious visual similarity. Typical signals from rain are shown in Fig. 11. The time-delayed correlation of these signals will yield the average time delay between them from which, with a knowledge of the slit separation, the average fall rate may be determined.

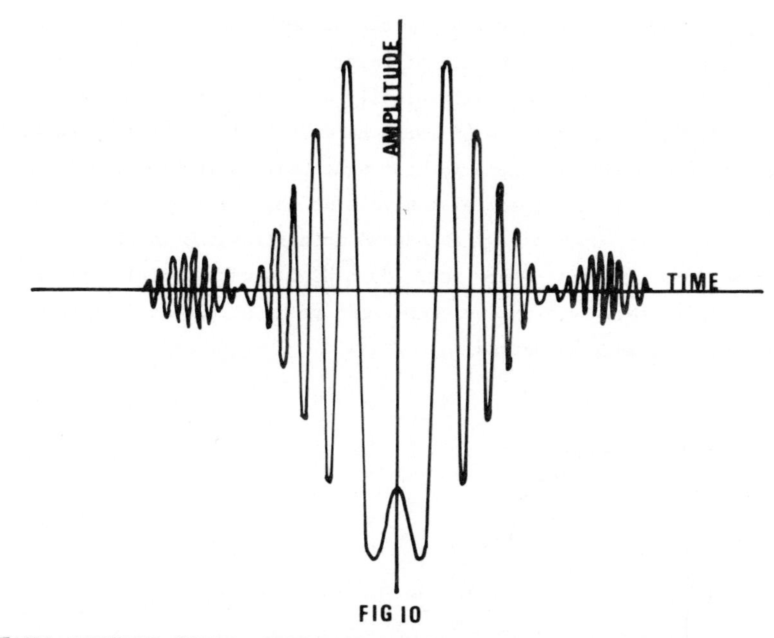

FIG 10

TIME VARYING SIGNAL DETECTED FROM A SINGLE 2MM DROP 50M AWAY FROM A THIN LINE DETECTOR (AFTER EARNSHAW)

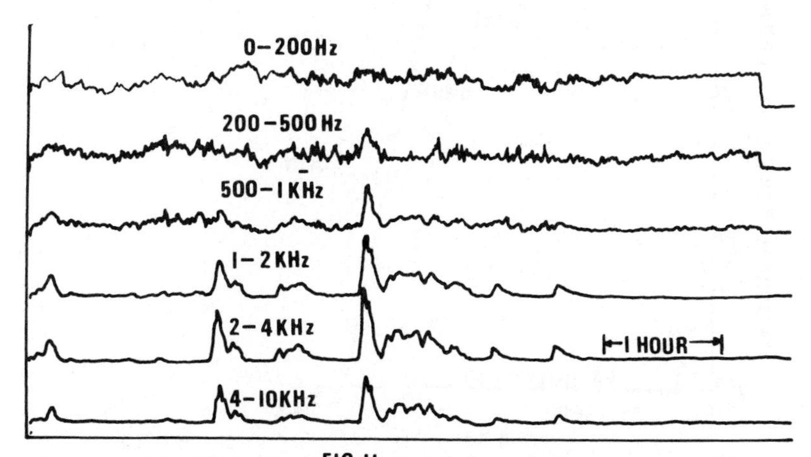

FIG 11

RAIN-INDUCED SCINTILLATION USING A HORIZONTAL LINE DETECTOR AND A 50M PATHLENGTH (AFTER EARNSHAW)

Furthermore, each raindrop acts as a short focal length lens and a fraction of the incident light is reflected from the inner surface of the drop and is back-scattered in a narrow cone at the rainbow back-scatter angle of 42°. This back-scatter can be detected and is an additional signature for the presence of rain.

The diffraction patterns and fall velocities of hail and snow are sufficiently different from those of rain and drizzle that these types of precipitation can, in principle be distinguished. A further aid to identification comes from a spectral analysis of the signals. The temporal power spectra of rain, snow and wind induced scintillations is shown in Fig. 12 taken from a recent paper (ref. 18). Wind induced scintillations represent noise in this type of device but can in principle be distinguished by the spectral analysis technique.

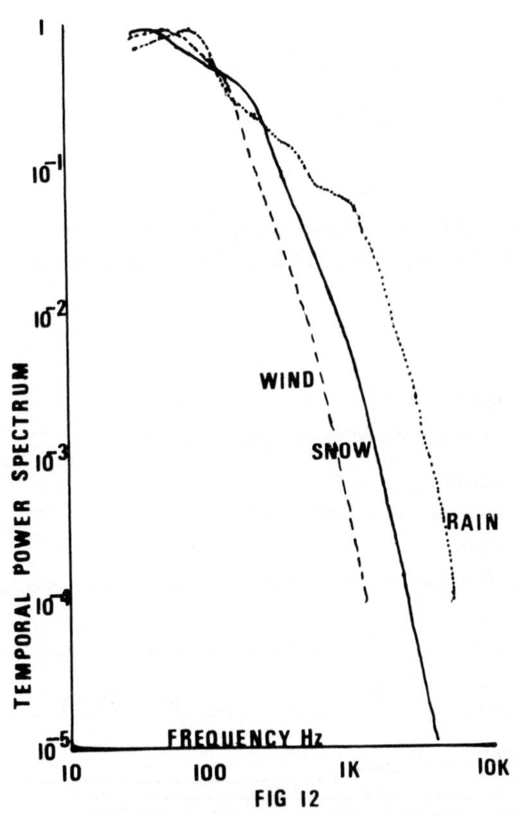

FIG 12
TEMPORAL POWER SPECTRUM OF WIND-, SNOW- AND RAIN- INDUCED SCINTILLATION (AFTER EARNSHAW)

There is no doubt that this interesting method works in well-defined precipitation situations – heavy showers, heavy snowfall, or hail. However, experience suggests that the majority of precipitation events are not really well defined and the method is likely to have great interpretational difficulties in such circumstances. Nevertheless, it is one of the few laser techniques which still appears to have potential to provide a really useful addition to the instrumental measurements of operational meteorology. The theory of the method has recently been extended by Ting-i-Wang et al (ref. 19) to cover the case of a divergent laser beam and this potentially improves the usefulness of the technique because the derived rain rates are less sensitive to the drop-size distribution than is the case with a collimated beam. However, a consequential disadvantage is that for this case the path-weighting function is biased strongly towards the receiver end whereas in the plane parallel beam case it is almost uniformly distributed along the path. Saturation effects limited the results obtained by this technique to path lengths shorter than 200 m for rainfall rates up to 140 mm hr^{-1} but as Fig. 13 shows, at low rain rates, good agreement was obtained over a 1000 m path with simultaneous measurements by tipping bucket raingauges.

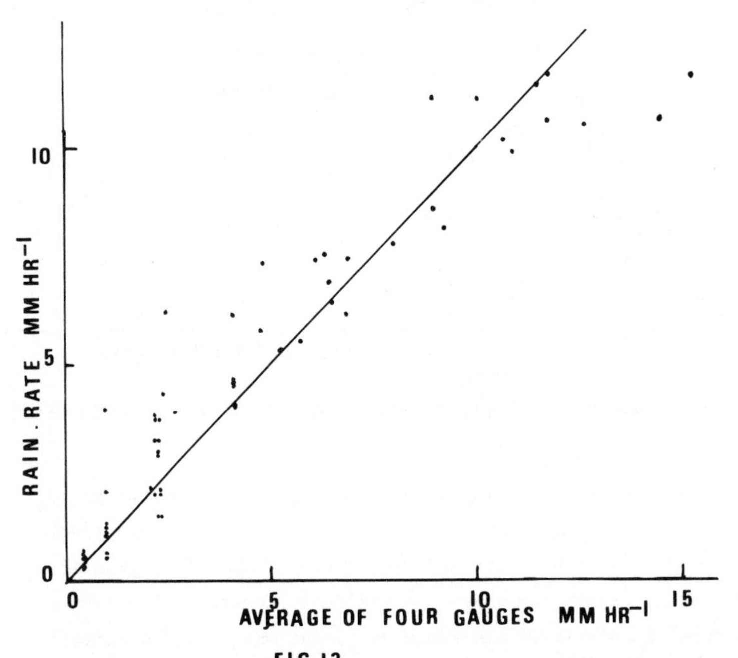

FIG 13
RAIN RATE FROM LASER SCINTILLATION OVER A 1000M PATH COMPARED TO VALUES FROM TIPPING-BUCKET RAIN GAUGES (AFTER WANG)

7. THE MEASUREMENT OF TEMPERATURE

There are at least two methods by which temperature measurements can be made using scattered laser light. The first of these, proposed by Cooney makes use of the fact that the shape of the rotational Raman spectrum of nitrogen is a sensitive function of temperature so that by comparing the intensities at two points in the spectrum, a measure of the gas temperature may be obtained. From Fig. 14 it can be seen that the detected wavelengths will be nearly equal so that problems associated with the wavelength dependence of the atmospheric transmission and the detection efficiency of the receiver are eliminated.

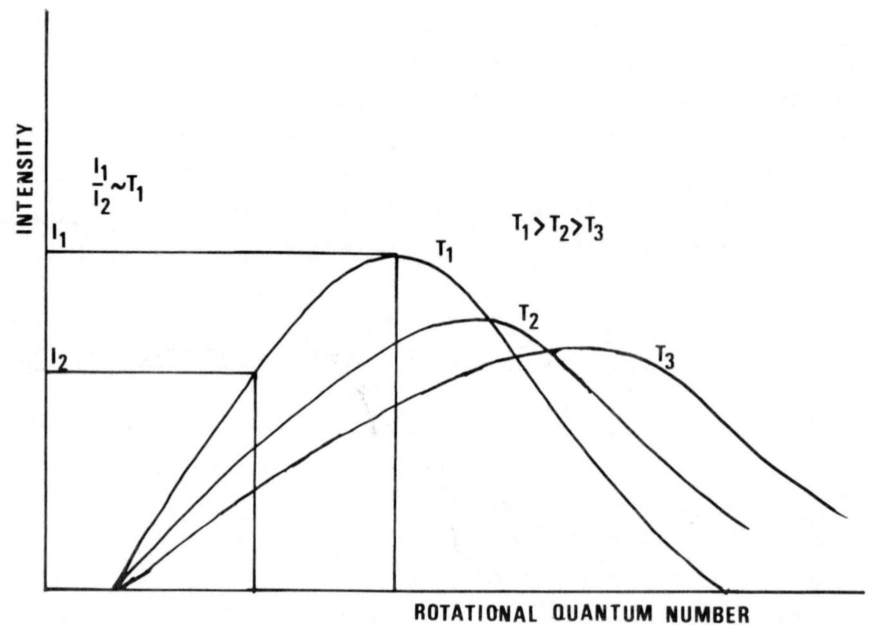

FIG 14
ROTATIONAL RAMAN SPECTRUM OF N_2 AS A FUNCTION OF TEMPERATURE

However, the detected wavelengths will be only a few Angstrom units away from the exciting wavelength for which there will be an elastically back-scattered signal between one and two orders of magnitude greater than the rotational Raman components we desire to measure. Furthermore, the temperature dependence of the Raman component whilst quite marked at high and low temperatures is at its least sensitive around 270°K which is, of course, the region of greatest interest, in the lowest levels of the atmosphere.

The second method for temperature, on the other hand, has been used with some degree of success at heights in the atmosphere up to 25 km. The method relies upon the vibration/rotational Raman effect. It is reasonable to propose that the use of the pure rotational spectrum in the same manner would result in the same measurements to greater accuracy because the cross-section for rotational Raman scattering is nearly two orders greater than for vibrational Raman scattering. However, the problem of the rejection of the nearby elastically scattered light has so far made the use of this component of the scattered light impractical.

However, for the vibrational/rotational case we may write the laser radar equation as

$$C(\lambda', Z) = \frac{P_0(\lambda)T(\lambda)T(\lambda')\ A\ Q(\lambda')\ \delta Z}{Z^2}\ N_{N_2}\frac{d\sigma}{d\Omega} \qquad (10)$$

where N_{N_2} is the number density of nitrogen molecules and all the other variables are as previously defined.

$C(\lambda', Z)$ is the experimental measurement in the absence of noise, which in laser radar experiments can often be measured independently by observing the photon count (or analogue voltage) in a range bin sufficiently high for the signal to be negligible, and N_{N_2} is the variable required from which the temperature must be found by an inversion technique.

For a particular experiment and a known height Z, the factor

$$P_0(\lambda)T(\lambda)T(\lambda')\ Q(\lambda')\ \frac{d\sigma}{d\Omega}$$

can be found by a normalization procedure for any level at which the atmospheric number density is known. The knowledge of the number density may be obtained from a concurrent radiosonde measurement at some height for which we have a scattering measurement. We assume for this that the number of photons detected from a given layer of the atmosphere of thickness, Z, is proportional to the average number density of scatterers through the layer. We may then write (10) as

$$C(\lambda' Z) = \frac{K}{Z^2}\int_{Z-\frac{\delta Z}{2}}^{Z+\frac{\delta Z}{2}} \frac{N(Z)}{Z}dZ \qquad (11)$$

Where K is the normalization factor given by

$$K = P_0(\lambda)T(\lambda)T(\lambda')\ AQ(\lambda')\ \frac{d\sigma}{d\Omega} \qquad (12)$$

7.1 Normalization procedure

To determine K we note that the function N(Z) can be closely approximated in the atmosphere below 100 km by

$$N(Z) = N_o e^{-(Z/H)} \tag{13}$$

where H is the number density scale height which is a slowly varying function of Z.

Thus

$$\frac{1}{\delta Z} \int_{Z-\frac{Z}{2}}^{Z+\frac{Z}{2}} N(Z)\, dZ = \frac{H}{\delta Z} \left(N(Z - \frac{\delta Z}{2}) - N(Z + \frac{\delta Z}{2})\right) \tag{14}$$

But from the gas equation:

$$N(Z) = \frac{p(Z)}{kT(Z)} \tag{15}$$

Where p(Z) is the pressure, T(Z) is the absolute temperature and k is Boltzman's constant. δZ is fixed in our experiment, it may range from a few metres in tropospheric experiments to a few kilometres in stratospheric experiments but must be small enough that H can be considered constant over it. p(Z) and T(Z) are obtained for the levels $Z \pm \frac{\delta Z}{2}$ from the independent radiosonde measurements and from (14) and (11), K may be found. It is worth noticing that because K contains T (λ) and T (λ') it is in fact valid only at the level for which it is determined and corrections are needed to account for additional transmission losses above the normalization level (or smaller losses below it). For stratospheric experiments these corrections are of the order of a few tenths of 1% but for tropospheric experiments they may be much larger than this. Having obtained K, N_{N_2} may be readily calculated for each height increment, δZ, in the experiment. Since the atmosphere is well mixed below about 90–100 km, N_{N_2} is a fixed fraction of the average total number density of the air in the layer δZ. Henceforth we assume that the value N_{N_2} has been adjusted to the number density for air molecules, N. We have then as our starting point for the recovery of temperature a set of average number densities, N_i, each belonging to a height increment δZ at a height Z_i.

7.2 Temperature recovery - method 1

For this method, we assume that each value of N_i can be ascribed to a height, Z_i, which is the geopotential height of the mid point of the layer δZ_i. (For an explanation of the geopotential height scale see any standard textbook in meteorology - it is a scale defined to enable the acceleration due to gravity, g, to be treated as constant with height. At low levels it is indistinguishable from geometric height but at high levels the differences become important.)

Let p_i, T_i be the pressure and absolute temperature at the height Z_i to which the number density N_i is ascribed. Then from the gas equation we have:-

$$M(p_i - p_{i+1}) = \rho_i RT_i - \rho_{i+1} RT_{i+1} \quad (16)$$

where M is the molecular weight of air $\rho_i = mN_i$ is the density of the air at Z_i if m is the mass of an air molecule, and R is the gas constant for dry air.

From (16) by rearrangement and substitution for ρ_i and ρ_{i+1} we obtain

$$T_{i+1} = \frac{mN_i - \frac{M}{R}(p_i - p_{i+1})}{mN_{i+1}} \quad (17)$$

If N_i, p_i and T_i are obtained from independent data for the level immediately below the lowest level of the laser sounding, an upward stepping calculation can be performed to obtain values of T at successively higher levels provided we have a knowledge of the pressure profile over the height range of the laser data.

This pressure profile may be obtained by a downward integration of the number density profile in which the pressure increment between two adjacent levels is equated to the weight of the molecules between the layers. From equation (13) we may write

$$H = \frac{Z_{i-1} - Z_i}{\ln(N_i/N_{i-1})} \quad (18)$$

and by integrating (13) the total number of molecules in the layer is

$$N_{tot} = H(N_i - N_{i-1}) \quad (19)$$

and by substitution from (18) for H we have

$$N_{tot} = \frac{(Z_{i-1} - Z_i)\ (N_i - N_{i-1})}{\ln(N_i/N_{i-1})}$$

or

$$N_{tot} = \frac{\delta Z\ (N_i - N_{i-1})}{\ln(N_i - N_{i-1})}$$

To obtain the pressure at the highest level of the sounding we may, provided that this level is sufficiently high (> 20 km), assume that the scale height, H, for the highest height increment applies also from the highest level of the sounding to the top of the atmosphere and the number of molecules in this layer can be obtained from

$$N_{tot} = H\ N_{top}$$

where N_{top} is the number density at the highest level of the sounding. The disadvantage of this method is that all the integration is performed downwards through the profile, which places the greatest reliance on the least accurate data. This is, however, partly offset by the fact that in the step calculation of successive pressure increments from equation (19), the first value adopted for N_i rapidly becomes of little significance.

7.3 Temperature recovery - method 2

A weakness of method 1 is that it treats the number densities derived from the laser measurements as point measurements at heights in the middle of each layer δZ. However, the number densities are strictly the average values of the number density through layers of thickness δZ and to take account of this a more sophisticated method for obtaining the pressure profile is required.

The pressure $p_{i+\frac{1}{2}}$ at the top of the layer for which N_i is the average number density can be found from the relation:-

$$p_{i+\frac{1}{2}} = p_{i-\frac{1}{2}} - \Delta p \qquad (20)$$

where $P = m\ \delta Z\ N_i\ g$ and g is the acceleration due to gravity which, as we have already remarked can be taken as constant if Z is expressed in geopotential units. N_i is, of course, the average number density in the layer from $Z_{i-\frac{1}{2}}$ to $Z_{i+\frac{1}{2}}$. As before we must obtain the first pressure from an independent measurement and the calculation then yields pressure at levels intermediate between those to which the N_i can be ascribed.

It is not necessarily sufficient to assume that the N_i can be placed at the mid point of the increments defined by the values $p_{i+\frac{1}{2}}$ because the function N (Z) is exponential rather than linear in Z. We may find the appropriate value of Z_i from the equation (12)

$$N_i = N_o \exp - (Z_i/H) \tag{21}$$

However, in this equation, N_o is unknown and will be different for each atmospheric layer. To avoid this difficulty we substitute for N_i from the equation:

$$N_i \,\delta Z = \int_{Z_{i-\frac{1}{2}}}^{Z_{i+\frac{1}{2}}} N_o \exp-(Z/H)dZ \tag{22}$$

so that we obtain

$$N_o \exp-(\frac{Z_i}{H}) = \frac{1}{\delta Z} \int_{Z_{i-\frac{1}{2}}}^{Z_{i+\frac{1}{2}}} N_o \exp-(Z/H)dZ \tag{23}$$

from which

$$Z_i = H \ln \left[\frac{-H}{Z} \left(\exp-(\frac{Z_{i+\frac{1}{2}}}{H}) - \exp(\frac{Z_{i-\frac{1}{2}}}{H})\right) \right] \tag{24}$$

H may be found from equation (18) for the two layers bounded by N_i, N_{i-1}, N_{i-2} and an average value of these two quantities is a sufficiently close approximation to the value of H for the layer containing N_i.

Having found Z_i, the value of the pressure, p_i, appropriate to N_i may be interpolated from the pressure/height data obtained from equation (20). The appropriate mean absolute temperature for the layer $Z_{i-\frac{1}{2}}$ can then be found from the gas equation in the form

$$T_i = \frac{p_i}{N_i k} \tag{25}$$

Results of this type of measurement have been published for stratospheric soundings (ref. 6) and for low level soundings (ref. 20) but the agreement with independent radiosonde measurements is not strikingly good in either case.

7.4 The required accuracy

It is worth considering here the accuracy required in the number density measurement for a given error in the derived temperature. The first point to notice is that the temperature is obtained from equation (25) in which the experimental errors enter twice. The error of each individual measurement of N_i appears in the denominator of (25) but the integrated errors of all the measurements of N_i from the highest level of the sounding down to the level for which T is being determined enter through the pressure, p_i.

Fig. 15, due to Minzer, (ref. 21) shows that for a height resolution, δZ, which is 40% of the scale height, H, a 1% uncertainty in number density yields a 3% error in absolute temperature. It is worthwhile, and is left as an exercise for the student, to compute the accuracy required in the measurement of number density to obtain, say, a $\pm 1^{\circ}$C uncertainty in the temperature of a layer of the atmosphere 100 m thick at a height of 2 km. A comparable measurement from a conventional radiosonde would be unlikely to have an error greater than $\pm 0.5^{\circ}$C and would more usually be within $\pm 0.2^{\circ}$C.

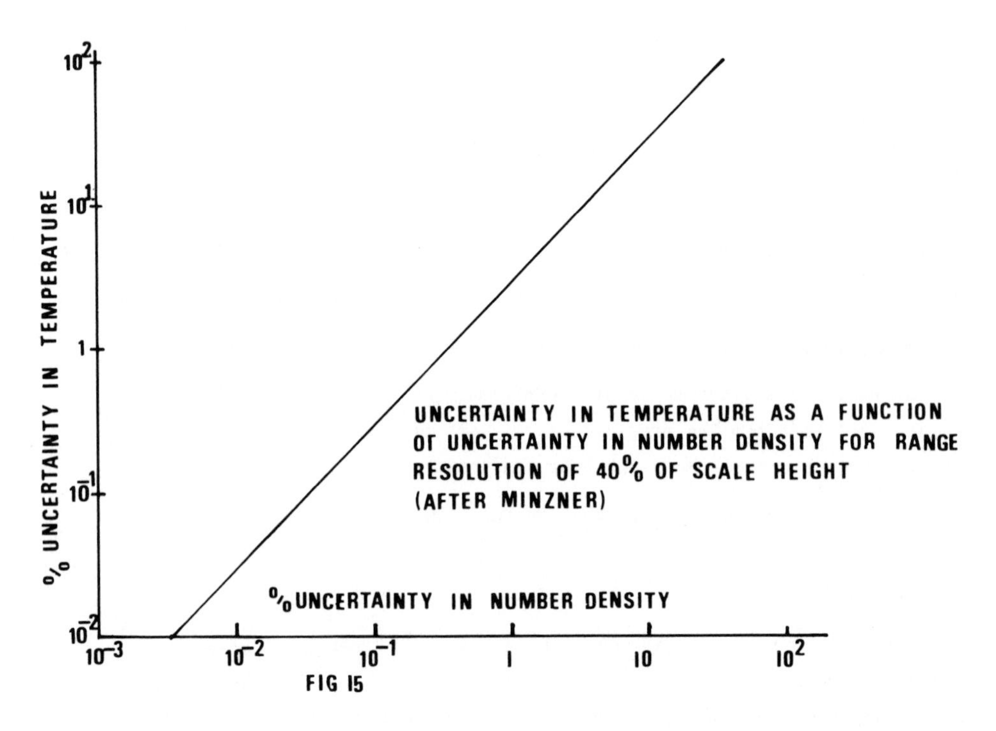

FIG 15

8. CONCLUSION

In a single chapter, it is not possible to cover all the possible techniques of optical remote sensing which have been applied to measurements in atmospheric physics. We have concentrated upon the measurements of meteorology and it must be recognized that some powerful techniques such as long path absorption (ref. 22), and differential absorption lidar (DIAL) (ref. 23) which have been widely used for such other atmospheric measurements as the concentration of atmospheric trace gases (including water vapour) have not been considered. They are covered elsewhere in these chapters. Moreover, even some methods which have been reported for the measurement of those variables which we have considered have had to be omitted, for example spectroscopic methods for simultaneous temperature and humidity (ref. 24). This is therefore, in no sense a complete review or survey of the subject but rather a brief and foreshortened introduction of some of the principal optical remote sensing methods which have been reported in recent years. As such, perhaps it can form a useful starting point for further study.

REFERENCES

1 Van der Hulst, Light Scattering by Small Particles, Wiley, New York, 1957.
2 M.Kerker, The Scattering of Light and Other Electromagnetic Radiation, Academic Press, New York, 1969.
3 H.A. Szymanski (ed), Raman Spectroscopy Theory and Practice, Plenum Press, New York, 1967.
4 L.G. Bird and N.E. Rider, The Met Mag: 97 (1968) pp. 107-115.
5 C.M. Platt, Jour. App Met. 18 (1979) pp 1130-1143.
6 R.E.W. Pettifer, G.J. Jenkins, P.G. Healey and J.H. Convery, Opt. Quan. Electr. 8 (1976) pp 409-423.
7 R.T.H. Collis and P.B. Russell, Topics in App Phys 14 Springer Verlag, New York, (1976). pp 118-132.
8 R.M. Huffaker, App Opt 9 (1970) p 1026.
9 R.M. Huffaker, App Opt 9 (1970) p 1026.
10 T.W. Tuer, Nichols Res Corpn. Rept NRC AA - 79 - LASBR - 002 15 Sept 1979 Ann Arbor Mich 48104. Vol 2.
11 W.L. Eberhard and R.M. Schotland, App Opt. 19 (1980) p 2967.
12 J.T. Sroga and E.W. Eloranta, Jour. App Met 19 (1980) p 598.
13 R.S. Lawrence, G.R. Ochs and S.F. Clifford, App Opt 11 (1972) pp 239-243.
14 V.E. Zvev, G.G. Matvienko and I.V. Samokhvalor, Proc 7th Int. Laser Radar Conf. Menlo Park, Calif. Nov 4-7 1975. p 53.
15 G. Benedetti Michelangeli, G. Congeduti and G Fiocco, Jour. Atm. Sci. 29 (1972) p 906.
16 J.C. Pourny, D. Renault and A. Orszag, App Opt 18 (1979) pp 1141-1148.
17 S.T. Shipley, E.W. Eloranta and J. Weinman, Jour. App Met 13 (1974) pp 800-807.
18 K.B. Earnshaw, Ting-I Wang, R.S. Lawrence and R.G. Greunke, Jour. App Met 17 (1978) pp 1476-1481.
19 Ting-I Wang, R.S. Lawrence and M.K. Tsay, App Opt 19 (1980) pp 3617-3621.
20 R. Gill, K. Geller, J. Farina and J. Cooney, Jour. App Met 18 (1979) pp 225-227.
21 R.A. Minzer, Private communication.
22 E.D. Hinkley, R.T. Ku and P.L. Kelley, Topics in App Phys. 14 Springer Verlag, New York, (1976) p 284 et seq.
23 R.M. Schotland, Jour. App Met O 13 (1974) p 71.
24 M. Endermann and R.L. Byer, Optics Lett. 5 (1980) pp 452-454.

Optical Remote Sensing of Air Pollution,
Lectures of a course held at the Joint Research Centre, Ispra, Italy, 12–15 April 1983,
P. Camagni and S. Sandroni (Eds). 123–142

THE PRINCIPLES OF LIDAR

R.H. VAREY

1. INTRODUCTION

Lidar provides a method of measuring properties of the atmosphere out to distances of many kilometres from the instrument itself. It has a wide variety of applications, military, scientific and industrial. It is employed as a range finder for military targets and in surveying and has been used to measure the distance of the moon from the earth. It has been used to measure sodium in the stratosphere and many other molecules are accessible to it in the troposphere. In meteorology it has applications in the study of water vapour, particulates and the depth and structure of the boundary layer. In the study of atmospheric pollution it is used to measure the rise and dispersion of plumes from industrial sources and in a multi-wavelength mode it can probe the vertical distribution of molecules such as SO_2, NO_2, O_3 and a variety of organic molecules. It can be operated from the ground, in aeroplanes and in satellites using natural targets, retroreflectors or the atmosphere itself to provide the signals necessary for its operation.

In this chapter the emphasis will be on applications for the study of atmospheric pollution using systems operating in the ultra-violet and visible regions of the spectrum with the atmosphere as the backscattering medium. The components which influence the strength of the backscatter signal will be discussed together with an analysis of the performance to be expected from a practical system. Some attention will be paid to the sources of noise in the signal and techniques of signal analysis and data presentation. Examples will be provided from measurements of the boundary layer and of plumes from power stations.

2. TRANSMISSION AND BACKSCATTER OF LASER RADIATION IN THE ATMOSPHERE

2.1 The lidar equation

The principle of lidar is that a laser pulse is fired into the atmosphere and as it proceeds along its path, radiation is continuously scattered back towards the laser where it is collected with a telescope and measured with a photomultiplier. The signal is analysed to provide information on the magnitude of backscatter and attenuation experienced by the pulse in its passage through the atmosphere. As with radar, the range from which the signal is being received is provided by the time interval between the firing of the pulse and the reception of the signal. By continuously monitoring the backscattered signal, information is received out to a range at which the signal becomes too weak to be detected, the range resolution obtainable being related to the length of the laser pulse. Lasers are particularly suitable for this application because they can conveniently provide intense pulses of short duration with a narrow bandwidth and small beam divergence. The narrow bandwidth is important because it makes it possible to measure the concentration of particular molecules along the path of the pulse by probing their absorption spectra. It also makes it possible to exclude a high proportion of the skylight radiation, which would otherwise swamp the receiving system, by including in the optics a narrow band filter which allows only radiation close to the wavelength of the laser to reach the detector. The small beam divergence is useful because it improves the lateral spatial resolution of the system and also reduces skylight radiation reaching the detector by permitting the use of a narrow aperture telescope.

The energy, dE, received at the telescope after backscatter from an element dr at range r along the path of the laser pulse is given by

$$dE = E_o TA \frac{\beta(r)dr}{r^2} \exp\left[-2 \int_o^r \alpha(r')\, dr'\right], \qquad \ldots (1)$$

where E_o is the initial laser pulse energy, T is related to the optical efficiency of the receiver system, A is the telescope area, $\beta(r)$ is the volume backscatter coefficient of the atmosphere at range r and $\alpha(r')$ is the volume extinction coefficient at range r'. $\beta(r)$ describes the radiation which is backscattered towards the laser, the factor $1/r^2$ arising from the usual inverse square law. The term involving $\alpha(r')$ describes the attenuation of the laser pulse on its path to and from range r. Both α and β are related to atmospheric visibility as described below.

The power received, P, is dE/dt where t is the interval between the start of the pulse from the laser and the return to the telescope of the radiation from range r. Since t = 2r/c where c is the speed of light then equation (1) becomes

$$P = \frac{E_o TAc\beta(r)}{2r^2} \exp\left[-2 \int_o^r \alpha(r')\, dr'\right] \qquad \ldots (2)$$

which is the so called lidar equation (see for example refs. 1 and 2).

2.2 The volume extinction coefficient

In travelling a distance r through the atmosphere light is attenuated according to the relationship

$$P = P_o \exp(-\alpha r), \qquad \ldots (3)$$

where if r is in metres the unit of α is m^{-1}. Our immediate experience of the extinction coefficient is through atmospheric visibility, we can see further on clear days when the extinction coefficient is low than on hazy days when it is high. Meteorologically, visibility is defined in terms of the contrast of an object seen against the background sky. It is a common experience in hilly country to see a series of distant ridges as successively lighter bands of grey merging more and more closely with the sky. It is found that the limit at which an object can be picked out against a background occurs when the difference between their apparent radiation is about 2%. Visual range V is defined (see for example ref. 3) as the distance at which this occurs. Hence

$$0.02 = \exp(-\alpha V)$$
$$\ln(0.02) = -\alpha V$$
$$\alpha = 3.91\ V^{-1} \qquad \ldots (4)$$

Clearly, the value of α depends upon the wavelength of the radiation being propagated and for visual range it is standard practice to use a wavelength λ of 550 nm. The problem is that for lidar operation we need to evaluate α at different wavelengths. If we neglect absorption, this will be dealt with in the chapter on differential lidar, and concentrate on elastic scattering then α has two components,

$$\alpha = \alpha_R + \alpha_M,$$

where α_R corresponds to Rayleigh scattering and α_M to Mie scattering. α_R applies to particles small compared with the wavelength of light being scattered, i.e. with a diameter less than about one tenth of the wavelength, and α_M to larger particles. The atmosphere contains molecules and particles covering a range of sizes and hence α has components of both α_R and α_M. As for the wavelength dependence of α, α_R presents no difficulties, it is composed principally of scattering by oxygen and nitrogen molecules and at the density at ground level we have to a good approximation,

$$\alpha_R = 1.1 \times 10^6\ \lambda^{-4}, \qquad \ldots (5)$$

where λ is in nm and α in m^{-1}. If molecules in the atmosphere were the only contributors to α then expression (5) could be used in equation (4). This gives a visual range at 550 nm in excess of 300 km. Even under exceptionally clear conditions horizontal visibility rarely exceeds 50 km so it is evident that we have to take into account the various liquid and solid particles which are present in the atmosphere.

The wavelength dependence of α_M is complicated, depending upon the size, shape and material of the particles. This problem has received considerable attention (see for example ref. 4) and it has emerged (ref. 5) that a good working relationship in the visible and near ultra-violet is

$$\alpha_M = 3.91\ V^{-1}\ (550/\lambda)^q, \qquad \ldots (6)$$

where α_M is in m^{-1}, V in m, λ in nm and where

$$q = 0.0585\ V^{1/3} \text{ for } V \leqslant 6000 \text{ m}$$

and $q = 1.3$ for 'average visibility'.

Equations (5) and (6) can be used to calculate the attenuation of the laser pulse to be expected for a range of wavelengths and meteorological conditions.

2.3 The volume backscatter coefficient

The extinction coefficient depends upon the total radiation scattered, no matter what the direction of scattering. The backscatter coefficient is a measure of that portion of the radiation which travels back along the path of the laser pulse. The two are clearly related with their ratio depending upon the angular distribution of the scattering. As with the extinction coefficient, β has two components, β_R and β_M. For Raleigh scattering, which occurs predominantly in a forward and backward direction, i.e. little radiation is scattered sideways, it can be shown (ref. 6) that

$$\beta_R = 1.5\ \alpha_R/4\pi, \qquad \ldots (7)$$

β being defined as the proportion of the incident radiation backscattered into unit solid angle (steradian) as the pulse travels unit distance through the atmosphere. In the system being used here the units of β are thus $m^{-1}\ Sr^{-1}$. Again for Mie scattering the situation is more complicated. For spherical particles of known size distribution, shape and refractive index the scattering distributions can be calculated. This cannot be done for particles of irregular shape or composition. In any case the size distribution of particles in the atmosphere is not in general known, so empirical relationships have to be used for β_M. A reasonably good approximation appears to be (ref. 7)

$$\beta_M = 0.5\ \alpha_M/4\pi, \qquad \ldots (8)$$

i.e. a smaller proportion of the scattered radiation travels back along the laser path than with Rayleigh scattering. However, the relationship varies with the shape and size distribution of the scattering particles and (8) is only an approximation.

TABLE 1

Backscatter and extinction coefficients at the ruby wavelength, 694.3 nm and at 300 nm. Scattering at ground level under normal conditions of pressure and temperature gives values for α_R and β_R as indicated at each wavelength. The notation a^{-n} indicates $a \times 10^{-n}$.

	λ = 694.3 nm, $\alpha_R = 4.7 \times 10^{-6}$ m^{-1}, $\beta_R = 5.7 \times 10^{-7}$ m^{-1} Sr^{-1}				λ = 300 nm, $\alpha_R = 1.4 \times 10^{-4}$ m^{-1}, $\beta_R = 1.6 \times 10^{-5}$ Sr^{-1}			
Visual Range (km)	α_M m^{-1}	α m^{-1}	β_M m^{-1} Sr^{-1}	β m^{-1} Sr^{-1}	α_M m^{-1}	α m^{-1}	β_M m^{-1} Sr^{-1}	β m^{-1} Sr^{-1}
0.2	1.8^{-2}	1.8^{-2}	7.2^{-4}	7.2^{-4}	2.4^{-2}	2.4^{-2}	9.6^{-4}	9.7^{-4}
0.5	7.0^{-3}	7.0^{-3}	2.8^{-4}	2.8^{-4}	1.0^{-2}	1.0^{-2}	4.1^{-4}	4.3^{-4}
1	3.4^{-3}	3.4^{-3}	1.4^{-4}	1.4^{-4}	5.6^{-3}	5.7^{-3}	2.2^{-4}	2.4^{-4}
2	1.6^{-3}	1.6^{-3}	6.4^{-5}	6.5^{-5}	3.1^{-3}	3.2^{-3}	1.2^{-4}	1.4^{-4}
5	6.2^{-4}	6.2^{-4}	2.5^{-5}	2.6^{-5}	1.4^{-3}	1.5^{-3}	5.7^{-5}	7.3^{-5}
10	2.9^{-4}	2.9^{-4}	1.2^{-5}	1.3^{-5}	8.4^{-4}	9.8^{-4}	3.3^{-5}	4.9^{-5}
20	1.4^{-4}	1.4^{-4}	5.6^{-6}	6.2^{-6}	5.1^{-4}	6.5^{-4}	2.0^{-5}	3.6^{-5}
50	5.8^{-5}	6.3^{-5}	2.3^{-6}	2.9^{-6}	2.9^{-4}	4.3^{-4}	1.1^{-5}	2.7^{-5}
100	2.9^{-5}	3.4^{-5}	1.2^{-6}	1.8^{-6}	2.0^{-4}	3.4^{-4}	8.1^{-6}	2.4^{-5}

Values for α and β with their Rayleigh and Mie components have been calculated for various values of visual range from equations (5-8) for the wavelength of a ruby laser, 694.3 nm, and for 300 nm, the wavelength at which some lasers operate in the ultra-violet. The results are shown in Table 1 where it can be seen that at the ruby wavelength Rayleigh scattering provides a minor contribution to α and β but that at 300 nm it is more significant. For example, at a visual range of 20 km Rayleigh processes contribute 25% of α and 45% of β at 300 nm whereas at 694.3 nm their contribution is less than 1%.

2.4 Pulse length and optical efficiency

It can be seen from equations (1) and (2) that at any instant signals reach the receiver from a finite length $c\tau/2$ of the atmosphere along the path of the pulse, where τ is the duration of the pulse. This defines the spatial resolution of the system since details of the structure within the length from which radiation is received simultaneously cannot be resolved, at least not without the use of deconvolution techniques. Clearly, for a given power the longer the pulse length the greater the received power. However, in practice with pulsed lasers it is not the instantaneous power which is fixed but the total energy in the pulse. This can, within limits, be used to give longer, lower power or shorter higher power pulses. There is thus nothing to be gained in using long pulses and it is best to use as short a pulse as possible because although this does not change the instantaneous power received it does give the best spatial resolution, provided of course that the electronics of the receiving system are fast enough to provide the time resolution τ.

The diameter of the receiving telescope can be up to several metres (see for example ref. 8) but it is more usual, particularly in mobile sytems, for it to be limited to half a metre or so. The collection efficiency of the receiver describes the proportion of the backscattered signal entering the telescope which actually reaches the detector. This is limited by the reflectivity or transmission of the optical components and by the geometrical design of the telescope. In fact the geometrical component of T is a function of the range from which the signal is being received. This is illustrated in Figure 1 for a system where the laser and receiving system are arranged side by side. Light only reaches the detector from the region where the telescope field of view and the path of the laser beam overlap. The overlap is small for short ranges, so the efficiency is low but for greater ranges the overlap becomes complete. It is normally arranged for this to occur in the region of the atmosphere of greatest interest. A similar effect occurs with coaxial systems where objects such as secondary mirrors in the field of view obstruct some of the light entering the system. The result of this effect is to produce a detected pulse of the form shown in Figure 2. To begin with the signal rises as the radiation

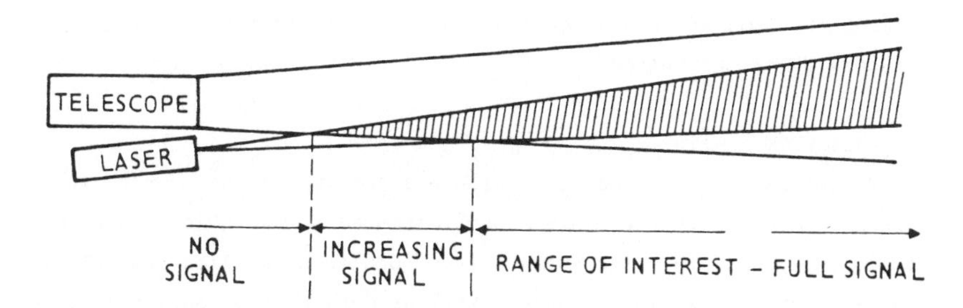

FIGURE 1 Overlap of telescope field of view and laser beam. The hatched region represents the volume from which backscattered radiation passes through the telescope system to reach the photomultiplier

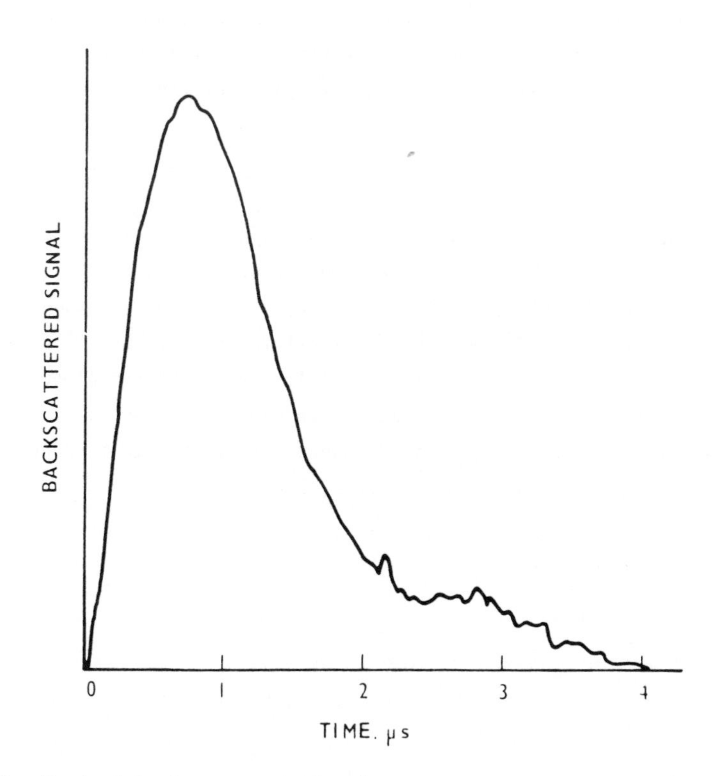

FIGURE 2 Typical backscattered signal

comes from regions of increasing overlap but then the $1/r^2$ term in equation (1) dominates and the signal decreases. This can often be used to advantage in avoiding problems of saturation and dynamic range in the receiving system, as discussed below in Section 4.

3. DETECTION LIMITS

To determine the range over which a lidar system can operate it is necessary to consider the minimum backscattered signal which can be detected. This is determined by noise which arises both from the lidar signal itself and from background radiation. Photomultiplier dark current and other internally generated noise are not usually important in lidar operation in the visible and ultra-violet. In order to examine the detection limit it is convenient to consider the photo-electron count rate, n, generated in the photomultiplier by the backscattered radiation. This is given by

$$n = Pq/h\nu, \qquad \ldots (9)$$

where q is the quantum efficiency of the photomultiplier, $h\nu$ is the energy of a photon at the laser wavelength and P is the power received at the photomultiplier. The signal to noise ratio is given by (ref. 9)

$$n_S\tau'/(n_S\tau' + n_B\tau')^{\frac{1}{2}}, \qquad \ldots (10)$$

where n_S and n_B are the count rates of the lidar signal and background radiation. τ' is the sampling time of the detection system, not to be confused with the laser pulse duration, τ. The choice of a value for τ' depends on two conflicting requirements. To improve the signal to noise ratio τ' should be large but this is at the expense of range resolution. A practical lower limit is $\tau' = \tau$ since no improvement in resolution is achieved for $\tau' < \tau$.

Returning to equation (10), if it is assumed that the limit of detection occurs when the signal to noise ratio is one, then

$$n_S\tau' = (n_S\tau' + n_B\tau')^{\frac{1}{2}}$$

$$(n_S\tau')^2 - n_S\tau' - n_B\tau' = 0$$

$$n_S\tau' = \frac{1}{2} + (\frac{1}{4} + n_B\tau')^{\frac{1}{2}}$$

$$n_S\tau' \simeq 1 + (n_B\tau')^{\frac{1}{2}}. \qquad \ldots (11)$$

Thus if the background radiation is zero then in principle only a single signal count is required for detection. In daylight conditions $n_B\tau'$ will generally be considerably greater than one and so then to a good approximation the detection limit becomes

$$n_S\tau' \geqslant (n_B\tau')^{\frac{1}{2}}. \qquad \ldots (12)$$

Values of $n_S\tau'$ can be calculated using equations (2) and (9) with values of α and β determined as in Table 1. This has been done for a ruby lidar with the following parameters: wavelength 694.3 nm; pulse energy, 0.2 J; telescope diameter, 0.5 m; quantum efficiency 0.03; sampling time, 50 ns. The field of view of the telescope has been assumed to include all the laser beam, a reasonable assumption since the minimum range considered is 200 m, and the optical efficiency has a value 0.2. For a given value of visual range, α and β are assumed to be constant, i.e. not varying with range, with values as in Table 1. The results are shown in Figure 3 for visual range having values between 0.2 and 50 km as indicated in the figure. It can be seen that as would be expected $n_S\tau'$ decreases with range for all values of visual range, the decrease being most rapid for poor visibility. However, the signal count from any given range does not increase monotonically as visual range increases, as can be seen from the figure there is an optimum value of V for which $n_S\tau'$ has a maximum value and beyond which it decreases as visibility improves. This occurs because the backscattered signal received at the telescope depends not only upon the attenuation of the signal but upon the strength of the backscatter, which in turn decreases as visibility improves. Nevertheless, when operating at the detection limit, even under bright daytime conditions, it will be shown below that for all practical conditions a system with the parameters detailed above has a maximum range which does increase with visual range.

The background signal depends upon the brightness of the background sky and the field of view and diameter of the telescope. In order to reduce the background signal, the telescope field of view should be reduced to a mimimum which is consistent with including all the laser beam, which typically has a divergence of a few milliradians. The field of view assumed in these calculations is 2×10^{-5} Sr which corresponds to a divergence of 5 milliradians. The background radiance varies with meteorological conditions, time of day and the wavelength under consideration. The values appropriate for these calculations range from about 10^{-1} Wm^{-2} Sr^{-1} nm^{-1} for a bright cloud background to 10^{-7} Wm^{-2} Sr^{-1} nm^{-1} in moonlight (refs. 6 and 10). The intensity reaching the photomultiplier depends also upon the bandwidth of the filter used to exclude wavelengths other than that of the laser, the value assumed here is 5.5 nm.

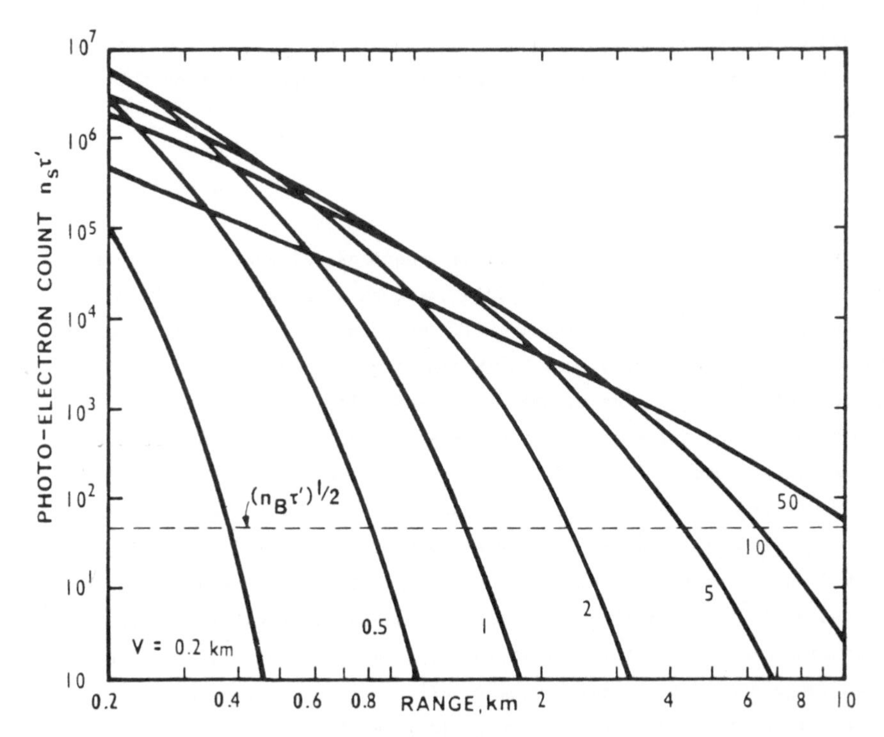

FIGURE 3 $n_S\tau'$ as a function of range for various values of visual range. The value of $(n_B\tau')^{1/2}$, dotted line, corresponds to a bright cloud background

In a system with the parameters described above, a bright cloud background corresponds to a signal count of about 2.2×10^3. As shown in equation (12) it is the square root of the background count which has to be compared with the lidar count. This is also shown in Figure 3 where it can be seen that in poor visibility, V = 0.2 km, detectable signals are received from ranges out to about 0.4 km whereas in moderate visibility, V = 10 km this increases to about 6 km. As the background radiance goes down the maximum range increases. In the limit of zero background radiance, i.e. night-time conditions, we have a signal to noise ratio of unity at a range of 0.45 km for V = 0.2 km and 11 km for V = 10 km.

If laser power is increased in order to increase range the effect is different for different values of V. If for example the laser pulse energy is increased by an order of magnitude to 2 J, thus also increasing $n_S\tau'$ by an order of magnitude, then with the bright cloud background as specified above the maximum range increases from 0.4 to 0.42 km with V = 0.2 km whereas with V = 10 km the increase is from about 6 to 9 km. The effect of increasing the telescope diameter is not so straightforward since as well as increasing the lidar return signal it also increases that due to background. There is an improvement, since it is the square root of the background which sets the detection limit but it is smaller than an equivalent gain from increasing laser power. For the sake of comparison take the unlikely possibility of being able to increase the telescope area by an order of magnitude. This increases $n_S\tau'$ by a factor of 10 and $(n_B\tau')^{\frac{1}{2}}$ by a factor of 3. Thus from Figure 3 with V = 0.2 km the improvement in range is from 0.4 to 0.41 km and with V = 10 km it is from 6 km to 8.2 km. Another possibility is to decrease the bandwidth of the filter, but care must be taken since a decrease in bandwidth is generally achieved at the expense of transmission at the required wavelength.

It is important to remember that these figures are only estimates of the maximum range from which detectable signals can be received. They are subject to the assumptions implicit in the calculation of α and β and to the assumption that a signal to noise ratio of one defines a detectable signal. In practice a larger signal to noise ratio may be required. Futhermore, it is generally some feature in the atmosphere which is to be detected and this itself in general contributes to both α and β. Nevertheless the figures give a general indication of the range of operation which is possible and the way in which it is affected by parameters such as laser power and telescope diameter.

4. SIGNAL ANALYSIS

The basic form of the lidar signal provided by atmospheric backscatter is that shown in Figure 2. For studies of meteorology and air pollution the objective of the analysis is to find the position in space of areas of enhanced backscatter caused by chimney plumes or features in the atmosphere such as clouds or haze layers. Repeated measurements over a period of time show the change and development of such features.

The most straightforward way of examining the return signals is to display signal strength against time on an oscilloscope screen. This gives a direct indication of the way backscatter varies with distance and gives an immediate measure of the range of features such as cloud height. In dense cloud α and β are very large so a substantial signal originates from the cloud base but the laser pulse penetrates only a short distance into the cloud. In this case the lidar is operating simply as a range finder in the same way as it would with

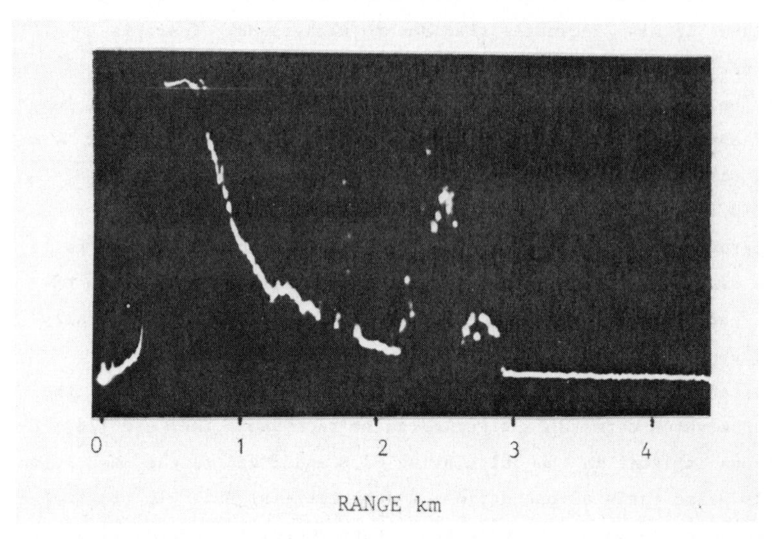

FIGURE 4 Backscattered signal with plume

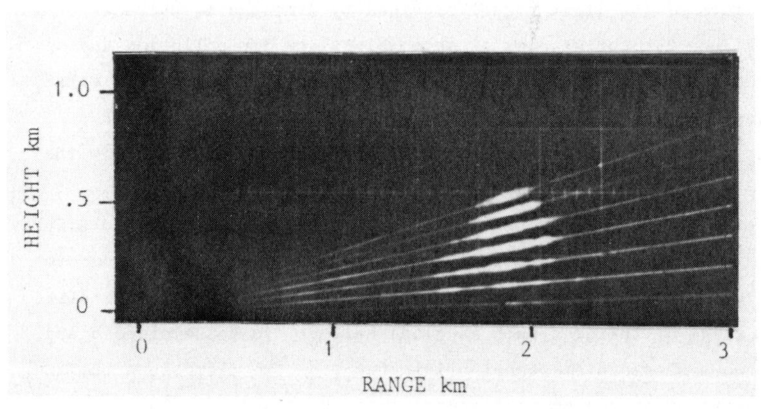

FIGURE 5 Oscilloscope display of plume cross-section

targets on the ground. On the other hand, less dense features such as chimney plumes and haze layers are traversed completely by the laser pulse and the oscilloscope display provides details of the dimensions and structure of the target. One such display (ref. 7) is shown in Figure 4 where the laser was fired through a power station plume. The signal exhibits the usual initial sharp rise and decay superimposed on which there is the backscatter from the plume. The first indication of the plume occurs at 1.2 km with a small perturbation to the signal followed by a much larger signal at 1.5 km. This section of the plume has a depth of about 200 m and is followed by a more substantial section starting at a range of 2.2 km and having a depth of 800 m. The far side of the plume has a clear cut boundary and it can be seen from the size of the signal that the beam has been signficantly attenuated in passing through it. The figure illustrates the structure of the plume which is not homogeneous but composed of irregular puffs. From such records, with measurements of the angle at which the laser was fired, the position of the plume features encountered along the path of the beam can be determined.

The same techniques can be used to scan an area of the sky. The laser is fired at a series of vertical angles and by noting the elevation of each shot and the range from which signals are received simple geometry gives the shape, dimensions and position of the area of enhanced backscatter. However, this soon becomes tedious if many shots are used and the first step in improving the analysis is to use the method illustrated in Figure 5 (ref. 7). Here the path of the pulse through the atmosphere is represented directly on the screen, a result achieved by coupling a potentiometer to the axis of elevation of the laser and using the voltage to provide the corresponding trajectory of the oscilloscope beam. The backscattered signal is used to modulate the intensity of the beam. Thus a bright region on the oscilloscope represents a cross-section of the area of enhanced backscatter in the atmosphere, in this case a cross-section of the plume. Displays such as this provide vivid, immediate information. A similar technique can be used with the laser always firing vertically upwards but with the signal from each successive shot displaced slightly along the x-axis, as illustrated by Figure 6. This shows measurements of the atmosphere under stable conditions in the morning where there are distinct layers of haze in the first 1500 m or so and above which there is relatively clear air. Later in the day as the earth is warmed by the sun the stability is disturbed by convection and the display shows the resulting turbulence in the haze layers.

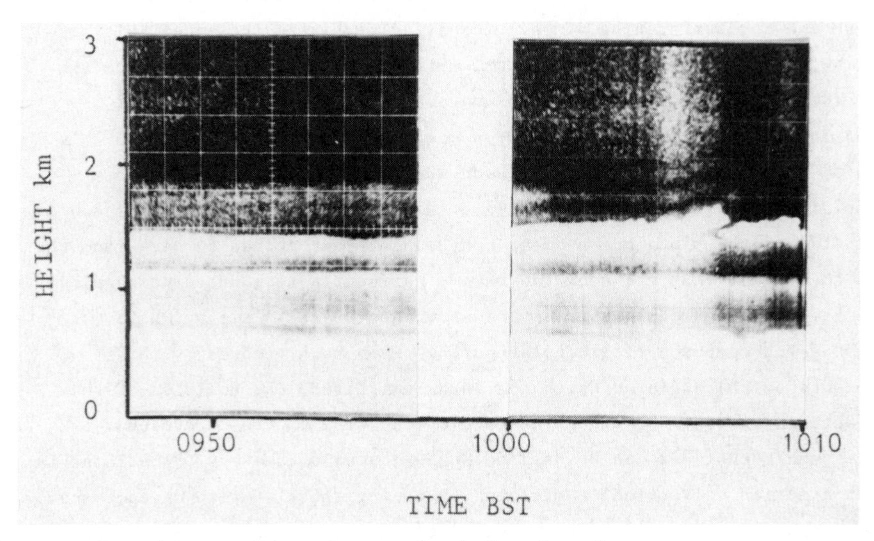

FIGURE 6 Lidar profiles of atmospheric boundary layer

Figures 2 and 4 illustrate one of the problems encountered in simple displays of this sort. It can be seen that after the initial rise, caused by the overlap of the telescope field of view and the laser beam, the signal rapidly decreases. This is due primarily to the $1/r^2$ term in equation (1) but also to the attenuation of the pulse as it travels through the atmosphere. The result is that the accuracy with which the effects of enhanced backscatter can be measured decreases rapidly with range. It also means that if backscatter from different ranges is to be compared then compensation has to be made for the effects of $1/r^2$ and attenuation. One method of doing this is to vary the gain of the photomultiplier. This is achieved by increasing the photomultiplier supply voltage in such a way that the gain increases as the square of the range. Further adjustment can be made to compensate for atmospheric attenuation depending upon the conditions which exist at the time of the experiment. In this way fairly constant signals can be obtained, typically for ranges between about 300 m and 3 km and it is upon this signal that the effects of enhanced backscatter are superimposed.

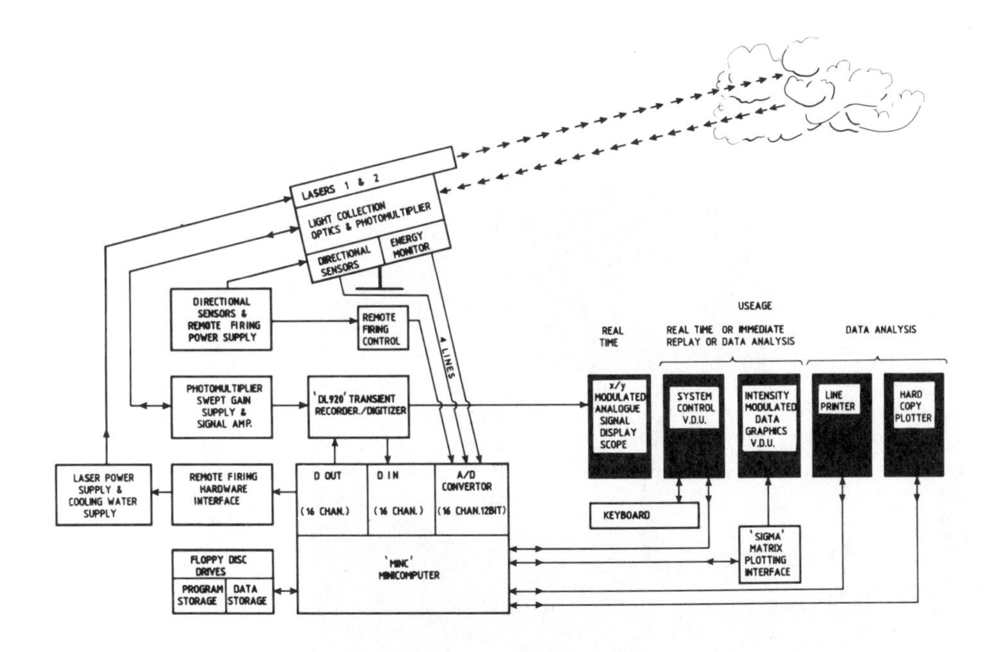

FIGURE 7 System diagram for CERL Ruby Lidar

The techniques described above provide excellent qualitative visualisation of return signals. For more detailed analysis it is clear that the examination of oscilloscope displays is unsatisfactory and that considerable advantage is to be gained from digitising the signal and coupling the system to a computer. An outline of the system used on the Electricity Generating Board ruby lidar, is shown in Figure 7, with a view of the vehicle in Figure 8. The photomultiplier is used with swept gain and the return signals are digitised in a 20 MHz, 8 bit transient recorder. The unprocessed signals are monitored by direct display on an oscilloscope screen and also stored on floppy disk in a DEC MINC computer. Azimuth and elevation of each laser pulse are also recorded together with the energy of each pulse as it leaves the laser. The signal is sampled every 50 ns, which is equivalent to a range resolution of about 7.5 m, and up to 1000 samples can be taken of each shot. The signals can be displayed immediately in an intensity modulated mode, either as an X-Y-signal strength display or as range-time-signal strengths. This is similar to the analogue displays used simply with an oscilloscope. Here however intensity modulation is achieved with a 6 level grey scale and a first step in processing is achieved by normalising each signal with respect to the energy in each pulse.

FIGURE 8 Central Electricity Generating Board Ruby Lidar

In terms of display this represents little advance over direct analogue displays, in fact in analyses of the mixing layer care must be exercised in identifying apparent structure introduced by the discrete nature of the grey scale quantisation. It does have the advantage of being able to recall the display at will without recourse to photographs but the great advantage lies in being able to do further analysis. Signals can be normalised and the results of many shots can be averaged. The statistical nature of plume dispersion can be investigated and corrections for variations in background attenuation can be incorporated in the analysis. Also the results are readily available for approaching the problems of making the results quantitative rather than just qualitative. This is difficult, the problems arising principally from the complexity of scattering from aerosols of unknown shape, composition and size distribution. This will be dealt with in more detail in the chapter on differential lidar and useful reviews will be found in refs. 1 and 11.

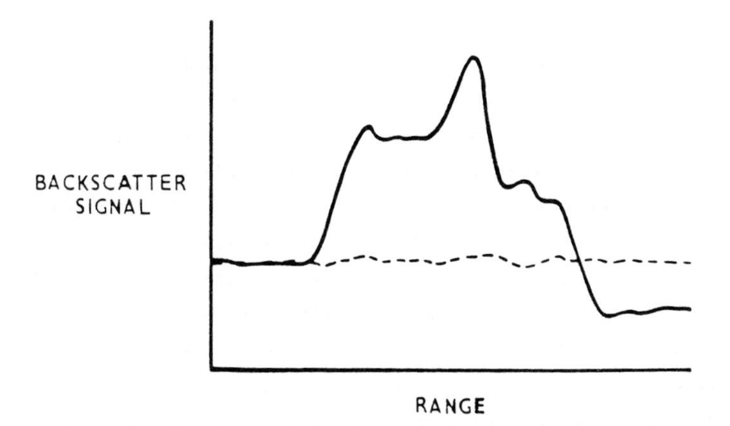

FIGURE 9 Diagram of attenuation caused by feature of interest. Dotted line shows backscatter in absence of smoke, adjusted by variable gain to be constant with range. Solid line shows backscatter with smoke. The attenuation is shown by the lower level of the signal at ranges beyond the plume

As an illustration of the use of a computer coupled lidar system in measurements where quantitative results can be obtained we can consider a study of dispersion in the wake of a large building. This was carried out at Oldbury Nuclear Power Station in Gloucestershire, England. The objective was to measure the density distribution of releases from the station as they were carried downwind and became entrained in the airflow around and into the wake of the building. A tracer gas, SF_6, was released from the roof at a known rate and measured with an array of ground level sensors. At the same time as SF_6, smoke was released from the same place. This was detected with the ruby lidar system described above arranged in such a way that it could make measurements of a vertical cross-section of the wake in a place coinciding with a line of ground level SF_6 sensors. If it is assumed that the smoke is released at a constant rate, that the size distribution and composition of particles are homogeneously distributed through the smoke and that the size distribution does not change as the smoke becomes entrained in the wake, then the size of the backscattered signal should be proportional to the density of the smoke particles. The results obtained at ground level, coinciding with the SF_6

monitors, were used to calibrate the system so that measurements over the complete cross-section could be interpreted in terms of a source dilution.

Atmospheric attenuation and the $1/r^2$ term were compensated for by firing the laser along a path at low elevations with no smoke present and adjusting the variable gain to give a signal as close as possible to being constant over the range where the wake was to be found, in this case 0.7-0.9 km. In general, attenuation due to the atmosphere varies greatly with angle of elevation, on a hazy day horizontal visibility is much less than vertical visibility, because the distribution of scattering centres has a marked vertical variation. In general therefore the correction for background attenuation will vary with the angle of elevation. In this study it was found that over the cross-section of the wake the correction was to a good approximation constant. In cases where complete compensation cannot be made with variable gain, corrections can be made to the signal in store as part of the processing procedure. A further check is required before the signals can be interpreted quantitatively, the attenuation caused by the smoke itself must be small or must be taken into account in the analysis. This is illustrated diagrammatically in Figure 9. The signal is clearly lower after the laser pulse has traversed the plume. The simplest way to allow for this is to assume a uniform distribution through the plume which would give an extinction coefficient which could be calculated from the measured width of the plume and the measured attenuation. In many cases it is clear from the backscattered signal that the distribution is far from being uniform. Then the procedure would be to measure the total attenuation caused by the plume, normalise it with respect to the integral of the backscatter signal across the plume and then in the analysis assume attenuation to be proportional to backscatter. The only relationship implied between extinction and backscatter being that the ratio is constant across the plume. In the measurements of smoke at the power station the attenuation was so small that no correction was necessary. Results are shown in Figure 10. In (a) are return signals from pulse trajectories about 17 m above ground level. The dotted line is the average of 7 shots obtained without smoke and with the variable gain adjusted to give a signal constant with range. The solid line is the average of 32 shots with smoke, normalised to the peak of the distribution of SF_6 measured by ground level samples, the result of which are indicated by crosses. The curves show that there is little attenuation caused by the smoke and that although the two distributions are slightly displaced the shapes are similar. The second figure, (b), shows a vertical cross-section of the plume obtained from the average of 30 cross-sections each comprising 15 individual shots. Averages and cross-sections such as these can be obtained relatively easily once the lidar is coupled to a computer.

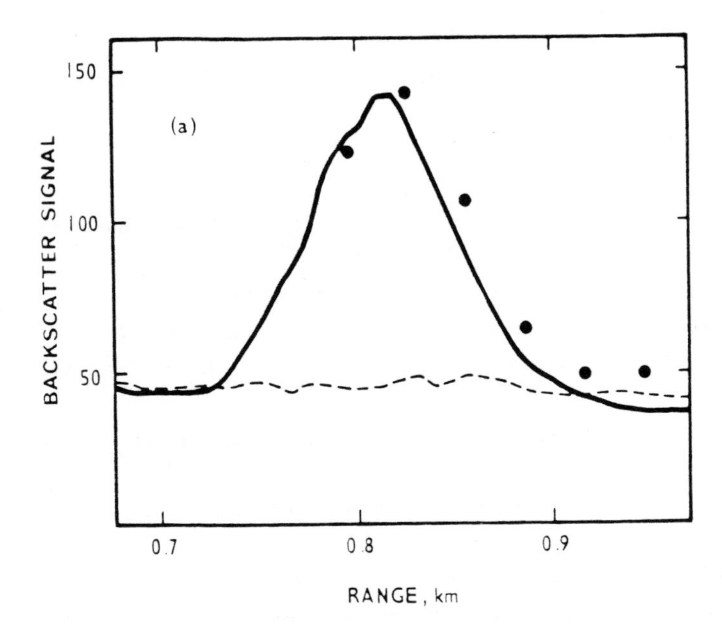

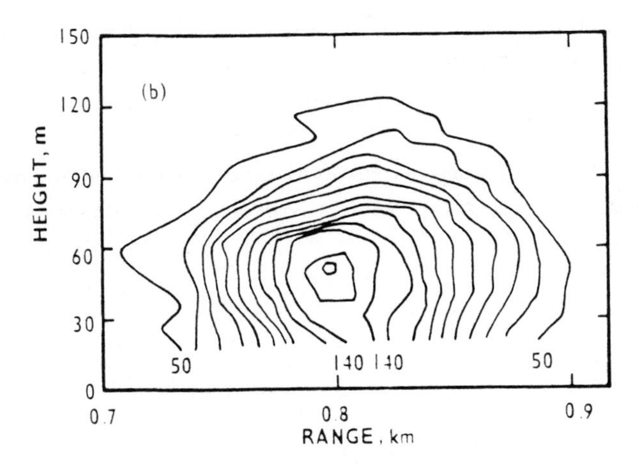

FIGURE 10 Lidar scans of smoke release 160 m downwind of power station. (a) shows lidar return signals 17 m above ground level. Dotted line is average of 7 shots without smoke, adjusted by variable gain to be constant with range. Solid line is average of 32 shots with smoke, normalised to peak value of distribution measured with ground level sensors, marked with dots. (b) shows vertical cross-section of plume, the average of 30 scans, each comprising 15 shots

ACKNOWLEDGEMENT

This work is published by permission of the Central Electricity Generating Board.

REFERENCES

1. R.T.H. Collis and P.B. Russel, in 'Laser Monitoring of the Atmosphere', E.D. Hindley. Ed. Springer, Berlin, 1976.
2. R.L. Byer and M. Garbuny, Appl. Opt. 12 (1973) 1496-1505.
3. E.J. McCartney, 'Optics of the Atmosphere', Wiley, New York, 1976.
4. Van de Hulst, 'Light Scattering by Small Particles', Wiley, New York, 1957.
5. P.W. Kruse, L.D. McGlaughlin and R.B. McQuistan, 'Elements of Infrared Technology', Wiley, New York, 1963.
6. W.E.K. Middleton, 'Vision Through the Atmosphere', University of Toronto Press, 1952.
7. P.M. Hamilton, Phil. Trans. Roy. Soc. 265 (1969) 153-172.
8. G.S. Kent, P. Sandland and R.W.H. Wright, J. Appl. Meteorol. 10 (1971) 443-452.
9. R.S. Adrain, D.J. Brassington, S. Sutton and R.H. Varey, Opt. and Quant. Electronics, 11 (1979) 253-264.
10. K. Ya Kondratyev, 'Radiation in the Atmosphere', Academic Press, New York, 1969.
11. H. Müller and H. Quenzel, 'Study on the Information Content of Ground Based Lidar Measurments', Final Report Euratom Contract 974-78-11 SISP D, 1980.

Optical Remote Sensing of Air Pollution,
Lectures of a course held at the Joint Research Centre, Ispra, Italy, 12—15 April 1983,
P. Camagni and S. Sandroni (Eds). 143—180

DIFFERENTIAL LIDAR AND ITS APPLICATIONS

D.J. BRASSINGTON

1 INTRODUCTION

Differential lidar is a technique in which the range resolved concentration of pollutants along the laser beam can be determined by measuring the optical absorption which they produce (ref. 1-6). The technique has many acronyms, the one now most commonly used being DIAL (DIfferential Absorption Lidar); in the past DAS and DASE have been used (Differential Absorption and Scattering [Equipment]). The term differential lidar is also sometimes applied, wrongly, to measurements in which the laser pulse is scattered back to the measuring site by a diffusely scattering target such as a tree or building. This produces a long-path absorption measurement with no range resolution and so is not a true DIAL measurement.

A DIAL system is similar to a LIDAR system except that the laser must be tunable. By analyzing the return signal with a fast digitizer (as in LIDAR) the radiation scattered back from any particular range along the beam can be measured. If we consider the size of the return signal at say 10 μs after the firing of the laser we know that this must correspond to radiation which has travelled to a range of 1.5 km ($= c \times 10^{-5}/2$) and been scattered back. The size of this signal will thus depend, apart from fixed geometric factors, on the amount of back scattering at 1.5 km range and on the absorption produced by the pollutant over the round trip to 1.5 km and back. The scattering is in general unknown but can be eliminated by repeating the measurement at a slightly different wavelength where the absorption coefficient of the pollutant is different but the scattering intensity is virtually the same. The change in signal between the two wavelengths thus corresponds entirely to the different amounts of absorption at the two wavelengths, and if the absorption coefficients of the pollutant at these two wavelengths are known the total burden of pollutant between the measuring site and 1.5 km range can immediately be deduced. (Total burden is the integrated concentration $\int_0^r n(r')dr'$ where n(r) is the concentration at range r.) Since the DIAL return signal is in fact continuous we can make this calculation at a whole series of ranges and determine the total burden as a function of range. By differentiating this total burden with respect to range one arrives at the pollutant concentration as a function of range.

This then is the principle of DIAL. A more precise treatment can be given starting with the LIDAR equations for the two wavelengths, which are traditionally known as the on-resonance and off-resonance wavelengths (on-resonance referring to the wavelength at which the absorption is largest - usually the peak of an absorption line or feature):

$$P_{on}(r) = \frac{E_{on}TA\beta_{on}(r)c}{2r^2} \exp\left[-\int_0^r 2\alpha_{on}(r')dr'\right] \quad (1)$$

$$P_{off}(r) = \frac{E_{off}TA\beta_{off}(r)c}{2r^2} \exp\left[-\int_0^r 2\alpha_{off}(r')dr'\right] \quad (2)$$

where the symbols have the same meaning as in the chapter on Principles of Lidar and the subscripts refer to the on and off resonance wavelengths. The atmospheric absorption coefficient can be expressed in terms of its components:

$$\alpha_{on}(r) = \alpha_M(r) + \alpha_R(r) + \sigma_{on}n(r) \quad (3)$$

$$\alpha_{off}(r) = \alpha_M(r) + \alpha_R(r) + \sigma_{off}n(r) \quad (4)$$

Here we are assuming that the Mie and Rayleigh scattering coefficients, α_M and α_R, change negligibly between the on and off resonance wavelengths. Even when the two wavelengths are so far apart that this is no longer true it is still sometimes the case that these two terms are small compared with $\sigma n(r)$ so that again the variation can be ignored. There are however cases (principally measuring ozone in the uv) where the variation of α_M and α_R is not negligible and corrections must then be made for this.

σ is the absorption cross-section of the pollutant (see Section 3) and n(r) the concentration of pollutant in molecules/unit volume. $\sigma n(r)$ thus represents the absorption produced by the pollutant. By dividing Eq. 2 by Eq. 1 we find:

$$\int_o^r n(r')dr' = \frac{1}{2(\sigma_{on} - \sigma_{off})} \log_e \left[\frac{P_{off}(r)}{P_{on}(r)} \cdot \frac{E_{on}}{E_{off}} \cdot \frac{\beta_{on}(r)}{\beta_{off}(r)} \right] \qquad (5)$$

and differentiating gives:

$$n(r) = \frac{1}{2(\sigma_{on} - \sigma_{off})} \frac{d}{dr} \log_e \left[\frac{P_{off}(r)}{P_{on}(r)} \right] \qquad (6)$$

To obtain (6) from (5) we have assumed that the ratio $\beta_{on}(r)/\beta_{off}(r)$ does not vary with range. We have not however needed to assumed that $\beta_{on}(r) = \beta_{off}(r)$. This is quite an important relaxation where λ_{on} and λ_{off} are widely separated since the ratio will only vary with range if the type of scatterer changes with range (e.g. if the aerosol size distribution changes).

Fig. 1 shows the variation with range of typical DIAL return signals for the case of a pollutant distribution consisting of a uniform concentration between R_1 and R_2. Also shown are the various stages in extracting the concentration from the measured returns.

There are various methods of determining the differential appearing in Eq. 6 from the experimental measurements but in the most straightforward scheme the return signals are sampled by a fast digitizer every τ seconds (where τ is of order 1 μs). τ determines the range resolution element Δr (= $c\tau/2$) and the concentration at r is then given in terms of two successive samples of P_{on} and P_{off} according to:

$$n(r) = \frac{1}{(\sigma_{on} - \sigma_{off})c\tau} \log_e \left[\frac{P_{on}(r)}{P_{off}(r)} \cdot \frac{P_{off}(r + \Delta r)}{P_{on}(r + \Delta r)} \right] \qquad (7)$$

We can use Eq. 7 to obtain an expression for the sensitivity, Δn, of a DIAL measurement giving:

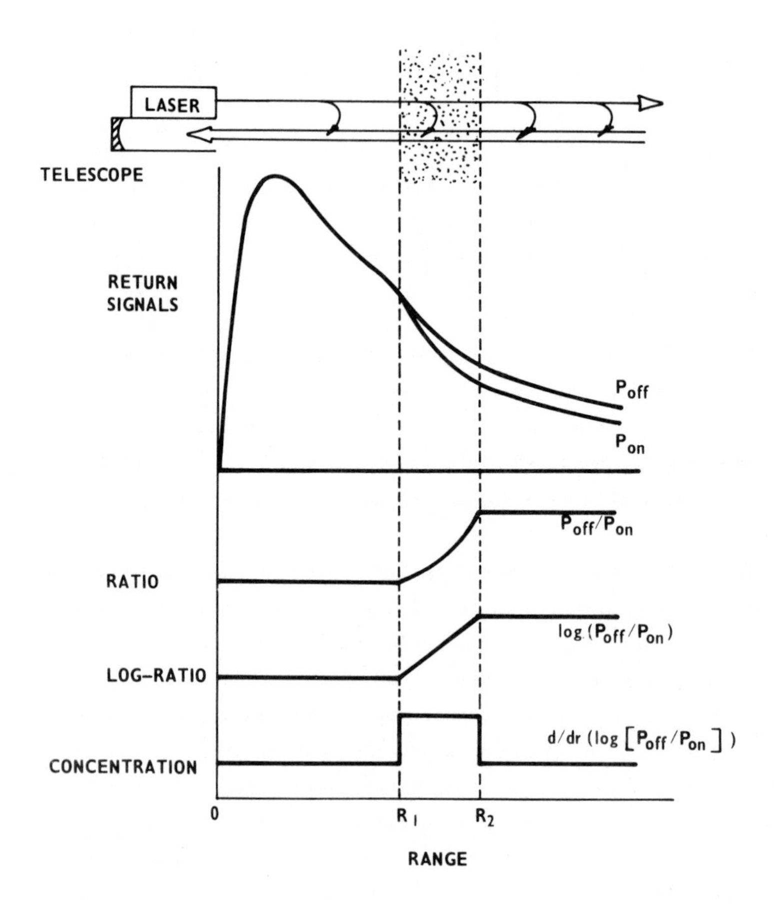

Fig. 1. Variation with range of the on and off resonance return powers and the stages in extracting from them the range-resolved pollutant concentration.

$$\Delta n = \frac{1}{(\sigma_{on} - \sigma_{off})c\tau} [\sum(\Delta P/P)^2]^{1/2} \tag{8}$$

where the summation is over the four measurements of P (on and off resonance, and at r and r + Δr). ΔP is the random error in each measurement of P and is given by:

$$\Delta P = \left[\frac{(P + P_b)\, h\nu}{q\tau} + \frac{\eta_D^{\,2}}{2\tau} \right]^{1/2} \qquad (9)$$

where P_b is the level of background radiation seen by the detector, q is the detector quantum efficiency, and η_D is the noise equivalent power (NEP) of the detector. The first term represents the shot noise and corresponds to the expression for signal to noise given in the chapter on Principles of Lidar; it is the dominant term for photomultiplier detection. The second term represents the inherent detector noise. For a photomultiplier this corresponds to the dark current and is negligible in DIAL measurements. For infrared DIAL systems where semiconductor detectors (e.g. indium antimonide or mercury cadmium telluride) must be used it is however the dominant term and the shot noise is usually negligible.

η_D is usually quoted as a noise power/$\sqrt{Hz}$. The factor $1/2\tau$ arises from assuming that this is integrated over a time τ in taking the sample. The correct expression will clearly depend on the precise details of the detection system following the detector (e.g. preamplifier pass-band).

Eq. 8 assumes a measurement comprising a single pulse pair. More usually in DIAL, sensitivity is improved by averaging over N pulse pairs. If the measurements are independent this gives a $\sqrt{N}$ reduction in Δn but often a smaller improvement than this is seen.

Examination of Eqs. 8 and 9 shows that detection sensitivity depends on $1/\tau^2$ so that reducing the range element by a factor 2 produces a fourfold decrease in sensitivity. In the infrared the sensitivity worsens by more than this because infrared detectors work less well at higher bandwidths. Thus a large penalty must be paid to achieve good range resolution and for most systems the limit is about 50 m corresponding to a sampling interval of 0.33 μs.

The above discussion relates entirely to random errors - for many systems, especially in the uv, the laser pulse energy is sufficient that systematic errors become important and we will consider these in later Sections.

2 MORE DETAILED DISCUSSION OF PARAMETERS IN THE DIAL EQUATION

In the analysis of the last Section some simplifying assumptions were made and we now examine these in more detail.

2.1 Range resolution

Although the question of sampling interval was dealt with in the previous Section it was assumed that this was a matter of experimental choice. In practice the laser pulse has a finite length and the detector has a time constant or bandwidth. The achievable range resolution will be determined by the longer of the laser pulse length and the detector time constant and there will be no point in sampling much faster than this. The effect of laser pulse-length can be computed by convolving the laser pulse shape with the ideal return signal given by Eqs. 1 and 2:

$$P'(t) = \frac{1}{E}\int_{0}^{\infty} P(t')\ \varepsilon(t' - t)dt'$$

where P' is the measured return signal and $\varepsilon(t)$ is the laser pulse shape: $(E = \int\varepsilon(t)dt)$. If the detector time constant is not negligible a further convolution must be performed with $\varepsilon(t)$ replaced by the response of the detector to a delta function input. This is correct assuming the detector is linear. (A non-linear detector would anyway be useless for DIAL measurements, because it would cause systematic errors).

When designing and building a DIAL system it is important to investigate the effect of these convolutions on simulated measurements because spurious results can sometimes result when the required resolution is comparable with the instrumental resolution.

2.2 The optical efficiency factor T

T represents the factor by which the amount of scattered radiation hitting the detector differs from the ideal value calculated from the solid angle subtended by the telescope primary mirror at the scattering site. T will normally be less than unity for several reasons:

(i) For the first hundred meters or so the scattered radiation will be out of focus on the detector. This out-of-focus image will often be larger than the detector so some radiation will be lost. This effect is worse for low f number telescopes and small detectors.

(ii) The detector area will normally be made sufficiently large that the field of view of the telescope system includes the whole of the laser beam. (Typically both laser beam divergence and telescope field of view would be ≈1 mrad.) However this is not always true and some radiation may be lost because of this.

(iii) Where the laser beam is not emitted coaxially with the telescope centre line there will be a certain initial range over which the laser beam is not fully within the telescope field of view, so reducing the signal at close range. It may also be desirable to purposely cross the laser beam and telescope centre line at a fixed range (rather than at ∞) resulting in a further loss of signal at longer ranges.

(iv) There will inevitably be some obscuration of the telescope primary mirror by, e.g. the secondary mirror. This will cause a reduction in T at all ranges but the effect will be greatest at close range.

(v) For a uv or visible system a narrow pass-band filter will be needed in front of the detector to reduce the value of P_b and thus the shot noise associated with it. This filter would typically have a pass band of 5 nm and a peak transmission of only 20%.

(vi) There will be reflection and transmission losses on all other optical components in the telescope system.

(vii) Atmospheric turbulence and associated variations of refractive index produce deviations in the path of both the laser beam and the scattered radiation. The effect is that not all the scattered light is brought properly to a focus on the detector causing a loss in signal which fluctuates both with range and with time.

Factors (i), (iii) and (iv) lead to a loss of signal at close range and produce the typical return signal which increases from zero, at zero range, to a peak and then decays with an approximate $1/r^2$ dependence. These three factors can actually be used to advantage in reducing the signal from close range (ref. 11). A large signal from this region would produce severe requirements on the dynamic range of the detection system.

The factor T is clearly strongly range dependent. This dependence does not in itself affect the derivation of the DIAL equations but it is most important that the range dependence of T does not vary between the on and off resonance wavelengths, i.e. $T_{on}(r)/T_{off}(r)$ must be constant. Any variation would lead to systematic errors in the pollutant measurement. This implies that where the two wavelengths are produced by two separate lasers their beams must be accurately aligned and the beam divergences must be the same. Where a single laser is used care must be taken that the beam direction does not change with wavelength - this can often occur as tuning elements within the cavity are adjusted.

2.3 Interference from other pollutants

In Eq. 3 and 4 the only absorption considered was that due to the pollutant being measured. However there will often also be absorption due to other atmospheric constituents or pollutants producing additional terms in Eq. 3:

$$\alpha_{on}(r) = \alpha_M(r) + \alpha_R(r) + \sigma_{on} n(r) + \sum_j \sigma'_{on,j}\, n'_j(r) \qquad (10)$$

Here $\sigma'_{on,j}$ represents the absorption coefficients, at the on-resonance wavelength, of other atmospheric constituents and $n'_j(r)$ represents their concentration. Similarly for Eq. 4. One would clearly try to choose λ_{on} and λ_{off} so that absorption by the unwanted atmospheric constituents was small but if this is not possible then there are three other approaches to the problem:

(i) If there is only one interfering species it is often possible to choose λ_{on} and λ_{off} so that $\sigma_{on} = \sigma_{off}$ for the interfering species whilst $\sigma_{on} > \sigma_{off}$ for the wanted species. The measurement is then insensitive to the presence of the interfering species.

(ii) If the concentration of the interfering species is known or can be measured easily by other means (as is the case for e.g. CO_2 or H_2O) then a correction can be applied.

(iii) By measuring the DIAL return at three wavelengths instead of two it is possible to determine the concentrations of two pollutants; similarly with n wavelengths, n-1 pollutants can be measured.

2.4 Effect of non-simultaneous measurement of P_{on} and P_{off}

When two lasers are used to produce λ_{on} and λ_{off} the two pulses can be emitted simultaneously. However when a single laser is used there will be a delay between λ_{on} and λ_{off} and this can cause both T and β effectively to change between wavelengths, whereas the derivation of the DIAL equation assumed they were constant with λ. T changes with time because of the atmospheric turbulence mentioned in Section 2.2. This turbulence has a spectrum which extends upto about 1 kHz (ref. 35), so provided the interval between P_{on} and P_{off} is $\ll 1$ ms. T will not change and no extra errors will be produced (ref. 12). β will change with time mainly in the case of plume measurements where the particulates in the plume produce a large contribution to β. In this case turbulence in the plume and the wind speed will cause β to vary. The time separation necessary to avoid effects due to this cause will depend on wind speed and the range of the plume but again a separation of <1 ms should be sufficient.

If the pulse separation cannot be reduced sufficiently to avoid these effects then an additional random error on the concentration measurement is introduced, but no systematic error. The variation of T due to turbulence is typically only a few percent but with dense plumes very large variations of β with time can occur.

When many pulse pairs are averaged to produce a measurement it is necessary to take the ratio of each pulse pair separately before averaging otherwise the advantage of quasi-simultaneous measurement of P_{on} and P_{off} is lost.

2.5 Effect of large separation of λ_{on} and λ_{off}

For pollutants whose absorption cross-section varies only slowly with wavelength it is necessary to use a large separation between λ_{on} and λ_{off} to attain adequate sensitivity (e.g. ozone measurement at 290 nm). We have already seen that a variation of $\beta(r)$ with λ does not cause an error provided the range dependence does not vary with λ. However if the size distribution of the scatterers changes with range this will no longer be true. A particular case where this occurs is in the uv where both molecular (Rayleigh) scattering and aerosol (Mie) scattering contribute. If the Mie scattering contribution varies with range (as it does as one traverses the mixing layer) then the range dependence of $\beta(r)$ varies with λ and systematic errors result (ref. 13).

The total scattering coefficient ($\alpha_R + \alpha_M$) also varies with λ and in this case <u>any</u> variation causes a systematic error since it is equivalent to differential absorption by the pollutant. Once again the error is greatest in the uv because Rayleigh scattering (which is only significant in the uv) shows a λ^{-4} wavelength dependence.

2.6 Non-monochromatic laser output

The spectral width of the laser output, although often very narrow, is always finite. Where $\sigma(\lambda)$ varies across the laser linewidth Eqs. 1-4 should be replaced with integrations across the laser linewidth and in this case the derivation of Eq. 5 no longer follows (essentially because the absorption of the laser output is no longer exponential). Assuming the usual DIAL equations then leads to errors (ref. 57).

3 ABSORPTION SPECTRA OF POLLUTANTS AND ATMOSPHERIC CONSTITUENTS

3.1 General

The transmission of monochromatic radiation through an absorbing gas is described by

$$I(x) = I_0 \exp(-\sigma n x) \qquad (11)$$

where I is the intensity after passing through a length x of the gas, I_0 is the incident intensity, σ is the absorption cross-section and n the concentration in molecules/unit volume. An alternative formulation of Eq. 11 is commonly used in which σn is replaced by αp where α is known as the absorption coefficient. The relation between α and σ (assuming a perfect gas) is:

$$\alpha = \sigma/kT$$

α is usually expressed in units of $atm^{-1}\ cm^{-1}$ and if σ is expressed in m^2 the numerical relationship at T = 20°C is

$$\alpha = 2.504 \times 10^{23}\ \sigma$$

The disadvantage of using α is its strong temperature dependence.

Absorption spectra normally result from the superimposition of absorption lines, each line corresponding to a particular electronic, vibrational, or rotational transition of the molecule. Electronic transitions produce absorption in the uv or at the blue end of the visible spectrum. Vibrational transitions produce absorption in the near and middle infrared, and rotational transitions produce absorption in the far infrared (>20 μm). Rotational transitions are not useful for DIAL because of the lack of tunable lasers in this region and also because of severe water vapour absorption.

3.2 The uv/visible region

Although many molecules show electronic absorption, in almost all cases the absorption occurs at <230 nm where the atmosphere is opaque because of oxygen absorption. The main exceptions are SO_2, O_3, and NO_2. Electronic spectra are extremely complicated with the absorption lines merging to form a continuous spectrum even when pressure broadening is reduced by working at low pressure.

The absorption spectrum of SO_2 in the 290-320 nm region is shown in Fig. 2 (ref. 14). It displays a series of well defined peaks about 2 nm apart - this is ideal for DIAL since λ_{on} and λ_{off} can be set to an adjacent peak and trough achieving maximum differential cross-section with minimum wavelength separation. The spectrum shown was measured with a 0.05 nm resolution. Higher resolution studies (ref. 15) show considerably more structure on the spectrum on a scale of ~.005 nm. This has two main consequences: (i) if the DIAL laser has narrow linewidth (<.01 nm) then the differential cross-section, and hence the calibration of the system, will be very sensitive to small changes in wavelength; (ii) for a DIAL laser of linewidth >.01 nm the cross-section will vary across the laser's linewidth leading to non-exponential absorption of the radiation. This in turn produces errors in the calculated SO_2 concentration although usually of only a few per cent (ref. 57).

The temperature dependence of the SO_2 cross-section has been measured (ref. 16) and a change of 20% in the differential cross-section between 20°C and 100°C was found. This must be taken into account when making measurements on plumes which may be above atmospheric temperature.

The ozone absorption spectrum (ref. 17) is shown in Fig. 3. It shows much less structure than the SO_2 spectrum and between 275 and 300 nm the absorption cross-section shows a monotonic decrease. In order to achieve adequate sensitivity a large wavelength separation of ~10 nm is needed.

NO_2 absorbs over a considerable region of the uv and visible but the region most suited to DIAL measurements is between 440 and 500 nm where the spectrum shows considerable structure and differential cross-sections of about 24×10^{-24} m^2 are obtainable (ref. 18).

3.3 The infrared region

Absorption in the infrared is due to fundamental, overtone and combination bands of vibration-rotation transitions. For diatomic molecules the spectra are quite simple with widely separated lines in a regular sequence. However, for larger molecules the spectra become more complex and the individual lines are often not resolvable at atmospheric pressure. They usually are resolvable, though, at reduced pressure (because pressure broadening is reduced). At wavelengths below 2.5 μm absorption is due to overtone and combination bands which are usually much weaker than the fundamentals so this region is not suited to DIAL measurements. However above 2.5 μm most molecules show strong absorptions and this makes the infrared a potentially more useful region than the uv where only SO_2, O_3, and NO_2 can be measured.

An absorption line in the infrared is characterized by its integrated intensity S:

$$S = \int_0^\infty \sigma(\nu)d\nu$$

where ν is the wavenumber ($\nu = 1/\lambda$). The width of the line is determined by the pressure. Below 10 mbar the line is said to be Doppler broadened - that is the linewidth is due the Doppler shift produced by the movement of the molecules. Above 10 mbar, pressure broadening has an effect and at atmospheric pressure it is dominant. The lineshape is then Lorentzian:

$$\sigma(\nu) = S\gamma_L/\pi[(\nu-\nu_0)^2 + \gamma_L{}^2]$$

where ν_0 is the centre frequency and γ_L is the pressure broadened half width at half maximum. γ_L varies with temperature and pressure approximately according to $p/T^{1/2}$. S also varies with temperature as the population of the rotational levels of the ground state changes. γ_L is typically between 0.05 and 0.1 cm^{-1}.

3.4 Atmospheric absorption

The atmosphere must be transparent at the DIAL wavelengths. Fig. 4 shows the atmospheric absorption as a function of wavelength for a 1 km path at sea level. Above the region of oxygen absorption there is little absorption in the uv/visible region. In the infrared however there are several bands of strong water vapour and CO_2 absorption and this restricts DIAL wavelenths to the various atmospheric windows i.e. 1.5-1.8 μm, 2-2.5 μm, 3.4-4.4 μm, 4.5-5.0 μm and 8-13 μm.

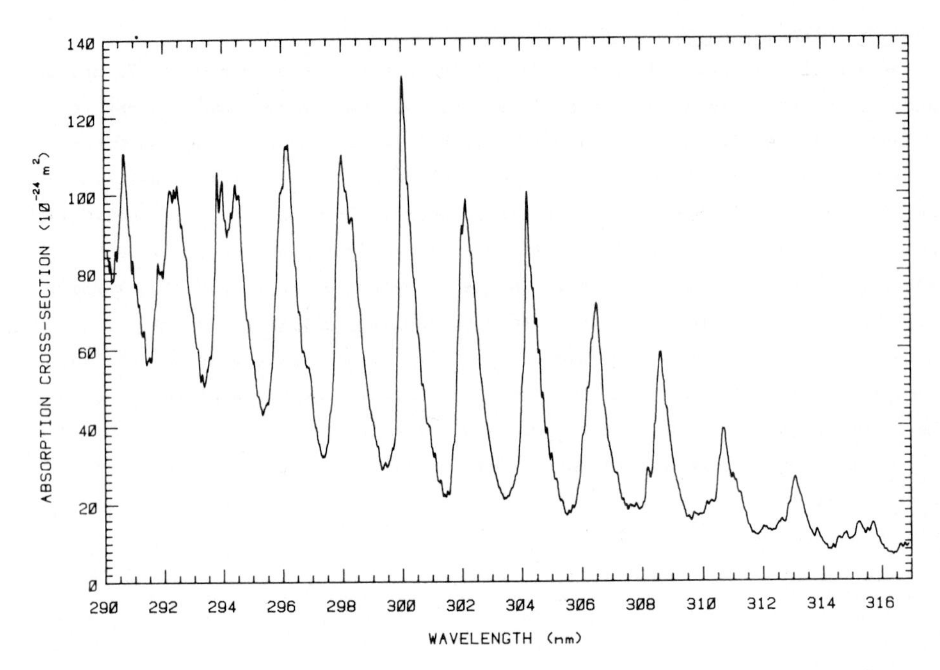

Fig. 2. SO_2 absorption spectrum.

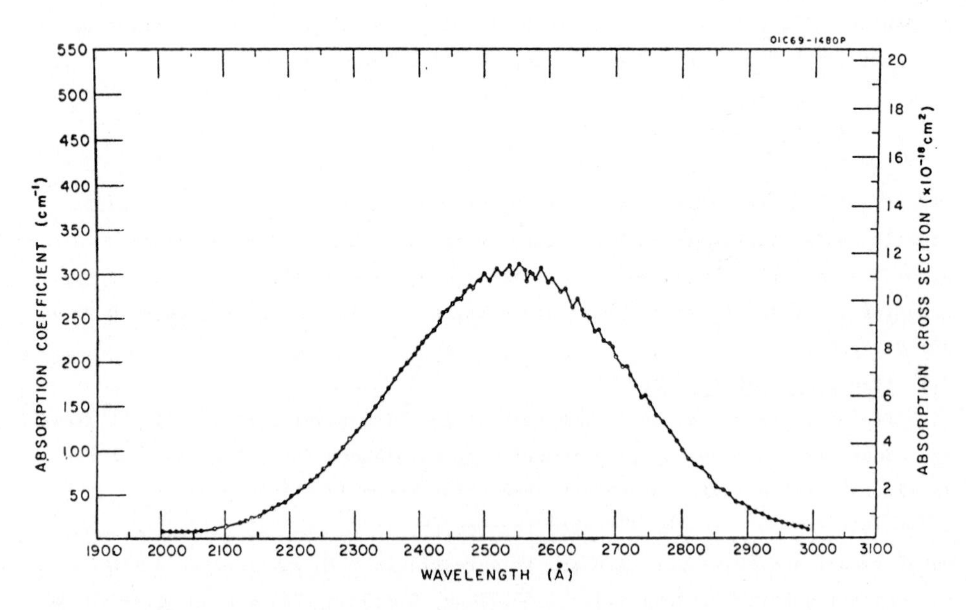

Fig. 3. Ozone absorption spectrum. Reprinted with permission from ref. 17. Copyright (1959) American Chemical Society.

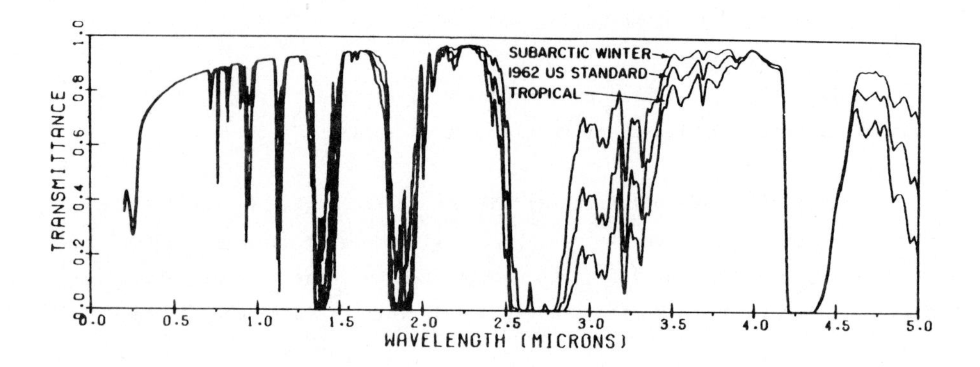

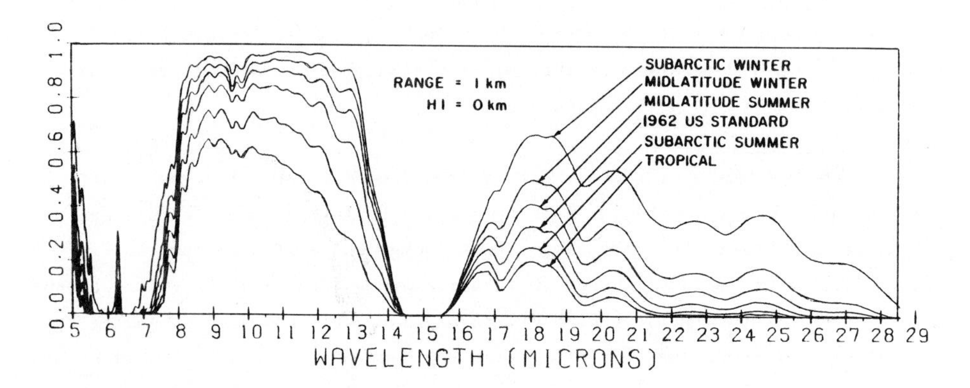

Fig. 4. Atmospheric transmittance for a 1 km path at sea-level for six model atmospheres (ref. 60).

3.5 Sources of information on absorption cross-sections

References for SO_2, O_3 and NO_2 have already been given. Additional data for the uv/visible region are contained in references 19-23.

In the infrared the main source of data is the AFGL line compilation (refs. 24,25). This comprises two data tapes, one for normal atmospheric constituents and one for pollutants, containing between them information on over 100 000 absorption lines. With the necessary computer program, absorption spectra can be generated for various atmospheric conditions and wavelength intervals. The data is also published in the form of atmospheric transmittance spectra (ref. 26) and an absorption lines atlas (ref. 27). Other sources of information are standard infrared absorption spectra compilations (refs. 28,29), and measurements on individual molecules published in the literature. Various measurements of absorption cross-sections for particular laser lines, e.g. CO_2 laser lines, are also available.

4 LASERS FOR DIAL MEASUREMENTS

4.1 General requirements

The laser is the most critical component of any DIAL system. The most important parameters in its specification are as follows:

(i) Pulse energy

The higher the laser pulse energy the better the signal to noise ratio for a given range or the greater the range for a given signal to noise ratio. In some systems the pulse energy is high enough that systematic errors are dominant in which case there is no advantage in further increasing pulse energy.

(ii) Pulse-length

This sets a limit to the range resolution. A 1 μs pulse-length is equivalent to a 150 m range resolution. For atmospheric measurements a pulse-length of 100 ns or less is adequate.

(iii) Beam quality

A beam divergence of 1 mr or less is required for the output beam. This is necessary for a uv system so that the field of view of the telescope can be minimized in order to reduce the amount of background light seen by the detector. For an infrared system it is necessary so that a small area detector may be used to reduce noise. Beam divergence can be improved by expanding the beam with a telescope and this also makes the beam more eye-safe. However this option is open only if the laser beam has a small diameter initially. Ideally a diffraction limited beam is required.

(iv) Repetition rate

A high repetition rate speeds the aquistion of the required number of pulses for averaging. Rates greater than 20 Hz would however cause problems for the computer in aquiring the data. Rates less than 1 Hz make the system difficult to align.

(v) Linewidth and wavelength stability

In the infrared the laser linewidth should be less than the absorption linewidth i.e. $<0.1\ cm^{-1}$ (ref. 58). For a continuously tunable infrared source a major problem is setting and maintaining the wavelength on the line centre. For a 10% accuracy the laser wavelength must be set and maintained to within $\sim 0.02\ cm^{-1}$ for a typical absorption line with a γ_L of $0.07\ cm^{-1}$ (ref. 30).

For the uv the problem is not so severe because the wavelength scale of gross structure in the spectrum is much larger. Usually a linewidth of 0.1 nm is sufficient.

(vi) Suitability for field use

The laser needs to be easy to operate, reliable, rugged, reasonably compact and have a low requirement on external services.

4.2 Lasers for uv and visible measurements

4.2.1 Dye lasers

In this region dye lasers are the almost universal choice. Various pumping options are available: Nd:YAG, flashlamp and excimer in order of current popularity. The Nd:YAG pumped dye laser is undoubtedly the current standard and Table 1 summarises typical parameters for such a system. Wavelengths in the 280-300 nm region for SO_2 and O_3 measurement are obtained by frequency doubling 560-600 nm output from a Rhodamine dye. Frequency doubled Nd:YAG radiation at 532 nm is used to pump the dye. For NO_2 measurements at around 480 nm tripled Nd:YAG radiation is used to pump usually a Coumarin dye. The Nd:YAG pumped dye laser offers almost ideal performance for DIAL with good beam quality, high output pulse energy, short pulse legth, and narrow linewidth.

Table 1: Typical Specifications for a Nd:YAG Pumped Dye Laser

Nd:YAG Pulse Energy	1.06 μ	700 mJ
	532 nm	225 mJ
	355 nm	125 mJ
Dye Laser Pulse Energy	600 nm	60 mJ
	300 nm	15 mJ
	480 nm	20 mJ
Repetition Rate		20 Hz
Beam Divergence		0.5 mrad
Beam Diameter		5 mm
Linewidth		0.1 cm^{-1}
Pulse Length		10 ns

Such a system is, however, expensive and sometimes the cheaper solution of flashlamp pumping is employed. The main differences are a longer pulse length (between 100 ns and 1 μs), slower repetition rate (~1 Hz) and lower reliability associated with the finite flashlamp lifetime. The longer pulse length limits range resolutiom and the pulse shape sometimes shows a long tail. This can be an embarassment since the tail will be producing a strong return from close range at the same time that the main part of the pulse is producing a weaker return from longer range. This is a case where it is particularly important to investigate the convolutions mentioned in Section 2.1. The slower repetition rate is often compensated by a high pulse energy (possibly 1 J in the visible).

4.2.2 Excimer lasers

Excimer pumping of dye lasers has yet to be used for DIAL but excimer lasers can be used on their own to produce uv output (ref. 32). Various output wavelengths are available, some having limited tunability e.g. the KrF laser can be tuned over the range 247.6-249.7 nm, although not continuously. Closer examination however shows that none of the excimer lasers produce wavelengths suitable for DIAL monitoring with the possible exception of XeCl, which can be tuned between 307.6 and 308.5 nm. This coincides with SO_2 absorption but the differential cross-section is low (ref. 14).

The main advantage of excimer lasers, if they produced suitable wavelenths, would be simplicity - a single laser compared with the complexity of the Nd:YAG pumped dye system.

A future possibility for the use of excimer lasers is in combination with Stimulated Raman Shifting. Here one focusses the beam into a high pressure cell of e.g. H_2, D_2, or CH_4 and the radiation is shifted by multiples of a fixed amount (4155 cm^{-1} for H_2) with high efficiency (~30%). This gives access to a large number of new wavelengths and some look promising for DIAL.

4.3 Lasers for infrared measurements

4.3.1 Line tunable infrared lasers

Since continuously tunable ir lasers are not yet well developed almost all ir DIAL work has used line tunable lasers. With these one selects, with an intra-cavity grating, one of a range of possible lasing lines of the active medium. Those most commonly used for DIAL are the DF laser, which can produce output between 2.5 and 4.1 μm, corresponding to the 3.4-4.4 μm atmospheric window, and the CO_2 laser covering the range 9.1-11.0 μm corresponding to the 8-13 μm atmospheric window.

With line tunable sources one relies on finding chance coincidences between laser lines and pollutant absorption lines. There is thus little freedom to choose the optimum absorption lines and often the wing of a line must be used where the absorption cross-section is not well known and is sensitive to changes in atmospheric conditions. Also the on and off resonance wavelengths may be widely separated leading to systematic errors. An advantage of line tunability however is that the laser wavelength is accurately known.

Typical specifications for DF and CO_2 lasers are shown in Table 2.

Table 2: Typical Specifications for DF and CO_2 TEA Lasers

	DF	CO_2	
Wavelength Coverage	3.5-4.1	9.1-11.0	μm
Typical Single Line Pulse Energy	10 mJ	200	mJ
Repetition Rate	3	3	Hz
Beam Divergence	1.0	0.6	mrad
Beam Diameter	30	30	mm
Pulse Length	500	100	ns

photomultipliers used for uv/visible systems but shot noise and background light are less of a problem.

The best method of optimising the wavelengths for DIAL is by means of a computer simulation because the number of factors involved is so large. Such programs are fairly simple to write and are well worth the effort involved.

6 DIAL SYSTEM COMPONENTS AND DESIGN

The most important component, the laser, has already been discussed; this Section deals with the remainder of the system.

6.1 Telescope and optics

One of the first decisions in designing a DIAL system is whether steerability is required. For many measurements a fixed pointing direction (usually vertical) is sufficient, particularly if the system is aircraft or van mounted so that pollutant mapping can be achieved by flying or driving around. However if steerability _is_ required e.g. to produce 3D maps of concentration from a fixed site, then three design options are available:

(i) Mount laser and telescope together and point the entire assembly in the required direction. This solution requires a compact laser of rugged construction that will operate in a range of orientations.

(ii) Keep the laser fixed and move the telescope. By using a system of mirrors to transmit the laser beam along the axes of rotation of the telescope, alignment can theoretically be maintained as the telescope moves. In practice it proves difficult to maintain the alignment of such a system and some adjustment is often required as the telescope is moved. Also reflexion losses through the chain of mirrors required in the laser beam path can be significant.

(iii) Keep laser and telescope fixed with the telescope and laser beam pointing vertically, and use a large plain mirror to steer both laser beam and telescope field of view in the required direction. The main disadvantage of this scheme is that the plain mirror must be larger than the telescope primary mirror and of equivalent optical quality. Another problem is that if elevations of more than about 60° are required the size of plain mirror becomes prohibitive. Vertical operation can always be achieved though by simply removing the plain mirror.

All three schemes are in use in DIAL systems around the world and the particular choice will depend on the detailed requirements of the system.

Telescope design for DIAL is now almost standard. A 0.5 m parabolic primary is used (the largest convenient size) of typically 2 m focal length. A Newtonian configuration is often employed but a Cassegrain or Gregorian design is more compact. For a uv system a variable aperture is placed at the telescope focus and this defines the field of view. A lens behind the aperture produces a parallel beam which passes through the background blocking filters onto the photomultiplier. It is necessary that the beam passing through the filters be parallel because interference filters have pass-bands which vary with angle. For an ir system the detector is usually placed at the focus of the telescope and the detector area then defines the field of view. For the infrared the best mirror coating is gold, although aluminium is satisfactory, whilst in the uv aluminium overcoated with magnesium fluoride works well.

Sometimes the output beam from the laser can be transmitted directly but usually an improvement in system performance will be obtained if a beam expanding telescope is used. Expanding the beam by a given factor will reduce beam divergence by the same factor. Either refractive or reflective designs of expanding telescope can be used.

With small laser beam divergence and telescope field of view (FOV) (i.e. <1 mr) an arrangement in which the laser beam is transmitted coaxially with the centre line of the telescope is usually needed to ensure that the telescope FOV includes the entire laser beam at all ranges of interest. With Newtonian telescopes this can often be achieved by coating both sides of the secondary mirror so that it can also be used to transmit the laser beam.

6.2 UV detectors and filters

A photomultiplier is universally used for the uv and visible regions. Quantum efficiency for the conversion of photons to photoelectrons is typically 20% and the signal to noise ratio of the signal is close to the theoretical shot noise limit of $\sqrt{n_e}$ where n_e is the number of photoelectrons generated on the cathode. The actual noise is somewhat higher than $\sqrt{n_e}$ because there is considerable random variation in the size of the anode current-pulse resulting from a single photo-electron. This additional noise source is known as multiplication noise and depends on the type of tube and the voltage. The increase in noise is typically only about 30%.

Linearity of response is extremely important and the peak anode current should be kept well below the specified maximum. The dynode resistor chain must also be designed for linear response to pulse inputs. The anode load would normally be 50 or 75 Ω and the signal across this can sometimes be fed directly into the digitizer. More usually a preamplifier is used and the performance requirements on this are stringent. The upper frequency limit should be high enough not to restrict range-resolution but should not be greater than half the digitization rate else extra noise will be introduced - a good combination is a 5 MHz preamplifier with a 10 MHz digitization rate. The lower frequency limit needs to be surprisingly low to avoid any distortion of the overall return signal shape. 100 Hz is usually sufficient; however even with a DC amplifier some undershoot following the peak of the signal can be experienced and this can result in large errors in measurements at long ranges.

In order to reduce background light falling on the photomultiplier it is necessary to use a narrow-band-pass filter. Ideally this would have a bandwidth just wide enough to encompass both λ_{on} and λ_{off}; however a narrow-band-pass filter inevitably has low peak transmission. The best available filters for SO_2 measurements have a 3 nm pass band with 25% transmission. More typically standard filters of 10 nm pass band and 20% transmission are used. The blocking filter is particularly important for low pulse energy lasers and in these cases it is sometimes found that the blocking factor away from the pass band (typically 10^{-4}) is insufficient to cut out all visible light. In this case a second visible blocking filter must be placed in series with the narrow-band filter.

6.3 Infrared detectors

To obtain the necessary sensitivity and speed for DIAL, liquid nitrogen cooled detectors must be used. Normally InSb is used for the 2-5.3 μm range and HgCdTe for the range upto 13 μm. These semiconductor detectors are characterised (ref. 33) by a responsivity, R, in volts/watt and a noise equivalent power (NEP) which is defined as the rms noise voltage divided by the responsivity, i.e. NEP is the noise expressed as an equivalent input power - NEP will depend on the bandwidth with which the detector output is monitored and it is usually quoted in units of watts/ $\sqrt{Hz}$. The inverse of the NEP is known as the detectivity, D, i.e. D = 1/NEP. For a given type of detector D is proportional to $a^{-1/2}$, where a is the detector area, so a parameter D^* is introduced:

$$D^* = Da^{1/2}$$

which is characteristic of a particular type of detector. D^* is usually quoted in units of cm $Hz^{1/2}/W$. D^* varies with wavelength and the peak value is the one normally given by manufacturers. The detectivity shows a sharp fall at longer wavelengths because the photon energy becomes less than the semiconductor band gap, but decreases slowly at shorter wavelengths. The composition of HgCdTe detectors can be varied to alter the position of peak D^*.

A typical value of D^* for an InSb detector is 10^{11} cm $\sqrt{Hz}$/watt. The equivalent noise power P_n for a 1 mm^2 detector and a bandwidth, Δf, of 1 MHz can be calculated from:

$$P_n = a^{1/2}\Delta f^{1/2}/D^*$$

which gives 10^{-9} watts. This is the effective sensitivity of the detector for a 1 MHz bandwidth. For a HgCdTe detector at 10 μm, D^* would typically be 3×10^{10} cm $\sqrt{Hz}$/watt.

Detector noise is mainly due to the background thermal radiation seen by the detector. By reducing the field of view with a liquid-nitrogen-cooled shield, background radiation and thus noise can be reduced. D^* values are normally quoted for a 60° field of view, and reducing this to 20° gives a theoretical improvement (= $1/\sin(\theta/2)$) of a factor of 3. In practice only about half this improvement is seen. For wavelengths greater than 3 μm the intensity of skylight corresponds closely to that of a 300 K black body so placing narrow band filters in front of the detector does not achieve much unless the filters are cooled. In practice this is usually not worth the expense or effort.

For infrared systems the limiting range resolution is usually determined by the response of the detector and its associated preamplifier. Both InSb and HgCdTe detectors can have rise times of about 100 ns but, at bandwidths of 1 MHz or greater, preamplifiers start to produce a significant noise contribution. For a 1 mm^2 InSb detector at 1 MHz bandwidth even the best preamplifier will produce at least as much noise as the detector (ref. 59). Note that a 1 MHz bandwidth corresponds to about a 300 ns rise time. Rise time probably gives a better indication of the range resolution (300 ns ≡ 45 m).

Detector noise is proportional to the detector element diameter so small detectors give better noise performance. However a small detector results in a small field of view of the telescope and requires a small laser beam divergence. A 1 mm diameter detector seems to give the best compromise when used in a 2 m focal length telescope.

6.4 Digitization of the signal

Two parameters of the digitizer are significant: sampling rate and number of bits. A 10 MHz sampling rate is adequate for DIAL and this is well within the capacity of current instruments. The number of bits required depends on the amount of noise on the return signal. Provided the digitizer bit size is less than the noise then no loss of resolution on the final averaged return is experienced. Thus an 8 bit digitizer can produce an averaged return signal with a resolution of 12 bits or better if several hundred returns are used for the average. If the noise is less than the bit size then the digitizer resolution will show through on the averaged return and it may be necessary to add noise to the initial return signals before digitizing. In doing this one obviously degrades system performance and it is better to use a higher resolution digitizer. In practice an 8 bit resolution is sufficient for low energy DIAL systems but when using a Nd:YAG pumped dye laser or a CO_2 laser source, for example, a 10 bit and even 12 bit resolution is necessary. 10 bit 10 MHz and 12 bit 5 MHz digitizers are currently available.

The problem of digitizer resolution is magnified by the large dynamic range of the return signal, i.e. a large peak at close range and a weak return from long range. As previously mentioned the peak can be reduced by utilizing obscuration and out of focus effects in the telescope but another technique is to use two digitizers in parallel, one set on a range suitable for the first, strong, part of the return, and the second set on a more sensitive range to record the tail. Problems can arise with this technique if the second digitizer does not recover sufficiently fast from the overload produced by the peak of the return signal.

When the on and off resonance wavelengths are produced by two separate lasers these are normally fired about 50 μs apart and the two return signals recorded in a single digitizer sweep. In this way only one digitizer is required and one does not have the problem of separating the two return signals by means of filters or etalons.

6.5 Calibration systems

If the absorption spectrum of the pollutant is known together with the values of the two transmitted wavelengths then the range-resolved pollutant concentration can be measured without need for any calibration. However with a continuously tunable laser the output wavelengths are not always easily determined and it is convenient to split off a small fraction of the output beam and pass this through a small cell containing the pollutant. One can then perform an online measurement of σ_{on} and σ_{off}. At the same time the output energy for each shot can be measured so that the return signals can be normalized. By monitoring the values of σ_{on} and σ_{off} with the control computer a warning can be issued to the operator if the wavelengths drift away from their intended values. It might alternatively be possible for the computer to adjust the wavelengths itself.

The calibration cell should contain the pollutant under the same conditions of temperature and pressure as in the atmosphere. In particular this implies that the pollutant should be mixed with a buffer gas at atmospheric pressure.

6.6 Computer control

Computer control is essential for any DIAL system. Among the tasks which can or must be undertaken by the computer are:

(i) Laser tuning, especially if many tuning elements require adjustment.

(ii) Laser firing; although sometimes the laser repetition rate is set manually and the rest of the system synchronized to it.

(iii) Control of the digitizer. This mainly involves reading the data from the digitizer store and re-arming it for the next shot.

(iv) Taking data from energy monitors and the calibration system.

(v) Overall sequence control.

(vi) Steering the telescope and laser beam to produce 3D maps of pollutant concentration.

(vii) Data analysis. This involves converting the raw data from the digitizer into range-resolved concentration profiles and is described in detail in the next Section.

A minicomputer similar to a PDP 11/23 or 11/34 is typically used although it may now be possible to use one of the new generation of 16 bit microcomputers. Programming is usually in a high level language such as Fortran with assembler language being used for time-critical parts of the program. Floppy disks are useful for mass storage, especially for mobile systems, since they continue to operate even under conditions of severe vibration.

Interfacing can be either by way of the computer's own interfaces, e.g. parallel input/output, A to D and D to A converters, or standard interfacing systems can be employed: CAMAC was widely used in the past but the IEEE interface-bus is now probably the most convenient system with many compatible instruments available.

7 DATA PROCESSING

The first task is to aquire the digitized returns from the digitizer. In most cases all N return signals for λ_{on} will be added together to provide a single averaged return and similarly for λ_{off}. However in low repetition rate systems it may be possible to store each return individually. In this case no information would be lost and a range of different analysis techniques could be applied subsequently.

For systems with two lasers fired quasi-simultaneously there is some disagreement about whether each pair of returns should be ratioed separately and then the average ratio taken, or whether the ratio should be taken after averaging the λ_{on} and λ_{off} returns as described above. That is, should one evaluate (see Eq. 6):

(i) $\log (\overline{P}_{off}/\overline{P}_{on})$

or

(ii) $\log (\overline{P_{off}/P_{on}})$

or even

(iii) $\overline{\log (P_{off}/P_{on})}$

Alternative (iii) is clearly not a good choice because P_{off}/P_{on} can easily be zero or negative due to noise and the logarithm could not then be evaluated. Alternative (ii) is probably a good technique if the main noise is correlated between the pair of returns, i.e. if the noise is due to atmospheric scintillations or variations of backscatter which remain constant for each pair of returns but vary between pairs. If the noise is mostly random, however, alternative (i) is the best technique and is the method most commonly employed.

By whatever method the averaged log-ratio is calculated it is necessary to subtract the zero level from $P_{on}(r)$ and $P_{off}(r)$ before taking the ratio. It is important that this is done accurately because any error will cause large systematic errors in the long-range measurements where the return is weak. The best method is to record, in addition to the on and off resonance returns, an equal number of returns with the laser beam blocked but with laser and all other electrical equipment running as normal. These background returns $P_b(r)$ can then be subtracted from $P_{on}(r)$ and $P_{off}(r)$ and this not only provides accurate zero subtraction but also removes any synchronous electrical pickup on the return signals (e.g. spikes associated with the firing of the laser).

Having determined the average value of log (P_{off}/P_{on}) Eq. 6 shows that the next step is to differentiate with respect to range. This can be accomplished by simply taking differences between successive range elements as indicated in Eq. 7 but it is more conveniently accomplished in conjunction with smoothing. Smoothing could be achieved by inserting a low pass filter in front of the digitizer but it is better practice to maintain maximum range resolution in the recorded signal and subsequently perform the smoothing numerically. The degree of smoothing required will be a compromise between noise on the measured concentration and range resolution. We have already noted that noise increases inversely with the square of the range resolution. Numerical smoothing is performed by convolving the log-ratio with a smoothing function (e.g. a Gaussian). If we denote the log-ratio by L(r), i.e.

$$L(r) = \log (\overline{P_{off}}(r)/\overline{P_{on}}(r))$$

and denote the smoothed log-ratio by $L_s(r)$ then:

$$L_s(r) = \int_0^{\infty} L(r') F(r' - r) dr'$$

where F(r) is the smoothing function. It is possible to show that differentiating $L_s(r)$ is equivalent to convolving L(r) with the differentiated smoothing function, i.e.

$$d(L_s(r))/dr = \int_0^{\infty} L(r') F'(r' - r) dr'$$

where $F'(r) = dF(r)/dr$. Thus smoothing and differentiating can be accomplished in one operation.

Having calculated dL_s/dr the concentration is then given by Eq. 6. The results can be presented as plots of concentration vs range or, if pollutant mapping is required, grey scale or contour plots can be produced from a series of one dimensional measurements.

8. DIAL APPLICATIONS AND CURRENT SYSTEMS

8.1 Applications

One of the principle uses of DIAL is in plume studies. Many national laboratories are studying DIAL as a possible policing technique for pollution emission legislation. A DIAL system can be used to identify the source of the pollution and provide a quantitative estimate of the emission rate. DIAL can also be used to study plume dispersion - a plume can often be detected by its SO_2 content long after the particulate content is too low to be detected with a lidar system. Plume chemistry and pollutant concentrations in the plume can also be studied. In the USA pollution legislation is often based on very conservative models for the relationship between ground level SO_2 concentrations and the total emission from the power station stack. DIAL measurements are being used to validate more accurate models and thus place less restrictions on the operation of the power station (ref. 36). DIAL systems mounted in aircraft are in use by NASA for studies of atmospheric chemistry during high pollution episodes. Studies can be made of ozone, water vapour, and SO_2 vertical profiles and correlated with mixing layer height (ref. 38).

Besides tropospheric studies DIAL can also be used for stratospheric measurements, chiefly of ozone. Balloon-borne (ref. 34) and ground based systems (refs. 32,56) have been used. One feature of stratospheric measurements in the uv is that because scattering is entirely Rayleigh all the terms in the basic lidar equation can be evaluated absolutely and a measurement at a single wavelength is sufficient.

8.2 UV and visible systems

A typical system for measuring SO_2 in the uv would use a 0.5 m telescope and a laser with 10 mJ output at 300 nm. Such a system has a sensitivity of about 20 ppb at a range of 1-2 km with a 50 m range resolution. These systems are normally adaptable to measuring O_3 with similar sensitivity and NO_2 with about three times lower sensitivity. We now consider a representative selection of such systems in more detail.

The system at the Central Electricity Research Laboratories is shown in Figs. 5-7. This employs a single Nd:YAG pump laser and two dye lasers. One dye laser is normally used for SO_2 measurements at 300 nm and the other for NO_2 at 480 nm. Each dye laser runs at 10 Hz and is switched on alternate pulses between the on and off-resonance wavelengths. Frequency doubling for the two SO_2 wavelengths is by two doubling crystals in series. The system is van mounted with the telescope fixed in a vertical position. Measurements are made with the van moving and a navigation system enables SO_2 vertical profiles to be plotted along the van's route. The system is used to study SO_2 dispersion from power stations. Fig. 8 shows some typical results.

SRI in California have a similar system also used for SO_2 plume studies (refs. 36, 37). This system features two separate Nd:YAG pumped dye lasers, one each for λ_{on} and λ_{off}. In this system the lasers are fixed and the telescope is steerable, the whole system being transportable but operated from a fixed site. Fig. 9 shows an SO_2 contour map produced with this system.

NASA Langley operate an airborne system, again with twin YAG/dye lasers (ref. 38) (Fig. 10). The telescope in this case looks vertically down giving vertical profiles from the aircraft, at a height of typically 4 km, to ground level. In this case the ground return can be used to provide a confirmatory total burden measurement. This system has studied SO_2 and O_3 profiles through the mixing layer and has also been used for water vapour measurements using absorption lines at about 720 nm. Output in the uv at 300 nm is unusually high at 25 mJ and, uniquely, every shot is recorded separately on magnetic tape for later analysis. Sensitivities of 10 ppb for ozone are obtained, with a 200 m resolution. Ozone profiles measured with this system are shown in Fig. 11.

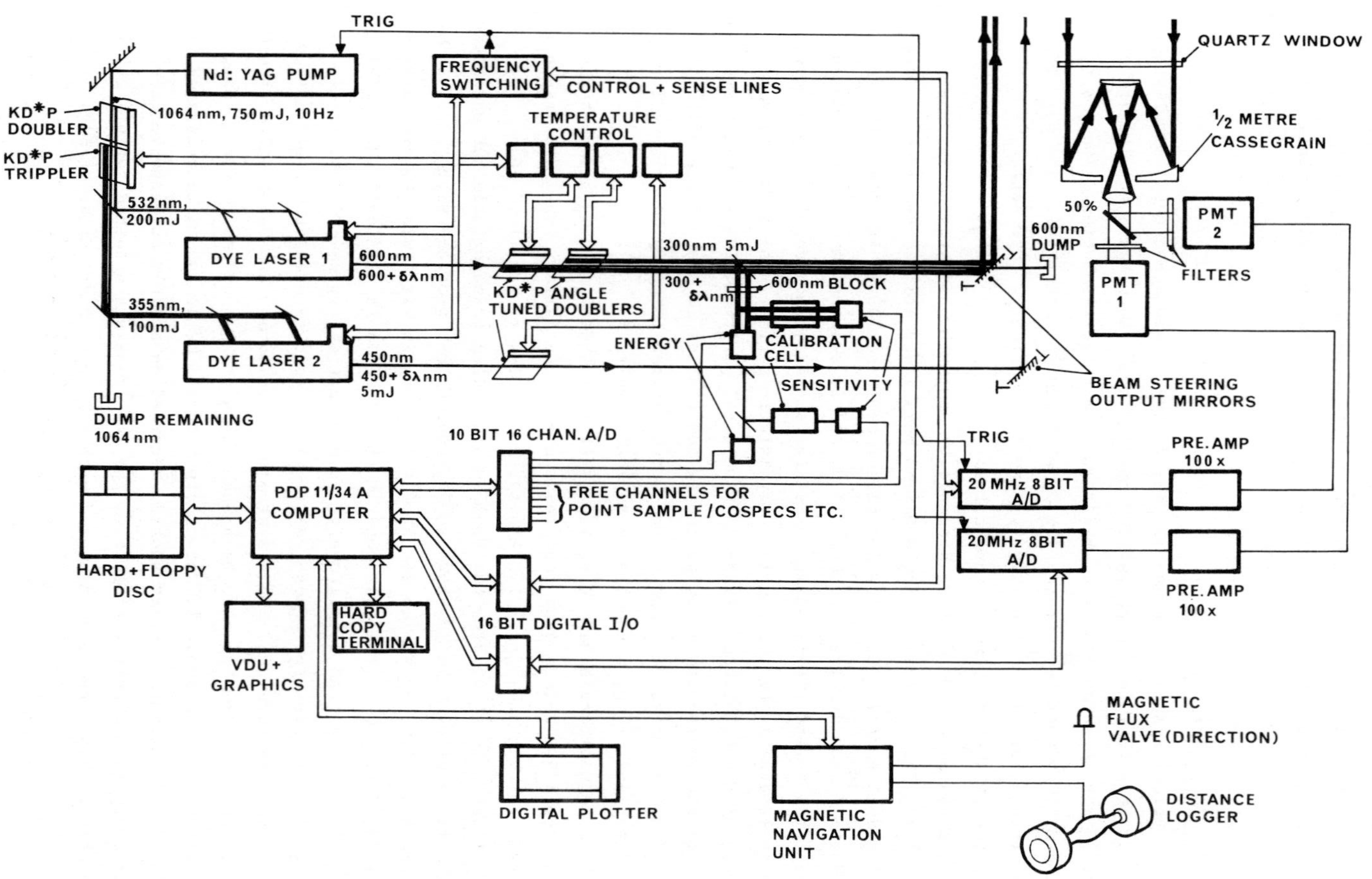

Fig. 5. Schematic of the CERL uv DIAL system.

Fig. 6. CERL uv DIAL van showing vertically mounted telescope.

Fig. 7. Interior of CERL DIAL van showing the two dye lasers (centre). The telescope can be seen through the window.

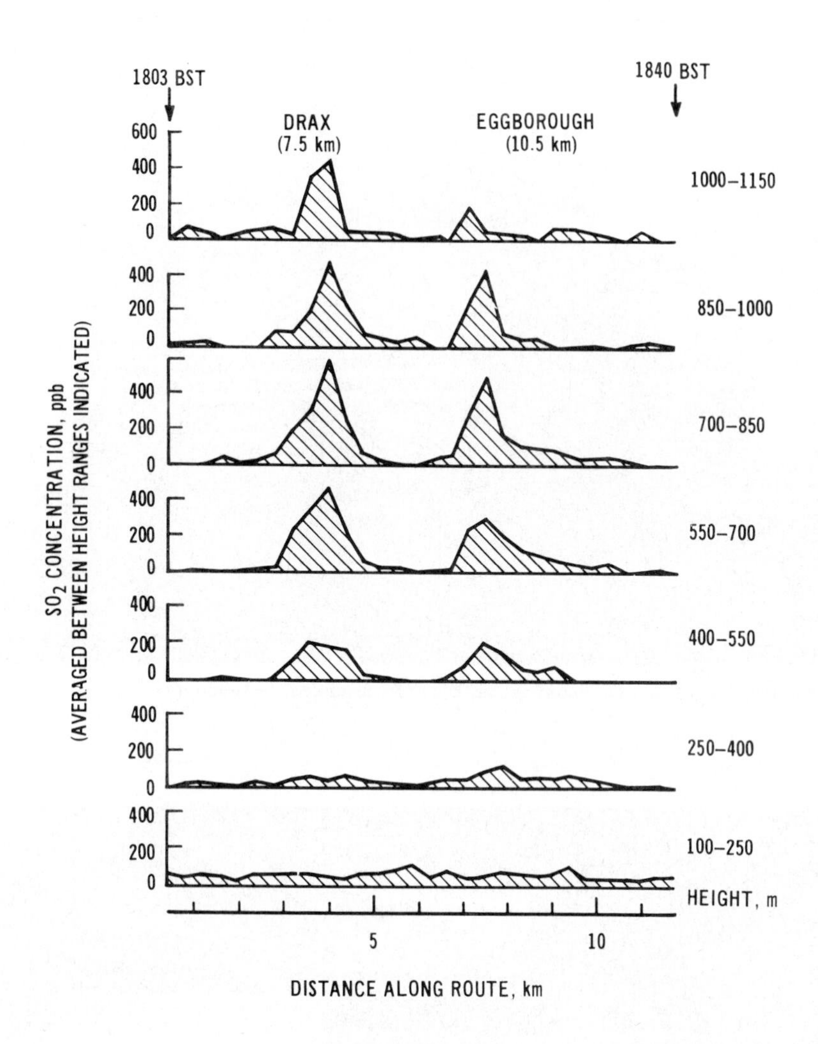

Fig. 8. Measurements of SO_2 concentration made by the CERL DIAL van when traversing beneath the plumes from two neighbouring power stations. The traverse was made downwind of the power stations and approximately perpendicular to the wind direction. The variation of SO_2 concentration along the traverse is shown for a series of height intervals.

Other YAG pumped dye-laser systems include those of the NPL in Britain and of Chalmers University, Sweden (ref. 40). Both of these are van-mounted and employ a large plane mirror for direction scanning. ENEL in Italy operate a system built for them by CISE. This again is van mounted but laser and telescope are mounted and steered as a unit. A system at the Haute Provence

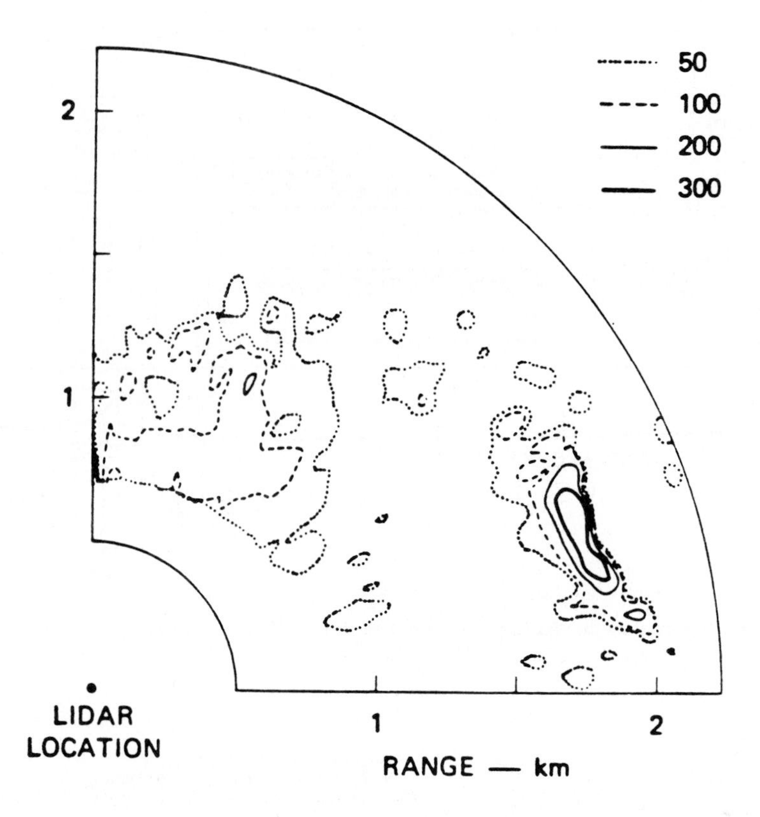

Fig. 9. Ground-level SO_2 contours measured with the SRI uv DIAL system (ref. 36). The system was scanned horizontally through a 90° arc. The power plant is 3 km to the right of the system. The figures are ppb.

Observatory in France is used to measure vertical profiles of H_2O in the troposphere and O_3 in the stratosphere. For the stratospheric measurements photon counting can be used. Not all systems use Nd:YAG pumping and GKSS in Germany have recently commissioned a system in which four flashlamp-pumped dye-lasers are used to provide the four wavelengths necessary for SO_2 and NO_2 measurements.

8.3 Infrared systems

Thanks to the availability of a near-ideal laser in the shape of the Nd:YAG pumped dye, uv DIAL systems are no longer built for their own sake but are in use in the field making valuable measurements. With one or two

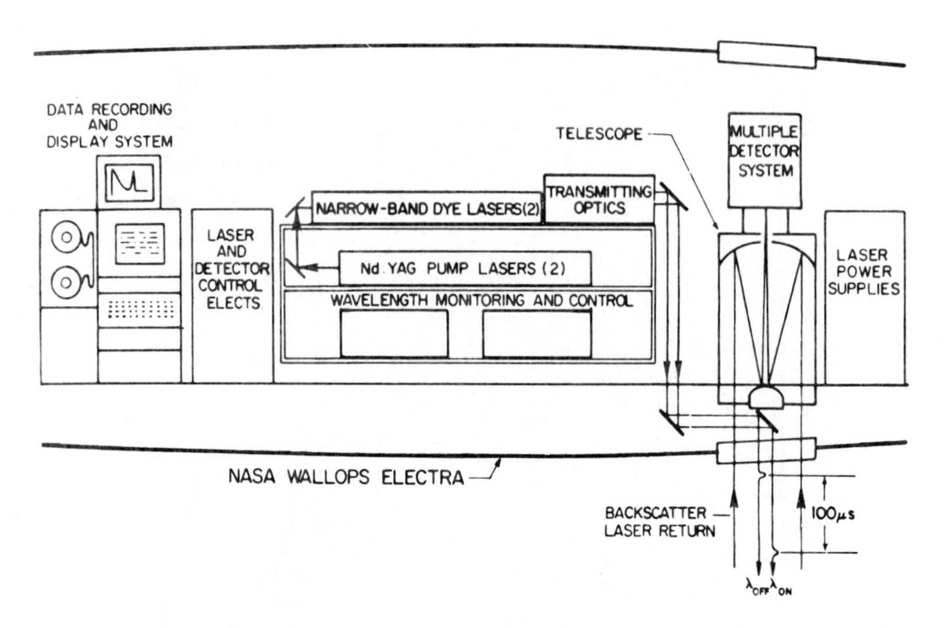

Fig. 10. NASA Langley airborne DIAL system (ref. 38).

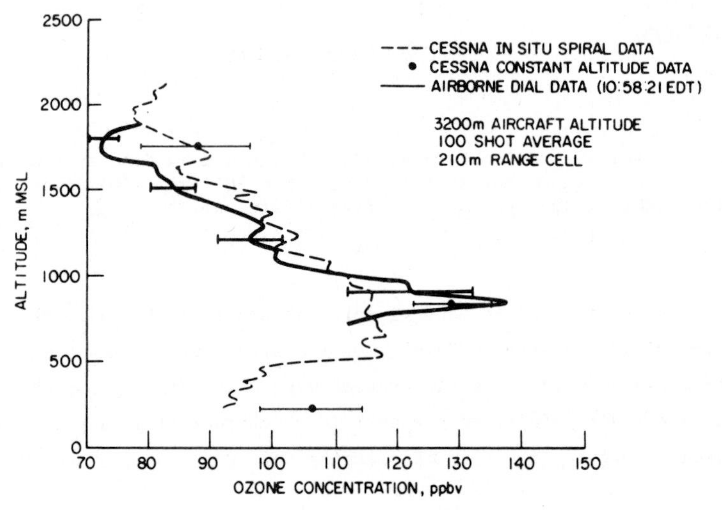

Ozone profile comparison of DIAL and *in situ* measurements on 29 May 1980.

Fig. 11. Vertical ozone profile obtained with the NASA airborne DIAL system (ref. 38).

exceptions this is not the case with infrared systems. The main problem is lack of suitable lasers, and ir systems in general suffer from low laser pulse energy and hence poor sensitivity and range, poor range resolution, and difficulty of operation. A few representative systems are now considered.

At the University of Munich a large CO_2 laser DIAL system has been constructed (ref. 40). Pulse energy is 1 J at a 70 Hz repetition rate. The laser and 0.6 m diameter telescope are mounted together in a massive framework, the whole assembly being moved to scan a plume. The system is not mobile but can be transported to a field site. It has been used to produce contour maps of ethylene concentrations above an oil refinery and water vapour concentrations above cooling towers (Fig. 12). For ethylene, sensitivity was about 10 ppb with 100 m resolution over a range of 500 m. Other CO_2 based systems include one for tropospheric ozone measurement at NASA Goddard (refs. 41,42) and a system at SRI California which has measured water vapour and ethylene (refs. 43,44).

Probably the only ir system which has made measurements of real value is the DF laser based system at GKSS in Germany which is used to study HCl emissions from an incinerator ship in the North Sea (refs. 45,46). To counteract the effect of ship motion the whole system is mounted on an inertial platform. The DF laser $P_2(3)$ line is used, as this coincides with the P(6) HCl line, and 30 mJ output is obtained. The sensitivity is only 500 ppb over a 800 m range but this still produces useful measurements on the high concentration of HCl in the plume. No other range-resolved measurements with DF lasers have been reported but there are several reports of long-path N_2O, CH_4 and HCl measurements using topographic-target backscatter (refs. 47,48).

All the above ir systems use line tunable lasers; only one DIAL measurement has used a continuously tunable source (ref. 30). This was an optical parametric oscillator based system with a pulse energy of only 1 mJ; H_2O measurements were made at 1.7 μm with a 200 m resolution and a 1 km range. Although the OPO is tunable out to 4.0 μm there are considerable problems in obtaining narrow line-width and high pulse energy in the 2.1-4.0 μm range, which is where most pollutants have strong absorptions.

9 FUTURE DEVELOPMENTS

Ultraviolet DIAL systems are probably now close to the peak of development. They are reaching the limit set by the difficulty of accurately measuring very small attenuations. This can be seen by realizing that 5 ppb of SO_2 produces an attenuation of only 1% over 1 km. If we take 1% attenuation as the minimum detectable then this corresponds to a sensitivity of 25 ppb SO_2 with 100 m range resolution, which is already achieved with current systems.

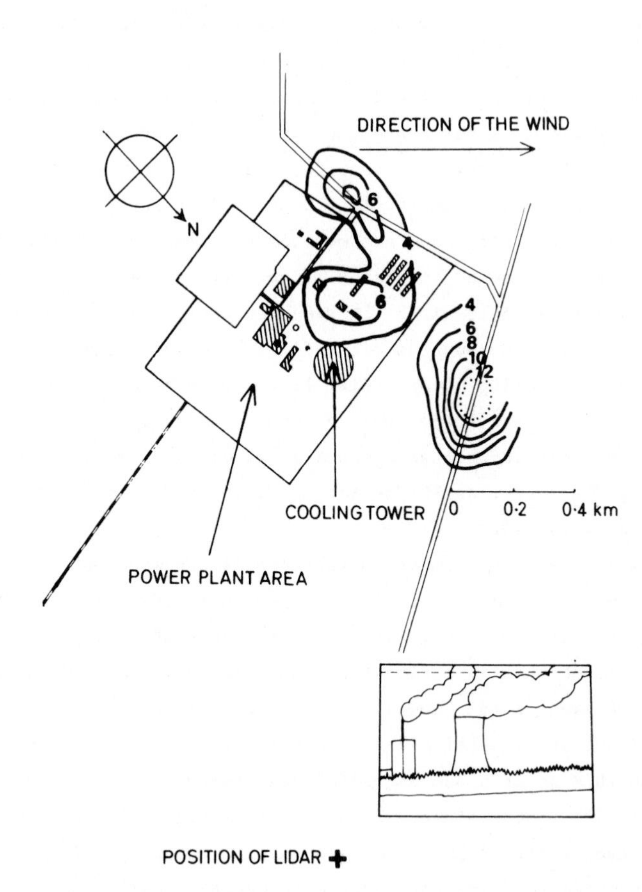

Fig. 12. Water vapour contours over a power plant measured with the Munich University CO_2 laser DIAL system (ref. 40). The numbers show the partial pressure in torr.

Improvements will be restricted to reducing the measurement time and increasing the range - although 1-4 km is already adequate for most purposes.

In the infrared improvements depend on better lasers becoming available. Unfortunately lasers which in the past appeared to show promise (e.g. the OPO) have now faded out and there are no obvious candidates for the future, except perhaps for the high pressure gas laser. There is however another area of improvement for infrared systems: the detection system. Heterodyne detection is already under development by the group at Hull University using a pulsed CO_2 laser with a cw CO_2 laser as the local oscillator (ref. 49). Heterodyne

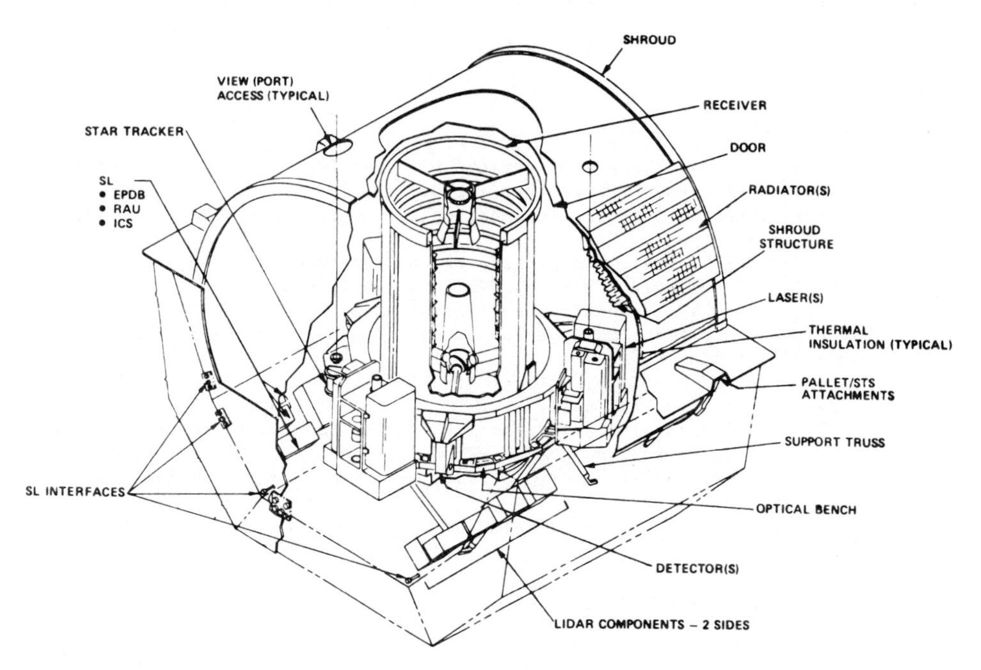

Fig. 13. Proposed shuttle lidar configuration (ref. 25).

detection theoretically allows an improvement of signal to noise by several orders of magnitude (refs. 50,51) but it requires very good wavefront matching between the local oscillator and the return radiation. There are also problems in interpreting the return signal. A further possibility for improved ir detection is up-conversion to a region where a photomultiplier can be used.

An area where future developments are likely is deployment. Already most DIAL systems are van mounted and several are flown in aircraft. The ultimate would be to fly a DIAL system in the Space Shuttle and proposals were made more than five years ago for such a Shuttle-borne lidar system (refs. 52-55). The original intention was that the system would fly in 1985 but this date has slipped considerably. The basic concept is shown in Fig. 13. For initial flights the system will operate as a lidar, being gradually upgraded to a DIAL system over several flights. Measurements of H_2O, CH_4, N_2O, O_3, and CO would be possible in the stratosphere and H_2O could also be measured in the troposphere. Unfortunately tropospheric SO_2 could not be detected because of beam attenuation by O_3 and scattering.

ACKNOWLEDGEMENTS

In preparing this chapter the author has drawn on the accumulated experience of several years work on DIAL systems at the Central Electricity Research Laboratories. He is particularly grateful to S. Sutton for his contributions concerning uv DIAL system operation. The work is published by permission of the Central Electricity Generating Board.

REFERENCES

General references

1 E.D. Hinkley (ed)., 'Laser Monitoring of the Atmosphere', Vol. 14, Topics in Applied Physics, Springer-Verlag, Berlin, 1976
2 R.L. Byer, 'Remote air pollution monitoring', Opt. Quant. Electron., 7, 147-177 (1975)
3 H. Kildal and R.L. Byer, 'Comparison of Laser Methods for the Remote Detection of Pollutants', Proc. IEEE, 59, 1644-1663 (1971)
4 R.M. Measures, 'Lidar equation analysis allowing for target lifetime, laser pulse duration and detector integration period', Appl. Opt., 16, 1092-1103 (1977)
5 R.M. Measures and G. Pilon, Opto-electronics, 4, 141-153 (1972)
6 R.L. Byer and M. Garbuny, Appl. Opt. 12, 1496-1505 (1973)
7 Conference abstracts of the International Laser Radar Conferences: ILRC8 Drexel University Philadelphia, 1977; ILRC9, Munich, 1979; ILRC10, Maryland, 1980; ILRC11, Madison, 1982; all organised by the American Meteorological Society

Specific references

11 J. Harms, Appl. Opt., 18, 1559-66, (1979)
12 N. Menyuk and D.K. Killinger, Opt. Lett., 6, 301-3, (1981)
13 S. Shipley, NASA Langley, private communication
14 D.J. Brassington, Appl. Opt., 20, 3774-9, (1981)
15 B.R. Marx, K.P. Birch, R.C. Felton, B.W. Jolliffe, W.R.C. Rowley and P.T. Woods, Opt. Commun., 33, 287, (1980)
16 P.T. Woods, B.W. Jolliffe and B.R. Marx, Opt. Commun., 33, 281, (1980)
17 E.C.Y. Inn and Y. Tanaka, Advances in Chem. No. 21, p 263, (1959)
18 P.T. Woods and B.W. Jolliffe, Opt. Laser Technol., 10, 25, (1978)
19 A.E.S. Green (ed)., 'The Middle Ultraviolet its Science and Technology', Wiley, New York (1966)
20 T.D. Wilkerson, B. Ercoli and F.S. Tomkins, 'Absorption Spectra of Atmospheric Gases', Technical Note BN 784, University of Maryland, Institute of Fluid Dynamics and Applied Mathematics, (1974)
21 E.D. Schultz, A.C. Holland and F.F. Marmo, 'A congeries of absorption cross-sections for wavelengths less than 3000 Å', NASA Contractor Report CR 15, (1963)
22 B.A. Thompson, P. Harteck and R.R. Reeves, Jr., J. Geophys. Res., 68, 6431-6, (1963)
23 K. Watanabe, M. Zelikoff and E.C.Y. Inn, 'Absorption coefficients of several atmospheric gases', AFCRC Technical Report, 53-23, (1953)
24 L.S. Rothman, Appl. Opt., 20, 791-5, (1981)
25 L.S. Rothman, A. Goldman, J.R. Gills, R.H. Tipping, L.R. Brown, J.S. Margolis, A.G. Maki and L.D.G. Young, Appl. Opt., 20, 1323-8, (1981)
26 R.A. McClatchey and J.E.A. Selby, 'Atmospheric attenuation of laser radiation from 0.76 to 31.25 μm', AFCRL-TR-74-0003, (1974) (available from National Technical Information Service)
27 J.H. Park, L.S. Rothman, C.P. Rinsland, M.A.H. Smith, D.J. Richardson and J.C. Larson, 'Atlas of Absorption Lines from 0-17900 cm^{-1}', NASA Reference Publication 1084, (1981)

28 D.G. Mureray and A. Goldman, 'Handbook of high resolution infrared laboratory spectra of atmospheric interest', CRC Press, Florida, (1981)
29 Sadtler, 'Gases and Vapours, High Resolution Infrared Spectra', Sadtler Research Laboratories Inc., Philadelphia
30 D.J. Brassington, Appl. Opt., 21, 4411-6, (1982)
31 P. Benner, Contract AF 61(052)-54, Technical Summary Report No. 1, (1960), (available from Armed Services Technical Information Agency as Report AD 261662)
32 O. Uchino, M. Maeda and M. Hirono, IEEE J. Quant. Electron., QE-15, 1094-1107
33 R.C. Jones, Proc. IRE, 47, 1495-1502, (1959)
34 W.S. Heaps, T.J. McGee, R.D. Hudson and L.O. Caudill, 'Balloon Borne Lidar Measurements of Stratospheric Hydroxyl and Ozone', NASA X-963-81-27, (1981)
35 N. Menyuck and D.K. Killinger, Opt. Lett., 6, 301-3, (1981)
36 J.G. Hawley, 'Ground based ultraviolet differential absorption lidar (DIAL) system and measurements', Technical Digest: Workshop on Optical and Laser Remote Sensing, Monterey California, US Army Research Office, (1982)
37 J.G. Hawley, EPRI Report EA-1267, Project 862-14, (1979)
38 E.V. Browell, A.F. Carter, S.T. Shipley, R.J. Allen, C.F. Butler, M.N. Mayo, J.H. Siviter, Jr. and W.M. Hall, Appl. Opt., 22, 522-534, (1983)
39 K. Fredriksson, B. Galle, K. Nyström and S. Svanberg, Appl. Opt., 20, 4181-9, (1981)
40 K.W. Rothe, Radio and Electronic Engineer, 50, 567-574, (1980)
41 R.W. Stewart and J.L. Bufton, Opt. Eng., 19, 503-7, (1980)
42 J.L. Bufton and R.W. Stewart, 'Measurement of ozone vertical profiles in the boundary layer with a CO_2 differential absorption lidar', Tenth International Laser Radar Conference, Abstracts, University of Maryland, (1980)
43 E.R. Murray, R.D. Hake, J.E. van der Laan and J.G. Hawley, Appl. Phys. Lett., 28, 542-3, (1976)
44 E.R. Murray and J.E. van der Laan, Appl. Opt., 17, 814-7, (1978)
45 C. Weitkamp, H.J. Heinrich and W. Herrmann, Laser and Electro-Optik (Germany), No. 3/1980, 23-27, (1980)
46 W. Herrmann, H.J. Heinrich, W. Michaelis and C. Weitkamp, 'DF lidar measurement of hydrogen chloride in the plume of incineration ships', Abstracts of 10th International Laser Radar Conference, University of Maryland, October, (1980)
47 J. Altmann, W. Lahmann, C. Weitkamp, Appl. Opt., 19, 3453-7, (1980)
48 E.R. Murray, Opt. Eng., 16, 284-290, (1977)
49 J.W. van Dyk, G.A. Greene, G.A. Hill, A. Layfield, J.C. Petherham, J.L. Pinto, B.J. Rye, S.E. Taylor, E.L. Thomas, F.J. Bryant, R.W. Miles and M.E. Uddin, 'A coherent laser radar for trace gas and meteorological measurements', Technical Digest: Topical Meeting on Coherent Laser Radar for Atmospheric Sensing, Aspen Colorado (NOAA and OSA), July (1980)
50 B.J. Rye, Appl. Opt., 17, 3862, (1978)
51 D.J. Killinger, N. Menyuk and W.D. DeFeo, Appl. Opt., 22, 682-689, (1983)
52 J.E. Harris and R.V. Greco, 'Atmospheric lidar multi-user system - a space lab payload', Conference abstracts, 9th International Laser Radar Conference, Munich (1979)
53 E.E. Remsberg and L.L. Gordley, Appl. Opt., 17, 624-630, (1978)
54 E.V. Browell, (ed), 'Shuttle Atmospheric Lidar Research Program', NASA SP-433, (1979)
55 R.V. Greco, (ed), 'Atmospheric Lidar Multi-user Instrument System Definition Study', NASA Contract Report 3303, (1980)

56 C. Cahen, J. Pelon, P. Flamant, J. Lefrere, M.L. Chanin and G. Megie, 'French lidar facility at the Haute Provence Observatory for tropospheric and stratospheric measurements', Conference Abstracts 10th International Laser Radar Conference, Maryland, (1980)

57 D.J. Brassington, 'Errors in Spectroscopic Measurements of SO_2 due to non-exponential absorption', CERL Note RD/L/N 168/80, Central Electricity Generating Board, (1981)

58 C. Cahen and G. Megie, J. Quant. Spectrosc. Radiat. Transf., 25, 151-7, (1980)

59 J. Altmann, S. Köhler and W. Lahmann, J. Phys. E: Sci. Instrum., 13, 1275-7, (1980)

60 J.E.A. Selby and R.A. McClatchey, 'Atmospheric Transmittance from 0.25 to 28.5 µm: Computer Code LOWTRAN 3', AFCRL-TR-75-0255, Air Force Geophysics Laboratory, Massachusetts, (1975)

Optical Remote Sensing of Air Pollution,
Lectures of a course held at the Joint Research Centre, Ispra, Italy, 12—15 April 1983,
P. Camagni and S. Sandroni (Eds). 181—204

RAMAN LIDAR AND ITS APPLICATIONS

R. CAPITINI*, D. RENAUT**, E. JOOS***

1 INTRODUCTION

Lasers have contributed as early as 1967 to the development of optical techniques for the acknowledge of the atmospheric parameters by sounding without probe. In the field of atmospheric remote sounding, Radar technique was well known so that the Lidar method derives from it. The laser electromagnetic waves generally extend from the ultra-violet to the infra-red so that the optical sounding range is limited, compared with the hertzien Radar; however, optical waves are transmitted without interferences and with a very good spatial resolution.

Lidar enables the remote measurement of aerosol particles and molecular species in the polluted as well as in the ordinary atmosphere. Information about physical and chemical parameters can be obtained from all the scattering phenomena. After a brief survey of the different Lidar methods, we focus in this paper on the Raman Lidar applications.

2 BACKGROUND INFORMATION ON LIDAR TECHNIQUES

2.1. LIDAR SYSTEM DESCRIPTION

The Lidar system is mainly composed with an emitter and a receiver.
The emitter is a laser source. The laser offers several advantages by comparison with incoherent sources:
- high power in a very short pulse (it is the pulse duration which gives the ran ge solution),
- spatial coherence (the very narrow laser beam allows one to use a narrow receiver field of view that reduces the background light),
- spectral purity giving a very high spectral brightness useful for spectroscopic measurements.

Some of the light in the transmitted pulse is scattered back to the receiver by the molecules and particles in the atmosphere so that backscattering acts as a distributed reflector. This method allows a range-resolved measurement.

Backscatter in the atmosphere can be replaced by a remote topographical target or a retroreflector to reflect energy back toward the receiver. The result is an integrated measurement along the optical path without depth resolution. The received radiation is collected by an optical system (mirror telescope) and focused on a photomultiplier. The electrical signal is amplified and displayed on an oscilloscope or analysed by a computer.

2.2. THE DIFFERENT KINDS OF LIDAR

Lidar systems can be used with fixed frequency or tunable laser sources as shown on Table 1.

Kind of Lidar	Scattering (or absorbing) Species	Cross-section (m^2)	Applications
. ELASTIC SCATTERING			
- Rayleigh	All the molecules	$\sim 10^{-31}$	High-altitude density
- Mie	Aerosols, droplets	$\sim 10^{-14}$	Air pollution/chemistry Atmospheric structure and motions Cloud and fog studies
. RAMAN SCATTERING	Molecules identified by λ_r	$\sim 10^{-34}$	Plume pollutants Boundary-Layer H_2O, Temperature
. RESONANCE RAMAN SCATTERING or FLUORESCENCE	Molecules identified by λ_r	$\sim 10^{-28}$ -10^{-32} $\sim 10^{-24}$ -10^{-28}	Minor species and pollutants
. DIFFERENTIAL ABSORPTION (DIAL)	Molecules identified by λ_{on}	$\sim 10^{-23}$	Low concentration or Long range minor species
. RESONANCE SCATTERING	Atoms identified by λ_o	$\sim 10^{-17}$ -10^{-21}	High-altitude atomic species (Na, K, Li,...)

Table 1 showing the different kinds of Lidar.

In the first case, one uses Rayleigh - Mie or Raman backscattering. When the scattered light is at the same wavelength as that of the transmitted beam, one deals with elastic scattering. This scattering has two components: the Rayleigh component which allows the whole molecular property of the environment to be probed and the Mie component which probes the aerosol concentration. The aerosols play an important role in atmospheric pollution and chemistry. They are a good tracer for atmospheric structure and motion, and they are at the origine of cloud and fog formation. However, the Lidar signal cannot easily be used to determine atmospheric density because it is difficult to separate the contributions of molecules and particles, except at high altitude. For the remote identification of molecular species, another kind of scattering is interesting which is shifted in wavelength with respect to the incident light. This type of interaction is called Ramanscattering. Raman lidar technique has great advantages: specificity (Raman shift is equal to a vibrational frequency of the observed molecule), proportionality (intensity directly proportional to the number density of corresponding molecules and independant of the others), simultaneous multiplicity of information (a lot of Raman lines can be observed at the same time). However, the low value of the molecular cross-section strongly limits the minimum detectable concentration and the range of the measurement. For this reason, Raman scattering has been chiefly used for boundary layer measurements and for pollutants plume measurements.

In the second part of the Table 1, the technique calls for a dye laser to adjust the laser wavelength to the sample absorption line.
Three methods are mainly used:

- Resonance Raman which has its efficiency increased by several orders of magnitude compared to the ordinary Raman intensity and fluorescence which is also usually much stronger than scattering. However fluorescence efficiency and spectral distribution are affected by collisions during the time where it is excited. As a result, it is often much harder to interpret quantitatively as we will see with the SO_2 example.
- DIAL which employs two emitted wavelengths respectively tuned on the molecule absorption line and beside it. In this case, concentration is deduced from the comparison of these two Lidar signals. This technique has been used for the detection of low level concentrations at long range, for example stratospheric ozone.
- Atomic resonant scattering concerns particularly photodissociated atomic species at high altitude. As an example, sodium, potassium and lithium profiles have been obtained by G. Megie, 1978, ref. 1, between 80 and 100 kilometers using a dye laser built in his laboratory.

2.3 CURRENT STUDIES WITH THE FRENCH LIDARS

Lidar equipments are chiefly used in research institutes for atmospheric investigations (meteorological or pollution studies, atmospheric physics...).In France we can point out: the C.N.R.S. Aeronomie Center with M.L. Chanin and G. Megie, 1981, ref. 2, who have developed Lidar techniques at the Haute Provence Observatory to explore the middle and the high atmosphere; E.D.F., with J.L. Lesne 1978, ref. 3 and C. Cahen have designed a mobile lidar station for aerosol measurements and plume dispersion studies; the E.E.R.M., which is the research department of the French Meteorological Office has developed instrumental techniques to explore the boundary layer:

radio sounding facilities, remotely piloted airplanes, sodar and lidar with D. Renaut and J.L. Gaumet, 1980, ref. 4; lastly, the C.E.A. with R. Capitini, 1980, ref. 5, has developed an ultra-violet Raman Lidar, for day and night-time remote sensing of pollutants and minor species in the atmosphere.

In the case of our C.E.A. Lidar unit, the design and the feasibility study are the result of an experimental program conducted in cooperation with the E.E.R.M. and Ecole Polytechnique to check the environmental and meteorological usefulness of the system.

3 OVERVIEW OF RAMAN LIDAR MEASUREMENTS

The Raman-Lidar studies have been conducted over the last decade or so. In 1967, D.A. Leonard, 1967, ref. 6, provided the first demonstration that Raman scattering could be observed by lidar. He used a pulsed N_2 laser to detect atmospheric O_2 and N_2. Since this pioneer work, a lot of studies have shown the interest of the Raman-Lidar in the meteorological field as well as for pollution studies.

3.1 METEOROLOGICAL APPLICATIONS

3.1.1 Water vapor profiles

Water vapor is abundant enough in the lower atmosphere (mixing ratio of about 1%) to give a strong Raman signal; moreover this species is the most important for meteorological knowledge (earth-air exchanges, boundary-layer thermo-dynamics...) For this reason, use of Raman-lidar to measure atmospheric H_2O profiles has been checked by a lot of authors. First observations have been carried out by J. Cooney, 1969, ref. 7, and Melfi, 1969, ref. 8, about ten years ago. The water vapor ratio profile r(z) is obtained by ratioing the H_2O and N_2 Raman signals.

$$r(z) = k.\frac{S_{H_2O}(z)}{S_{N_2}(z)}$$

The constant k is generally deduced from comparisons with in-situ sensors. Using a ground-based Lidar, S. Melfi, 1972, ref. 9, measured the vertical profiles of water vapor up to 2 or 3 km. The 347,2 nm wavelength of a doubled frequency ruby laser and a 40 cm Newtonian telescope were employed to obtain these data. There was a fair agreement between Raman-lidar and radiosonde profiles. J. Cooney,

1971, ref. 10, also measured vertical profiles of atmospheric humidity with a frequency-doubled ruby laser. The comparison with data obtained by radiosondes and helicopters showed a 10% relative accuracy and a 13% absolute accuracy. More recently, similar experiments were attempted in France. After a feasibility study performed at the E.N.S.T.A. J.C. Pourny, 1979, ref. 11, a Raman-Lidar was designed at the E.E.R.M. for the measurement of water vapor and aerosols vertical profiles Renaut, 1981 and 1983, ref. 12, 13. The Lidar is built with a doubled-frequency ruby laser and a 45 cm-Cassegrain telescope; a double path focal assembly allows simultaneous detection of H_2O and N_2 Raman signals. Vertical profiles of water vapor are measured at night-time with a 2 km-range for a 30 m-range resolution and a 10 mn-sounding duration. Figure 1 shows a comparison of the lidar profiles with simultaneous aerological soundings; the two instruments show a strong inversion at a height of 500 meters.

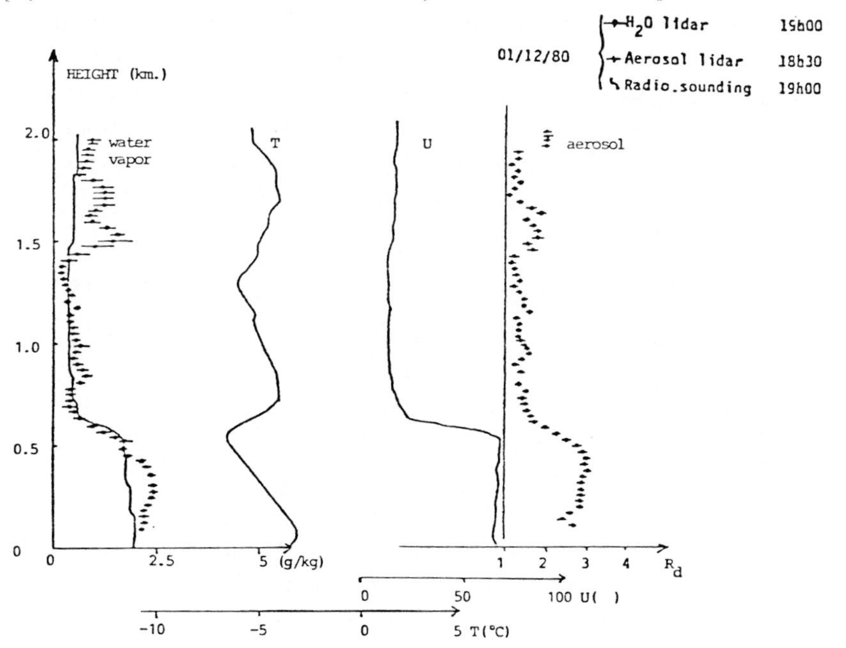

FIGURE 1 - SHOWING A COMPARISON BETWEEN RADIO SOUNDING AND AEROSOL AND H_2O LIDAR PROFILES

3.1.2 Temperature measurements

Remote and instantaneous determination of temperature profiles in the atmosphere is a very attractive possibility for the meteorologists. For this reason, a lot of work has been done for Raman-Lidar methods; up to now, studies are not as advanced as for water vapor measurements. Two Raman techniques have been proposed for temperature measurement:

* determination from N_2 density measurements

This method, suggested by R. G. Strauch, 1971, ref. 14, is based on the observed correlation between the Raman vibrational N_2 signal and the temperature variations. This correlation originates from the ideal gas law; the signal received by a lidar system measuring the backscattering from N_2 would indicate

$$\frac{\Delta S_{N_2}(z)}{S_{N_2}(z)} = \frac{-\Delta T(z)}{T(z)}$$

In a feasibility study, Raman signals were compared by R.G. Strauch with temperature data obtained from thermistors on a 30 m height tower. In a more recent study, D. Renaut, 1977, ref. 15, attempted to apply this method for the determination of vertical temperature profiles. Following conclusions were pointed out: the influence of aerosols is not negligible; it can be corrected by using two emitted wavelengths and by analysing the corresponding Rayleigh-Mie and N_2 Raman signals; an additional hypothesis on the optical fonctions of aerosol is necessary. Uncertainties in the temperature determination resulting from noise in the Raman signal are estimated for the E.E.R.M. Lidar and given in Table 2.

Table 2. Estimated uncertainty in temperatures obtained using Raman signals

Altitude (meters)	Density method	Raman Spectra
100	0.2°C	0.2°C
500	1.0	1.0
1000	2.0	2.1
2000	5.0	4.4
3000	10.0	6.8

* determination from rotational Raman spectra

This method, suggested by J. Cooney, 1972, ref. 16, and J.A. Salzman, 1972, ref. 17, uses the sensitivity of the Raman rotational spectrum of air ($N_2 + O_2$) to the temperature variations. The shape dependence of the spectrum with the temperature has been shown. The ratio of two backscattered signals, at the Raman wavelengths selected by two filters centered in the Anti-Stokes wing, provides a measure of the scattering volume temperature. A feasibility study of the method was made by A. Petitpa, 1977, ref. 18; he recommended the use of a ruby laser (694,3 nm) and of the Anti-Stokes part of the spectrum. Optimal wavelengths for the filters were determined. Finally, the uncertainties in temperature due to noise in the Raman signals were estimated for the E.E.R.M. Lidar (Table 2, second column). Experimental checking of the method was attempted by J.A. Salzman, 1973, ref. 1 and Y.F. Arshinov, 1981, ref. 20, 9, on short horizontal paths; standard errors of ± 3°C and ± 1°C, respectively, were obtained. Measurements of atmospheric profiles were attempted by R. Gill, les 1979, ref. 21, up to 2 km with a ruby laser; compari with a radio-sonde showed son an average absolute temperature of 0.85°C.

3.1.3 Density measurements

At high altitude where the aerosol attenuation is negligible, the N_2 Raman signal can be used for molecular density measurements. With a giant Lidar system using a pulsed ruby laser and a large collecting area of 16 m^2, M.J. Garvey, 1974, ref. 22, was able to detect N_2 up to 40 km height. Close agreement was noticed between the observed density profiles and data from local radiosonde measurements and the US Standard Atmosphere. Nevertheless, because of the giantism of the Lidar, this kind of measurement seems to be marginally effective.

3.2 POLLUTION MEASUREMENTS

As early as the first Raman Lidar measurements in the atmosphere were made, a great need for identifying the pollutants in the air around cities or factories was felt.

H. Inaba and T. Kobayasi, 1969, ref. 23, provided the first demonstration that chemical analysis of polluted air could be carried out by Raman Lidar. Other experimental studies have followed for the measurement of the atmospheric pollution as well as for the analysis of various molecular constituents in tanks.

A brief survey on these experimental studies is given now.

3.2.1 In the polluted atmosphere

Spectroscopic analysis of Raman backscattering from SO_2 and CO_2 molecules was performed by T. Kobayasi and H. Inaba, 1970, ref. 24, with a Q switched ruby laser with 5 to 10 MW peak power and 30 ns width at 6943 Å. In these measurements, SO_2 concentration was artificially enriched. The same group, H. Inaba, 1972, ref. 25, succeeded in measuring the pollution in a smoke plume settled at 30 m from the Lidar. The intensity peaks were observed at the Raman wavelengths corresponding to SO_2, C_2H_4, HCO, NO, CO, H_2S, CH_4. In their article, a spectral distribution of Raman shifted components from various molecules in an automobile exhaust gas was also shown. The Raman lines due to the presence of C_2H_4, NO, CO and liquid H_2O were identified in addition to the ordinary atmospheric components. With a frequency doubled Nd-Yag laser and a 50 cm-Cassegrain telescope mounted on a mobile Lidar unit, S. Nakahara, 1972, ref. 26, has demonstrated successfully the remote measurement of trace amounts of SO_2 in a smoke plume. The plume originating from the 150 m high stack of an operating power plant was located at about 200 m from the Lidar. SO_2 concentration in this plume has been estimated at 1850 ppm.

Similarly Lidar measurements have been performed by S. Melfi, 1973, ref. 27, using a ruby laser with 1.5 J energy and a 61 cm-Newton telescope. The chimney was 119 m high and settled at 210 m in front of the Lidar.
SO_2 concentration has been calculated at 800 ppm, in good agreement with in situ measurements.

Whereas all previous measurements were performed at night, T. Hirschfeld, 1973, ref. 28, published the first day time Raman measurements. The lidar unit was equipped with a frequency doubled ruby laser delivering 160 mJ with a 2 Hz repetition rate, and a 90 cm Cassegrain telescope. The unit was tested at ranges of several hundred meters in full day-light under both clear and foggy conditions. A good signal to noise ratio was obtained for the detection of 30 ppm of $SO_2$and a 1.7 ppm of kerosine.

A mobile UV lidar was developed by V.M. Zakharov, 1978, ref. 29, for the 24 hour measurement of gaseous and aerosol atmospheric composition by Raman scattering and fluorescence spectra. The third and fourth harmonics of a Nd-Yag laser have been used with a 42 cm Cassegrain objective. Later, Raman detection of CO, CO_2, C_2H_4, C_6H_6, H_2CO, H_2S, CH_4, NO and NO_2 was performed in an artificially contamined atmosphere, in an automobile exhaust and in a automobile traffic area. Strong fluorescence from vapors and aerosols of diesel fuel, soil dust and industrial smoke was identified in the atmosphere from distances of 300 -500 m at a sensitivity of $10^{-4} g/m^3$.

Y.G. Vainer, 1979, ref. 30, also designed a day-time Raman Lidar unit. He used the fourth harmonic of a Yag-laser emitting 1 mJ at 50 pulses per second. With a 80 cm diameter-telescope, CO_2, N_2, O_2, SO_2 have been detected in a plume located at 200 m from the Lidar unit with a minimum detectable concentration of 50 ppm and 50 % of accuracy. In these experiments, the integration time for the detection system was set to be about 3 minutes (or 1000 laser shots).

3.2.2 In a tank

S.K. Poultney, 1977, ref. 31, performed night-time measurement of about 1000 ppm SO_2 inside a tank with a 12% accuracy at a distance of 300 m. He used a 1.5 J ruby laser at 30 pulses per minute, 6 m range resolution and an observation time of 15 minutes.

4 OUR WORK

As was just shown, most of the Raman Lidar experiments are made with visible or near ultra-violet lasers (λ >300 nm). Consequently, probing is generally limited to night-time and, as only ordinary Raman scattering occurs in this part of the spectrum, the sensitivity of the measurements is restricted to a few ppm at a range of a few hundred meters. Application is then limited to smoke stacks or combustion exhausts monitoring. Using a U.V.Raman-Lidar allows one to work in full day-time and to benefit of resonance enhanced cross-sections on some absorbing species. In this prospect, a U.V. Raman-lidar system has been designed for meteorological studies and pollution measurements.

4.1 THE U.V. RAMAN LIDAR SYSTEM (Table 3) The U.V. Raman-lidar system is assembled with a frequency-quadrupled Nd-Yag laser emitting 80 mJ pulses at 266 nm. The repetition rate is 10 Hz and the pulse duration is 15 ns. The full angle beam divergence is 0,5 mrad. The laser beam is emitted coaxially with the

LIDAR CONFIGURATION	COAXIAL
EMITTER	
LASER	QUADRUPLED Nd-YAG 266 nm
LASER ENERGY	80 mJ
PULSE DURATION	15 ns
REPETITION RATE	10 Hz
RECEIVER	
CASSEGRAIN TELESCOPE	Φ 60 cm (0,22 m^2)
FIELD OF VIEW	1.8. x 2.7. mrad (full angle)
OPTICAL FILTER, ONE PATH, DOUBLE MONOCHROMATOR	
OPTICAL BANDWIDTH (Slitwidth 2 mm)	0.4 nm
OPTICAL TRANSMISSION	15 %
PM EFFICIENCY	25 %
COMPUTER ACQUISITION	
ACQUISITION SYSTEM	TEKTRONIX 7612 D (8 bits, 2 x 5 ns)
PROCESSING SYSTEM	TEXTRONIX CP 4165 (32 K words - 16 bits)
LIDAR SCANNING RATE	
ELEVATION (0 - 45°)	0.4 °/s
AZIMUTH (0 - 90°)	1.5 °/s

Table 3 - Technical specifications of the Raman-Lidar system telescope. The laser energy is controlled by means of a pyroelectric detector. The receiver is a Cassegrain telescope with 60 cm diameter. The full angle of the receiver field of view is 1.8 x 2.7 mrad. The telescope is coupled to a monochromator made of 2 holographic gratings with 3600 rules/mm, giving a linear dispersion of 1.75 $A.mm^{-1}$. The transfer function of the monochromator has been measured for different slitwidths (for example, 10^{-7} rejection factor is obtained at 400 cm^{-1} with 2 mm slitwidth). The optical transmission of the receiver is about 15%. A photomultiplier localized at the monochromator output detects the Raman light; its efficiency is 25%. The whole acquisition computing system is a TEKTRONIX unit. The programmable digitizer is a 7612 D transient recorder with 2 separate channels, 8 bit-amplitude and a 200 MHz maximum sampling rate. The processing system includes a computer and a floppy disk unit. This computer allows to average 2000 signal samples at the rate of 10 Hz.

The emitter and the receiver are mounted on a pedastal which provides scanning in both elevation (0 - 45°) and azimuth angle (- 45°, + 45°). The whole system is fixed on a rail-waggon allowing to pull the Lidar outside.

4.2 U.V. BACKGROUND LIGHT AND ATMOSPHERIC ATTENUATION

Day-time atmospheric Raman detection is made difficult by the strong sky radiance. To avoid this problem, two ways are possible :

- when using a laser wavelength greater than 300 nm one has to reduce the optical bandwidth to the Raman Q branch width ($\sim$ 10 cm^{-1}) and to reduce the receiver field of view (minimal value being called for by the laser divergence). Even with these constraints some background light can still be present. Moreover, optical adjustments become very critical.
- another method for avoiding background light is to operate in the 250 - 300 nm spectral range in which stratospheric ozone absorbs the incoming solar radiation. Our experimental results show that the diurnal sky signal decreases of 6 orders of magnitude when the wavelength is tuned from 320 to 295 nm. So, it is more interesting to use the 266 nm laser wavelength even if it is at the expense of laser energy losses and atmospheric attenuation.

Taking into account Mie and Rayleigh scattering and SO_2 and O_3 absorption, we have calculated, ref. 32, the atmospheric attenuation as a function of wavelength between 250 and 310 nm for a rural (clear) and an urban (polluted) atmosphere (Fig. 2a).

As can be seen, O_3 attenuation is the dominant factor up to 290 nm. Moreover, this absorption is strongly wavelength-dependent so that it has to be considered for moderate range measurements (water vapor profiles for example). From these data, atmospheric transmission has been computed for three different horizontal paths (Fig 2b).

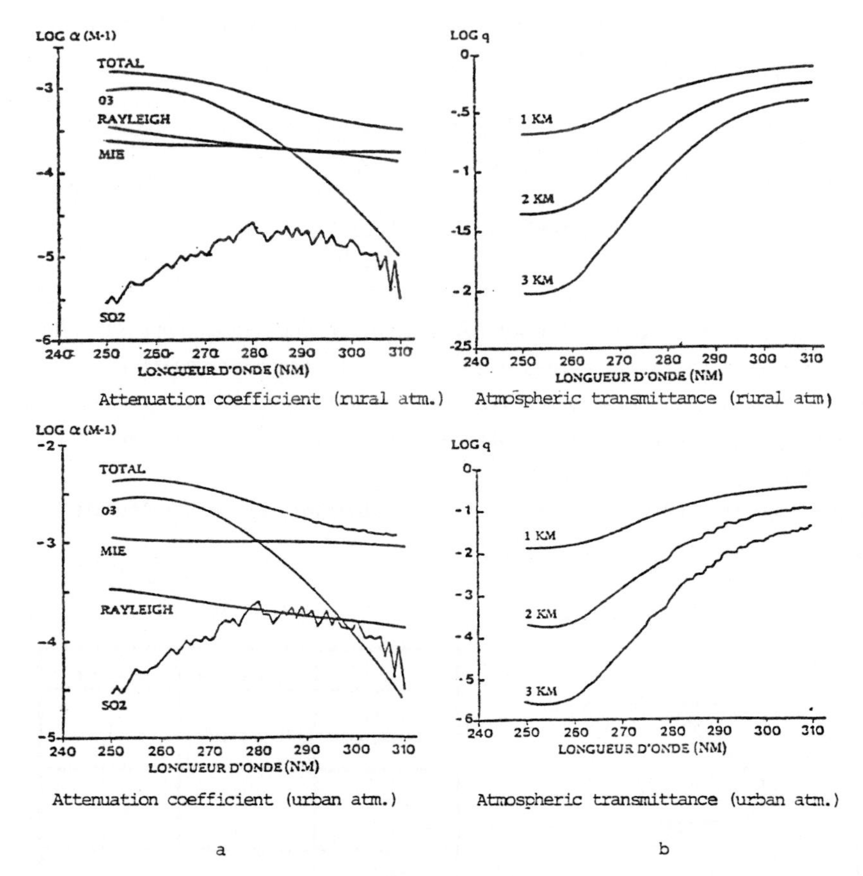

Figure 2 - SHOWING ATMOSPHERIC ATTENUATION (a) AND TRANSMITTANCE (b) IN THE 250-310 nm. RANGE

We can notice that:

- transmission becomes lower when the wavelength moves to U.V.,
- attenuation is not too severe for the rural case (10% for a 3 kilometer-range at 280 nm),
- transmission is much degraded for the urban case (10% for a 1 kilometer-range at 280 nm).

4.3 WATER VAPOR MEASUREMENTS

The water vapor mixing ratio r is determined with the help of the three Raman signals for O_2, N_2, H_2O, ref. 33

$$r = K \cdot \left[\frac{S_{O2}(z)}{S_{N2}(z)}\right]^{\gamma} \cdot \frac{S_{H2O}(z)}{S_{N2}(z)}$$

where
$$\gamma = \frac{\Delta\sigma\,(H_2O,\ N_2)}{\Delta\sigma\,(N_2,\ O_2)}$$

γ is the ratio of the ozone differential absorption cross-sections. Ratioing oxygene and nitrogen signals allows one to eliminate ozone attenuation . The calibration constant is determined by comparison with in-situ measurements. In this equation, the other factors of attenuation (Rayleigh, aerosols and SO_2) are neglected.

The error on the H_2O profiles induced by attenuation has been calculated for the 2 previous atmospheric models (Fig 3).

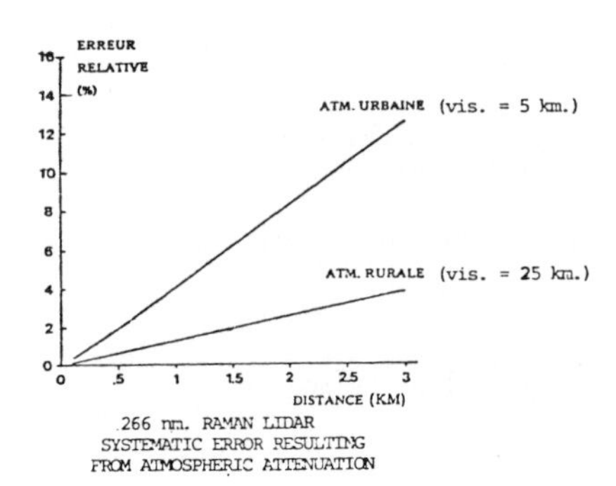

266 nm. RAMAN LIDAR
SYSTEMATIC ERROR RESULTING FROM ATMOSPHERIC ATTENUATION

	rural atmosphere	urban atmosphere
Rayleigh attenuation	+ 3 %	+ 3 %
Mie attenuation	+ 1 %	+ 4 %
SO_2 attenuation	+ 1 %	+ 6 %

DETAILED CAUSES OF ERROR AT A RANGE OF 3 KM

FIGURE 3 -

Maximum error which occurs in an urban atmosphere is about 13% at a range of 3 km; this error being due to Rayleigh (3%), aerosol (4%) and SO_2 (6%) extinction.

Experimental results have been presented at the 11° I.L.R.C., ref. 34.

Two examples of comparison with radio-soundings are shown Fig 4. The

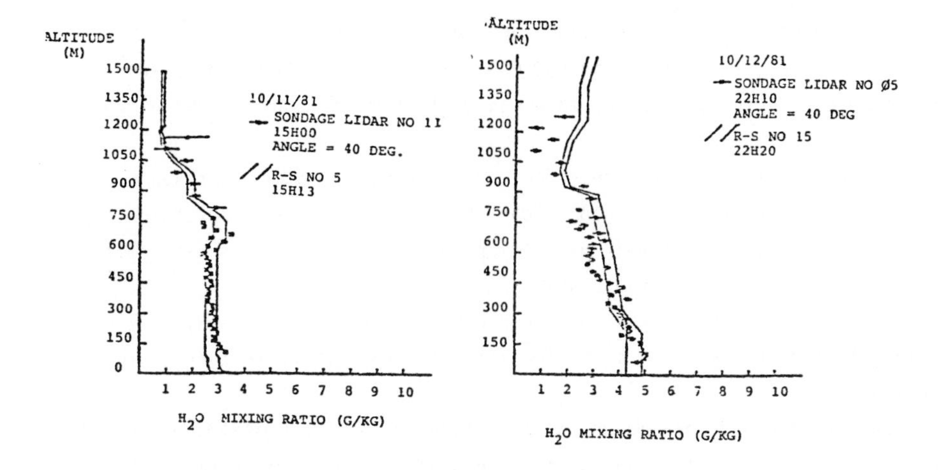

FIGURE 4 - 266 nm RAMAN LIDAR COMPARISONS WITH RADIO.SOUNDINGS

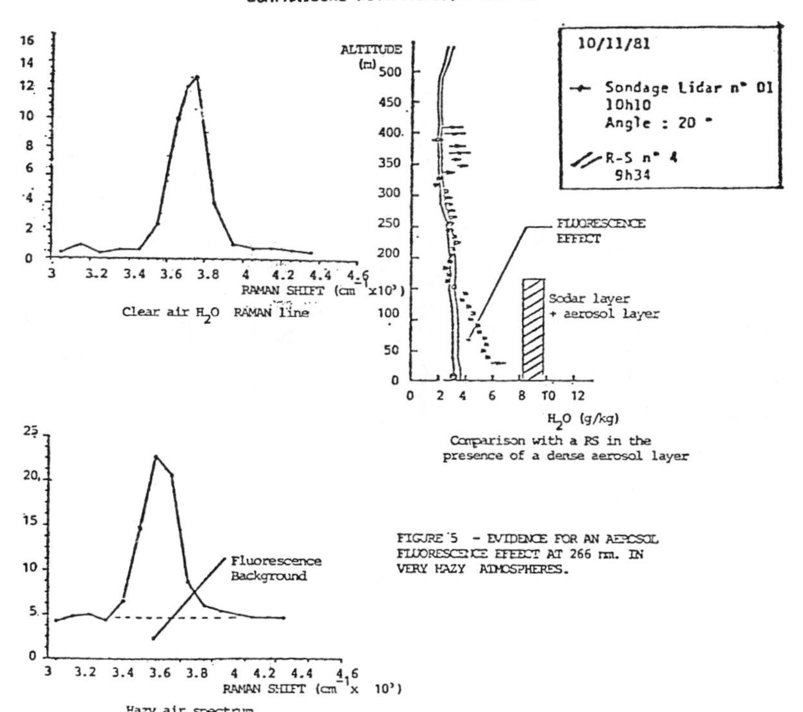

FIGURE 5 - EVIDENCE FOR AN AEROSOL FLUORESCENCE EFFECT AT 266 nm. IN VERY HAZY ATMOSPHERES.

first one gives a very good agreement. The second one shows a good general agreement with some differences; up to now, these discrepancies are attributed to the sequential measurement of the 3 Raman lines and to the non-vertical sounding of the Lidar.

A special aerosol broadband fluorescence effect has been found in very hazy atmospheres. To illustrate that point two Raman spectra are shown Fig 5. The first is recorded in clear conditions and the second in heavy haze. In the second case, one can see that the H_2O Raman lines is superposed on a large fluorescence background. Fluorescence was equivalent to about 1 g/kg of water vapor for a 2 nm bandwidth.

An other experiment shows the effect of fluorescence on a water vapor profile; the sodar and the aerosol Lidar signal were giving an inversion layer up to about 180 m; in this very hazy layer, H_2O Lidar shows an excess of 1 to 2g/kg when compared to the radio-sounding.

This aerosol fluorescence effect should be studied more in detail; it could be strongly reduced by using a smaller optical bandwidth or removed by subtraction.

To come to the end, are shown on Fig 6 four successive profiles of water vapor in the presence of a subsidence inversion. Under the inversion, the mixing ratio is between 3 and 4 g/kg; above 2 g/kg or less. The lowering of the inversion is marked by the continuous line.

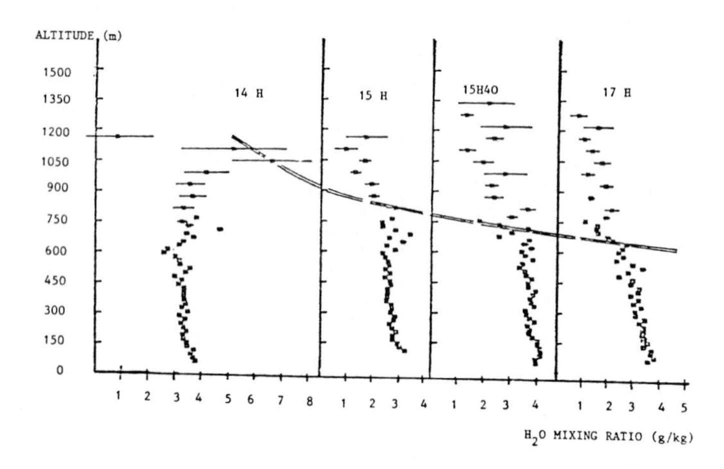

figure 6- 266 nm RAMAN LIDAR
4 successive profiles showing the lowering of a subsidence inversion

The synoptic situation shows for this day a cyclonic cell over France; this cell is vanishing with the arrival of atlantic disturbances. For this reason, the cold anticyclonic pellicle gets lower as warmer maritime air is coming at high altitude. The lowering of the subsidence inversion is seen by Lidar and radio-soundings. Good agreement between the two measurements is noticed (Fig. 7).

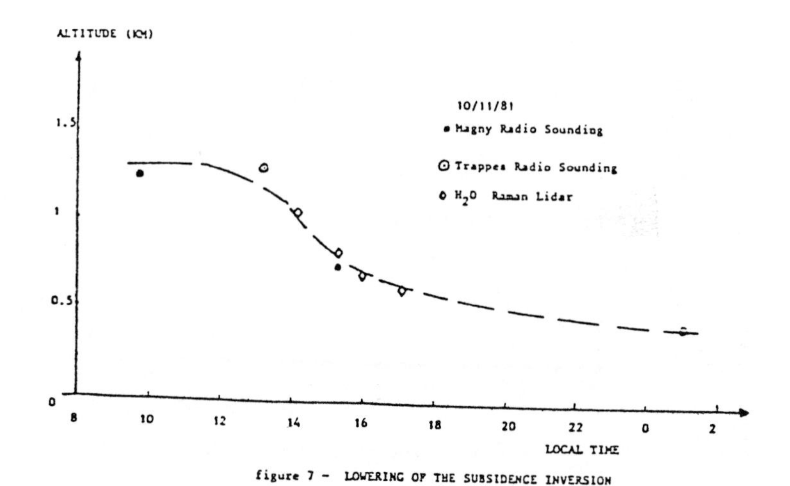

figure 7 - LOWERING OF THE SUBSIDENCE INVERSION

4.4 MINOR SPECIES MEASUREMENTS

Experimental results have been presented at the 3° I.A.M.A.P. Scientific Assembly, ref. 35.

4.4.1 In the atmosphere

For the measurements on minor species, the monochromator slitwidth is 2 mm which gives a good spectral resolution and a good rejection factor without severe degrading of the detected light intensity.

As said upper, ozone has an intense absorption continuum between 200 and 300 nm. We have made laboratory measurements to get ozone Raman cross section for a 266 nm excitation and a ratio σ ozone on σ oxygene equals 1200 was determined.

The Fig 8 shows a remote atmospheric spectrum obtained with a 45 cm^{-1} spectral resolution.

The vertical axis has a logarithm scale. The three Raman lines for oxygene, carbon dioxyde and ozone are apparent, the CO_2 ν_1 line being masked by the oxygene wing.
Atomospheric concentrations were not measured at this time; nevertheless, one can observe the strong enhancement for ozone scattering (about 1000) which agrees with laboratory measurements. For CO_2, only ordinary Raman scattering occurs and, assuming a 350 ppm mixing ratio, one gets σCO_2 on σO_2 equals 0.6.

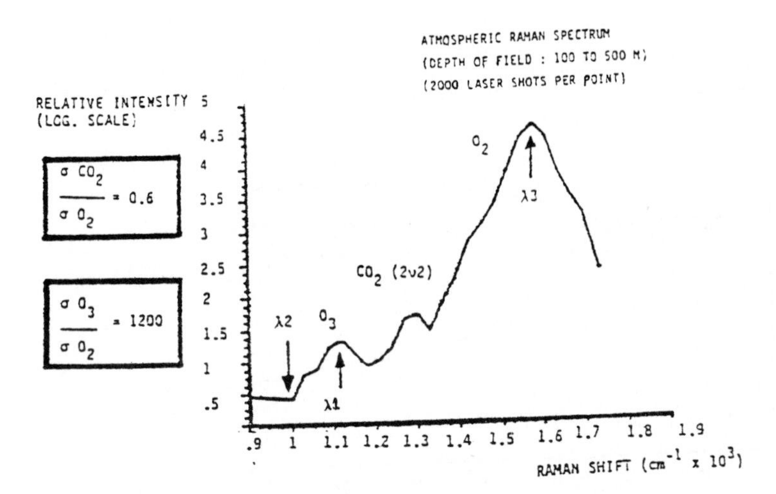

FIGURE 8 - ATMOSPHERIC RAMAN SPECTRUM

The profile of O_3 mixing ratio is obtained by measuring O_2 and O_3 signals, and noise including spurious light in the monochromator and unrejected wings of major components band. We present Fig 9 an atmospheric ozone profile. The constant of calibration k is calculated with the cross-sections ratio.

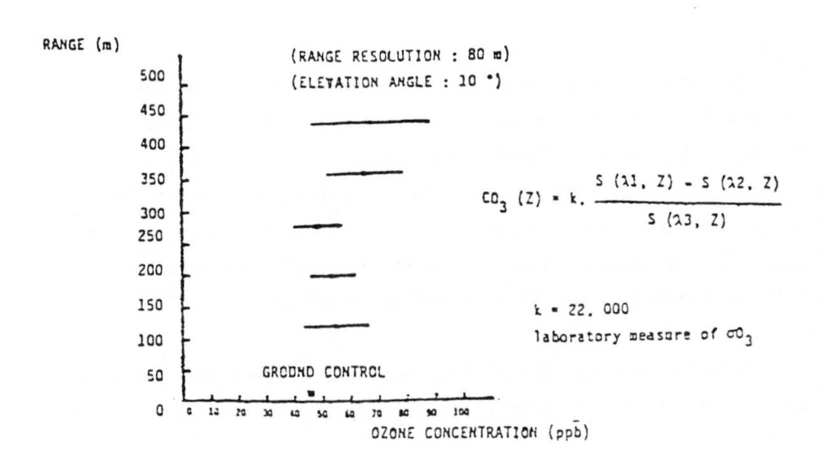

FIGURE 9 - ATMOSPHERIC OZONE PROFILE

Two thousand laser shots are averaged for each line. A good agreement can be noticed with the ground value measured by a chemiluminescence ozonometer. The accuracy and the range of the measurement were limited by the lack of laser energy and by the difficulty of recording the very low ozone and noise signals. It is planned to make this recording easier with the help of a photo-electron-pulse shaping circuity.

4.4.2 In a tank

In order to simulate a polluted air, a measurement tank was settled at about one hundred meters in front of the Lidar. This tank is a 2 meters long inox cylinder with a 15 cm diameter. Both ends are closed with quartz windows at Brewster's angle. In a first step, SF_6 was chosen as non resonant scattering gas at 266 nm. SF_6 was mixed with nitrogen inside the tank; nitrogen is the reference gas. Different concentrations were obtained by varying the relative flow rates of the two gases and controlled by sampling the outgoing gas in a cell. This cell was analysed with a mass spectrometer. The SF_6 Raman line is shown Fig 10. The comparison of the ratio of the SF_6 and

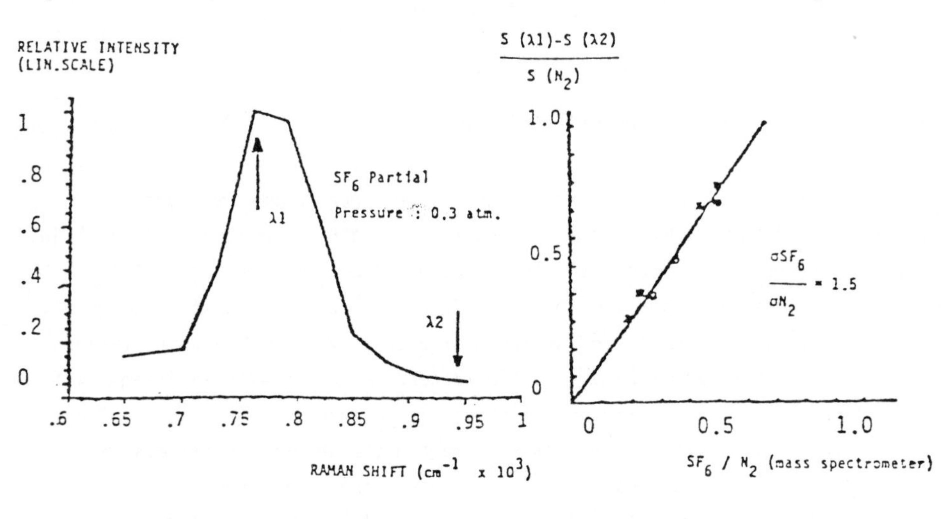

SF_6 RAMAN LINE
(DEPTH OF FIELD : 87.5 to 89.5 m)
(100 LASER SHOTS PER POINT)

RAMAN-LIDAR MEASUREMENT OF SF_6
IN A CALIBRATION TANK
COMPARISON WITH MASS SPECTROMETRY

FIGURE 10

nitrogen signals, measured by Raman Lidar, and the volume mixing ratio of SF_6 in nitrogen, measured by mass spectrometry, was made. The linearity is good and the least-squares-fitted adjustment gives the 1.5 cross-section ratio.

In the same way CO_2 has been detected. The Fig. 11 shows carbon dioxide Raman spectrum with the $2\nu_2$ and ν_1 bands; the ratio of the two lines cross-section is 1.55.

SO_2 presents a strong absorption in 266 nm range, so that resonance Raman or fluorescence is to be expected. Due to a strong absorption in the tank,

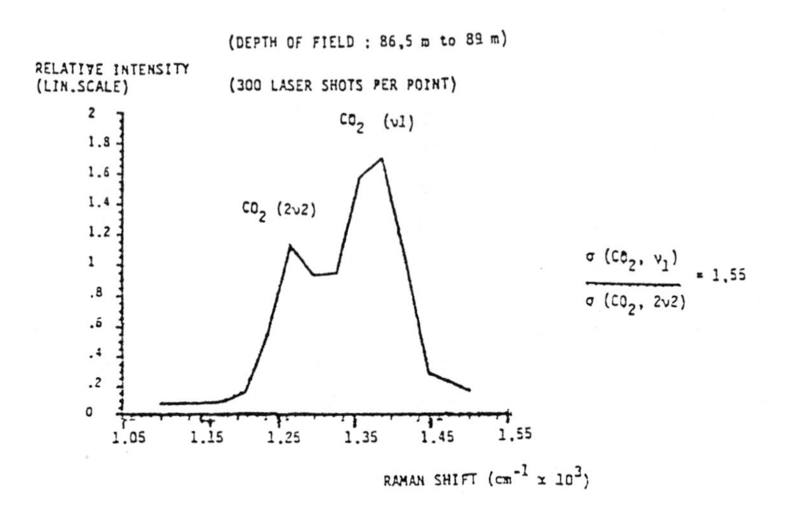

FIGURE 11 - CO_2 RAMAN LINES LIDAR

SO_2 backscattered signal was inferior to fluorescence echoes from the windows. So we have made a laboratory experiment with a smaller quartz cell in order to reduce the optical path. The gases are flowing in the cell at the atmospheric pressure. First, we have obtained a nitrogen Raman spectrum on a fluorescence background induced by the quartz cell. With a 5% SO_2 in nitrogen mixture, we have obtained Fig. 12 a fluorescence spectrum with an intensity 5 to 10 times greater than nitrogen spectrum. To measure the SO_2 cross-section some energy corrections are necessary to remove the absorption effects. We obtain a ratio σ SO_2 fluorescence on σ nitrogen Raman between 115 and 125 at 2330 cm^{-1} for different concentrations. We tried to measure the fluorescence decay time of the SO_2 signal. In fact it was inferior to the half of the 15 ns laser pulse duration. Fluorescence cross-sections depend largely upon quenching factors, so that it seems difficult to do a SO_2 quantitative measurement at 266 nm.

Starting from the experimental results obtained on minor species, the detection capabilities of the U.V. Raman Lidar for pollution measurements have been calculated.

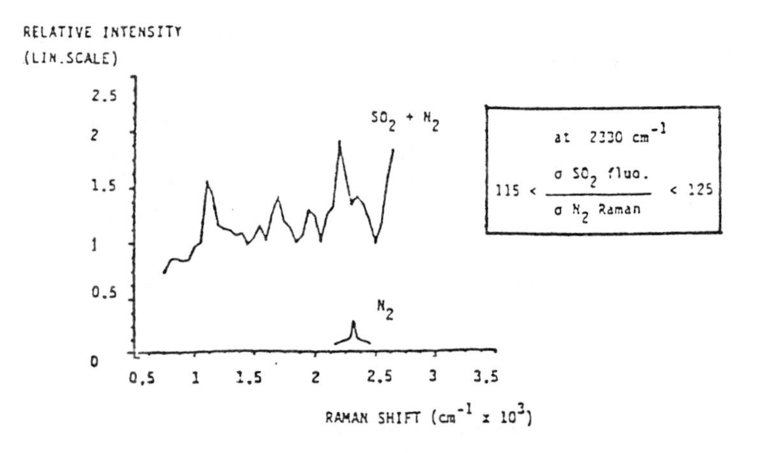

FIGURE 12 FLUORENCE SPECTRUM OF SO_2 IN CELL EXCITED BY 266 nm LASER LINE

4.5 SPECTROSCOPIC SIMULATIONS AND S.N.R. CALCULATIONS.

4.5.1 Simulation in a polluted atmosphere

Starting from the measured cross-sections or litterature data and from the monochromator transfer function, calculations have been made to establish the spectroscopic feasibility of the detection in a clear as well as in a polluted atmosphere, ref. 36.

For these computations, we used the data shown in the Tables 4 or 5.

As example of these computed Raman spectra is shown Fig. 13. The computed Raman atmospheric spectrum agrees quite well with the experimental spectrum. In a polluted atmosphere such as an urban environment, only ozone and carbon dioxide can be detected as pollutant gases Fig. 14. With the features mentioned here, we find a signal to noise ratio of about 10 for CO_2 and ozone so that Lidar measurements are possible.

4.5.2 Simulation in a plant plume

Calculations require pollutant concentrations inside a plant plume at a few hundred meters from the stack. Spatial distribution of concentrations of gaseous pollutants and aerosols in a plume are calculated using the classical Pasquill-Gifford plume dispersion model and the Turbigo plant data.

These data are given for a distance of 250 meters from the chimney and at the

SCATTERING SPECIE , X.	$\sigma X / \sigma N_2$		Nx ppm
O_2	2.1	(*)	21 10^4
O_3	2500	(*)	3.5 10^{-2}
CO_2	$2\nu_2$ 1.2	(*)	330
	ν_1 1.9	(*)	
RAYLEIGH SCATTERING	1240	(**)	10^6

(*) measured
(**) calculated

Table 4 - Used data for computation of the atmospheric spectrum

=-=-=-=-=-=-=-=-=-=-=-=-=

SCATTERING SPECIE, X	$\sigma X / \sigma N_2$	X CONCENTRATION IN THE PLUME m^{-3}	X CONCENTRATION URBAN ENVIRONMENT m^{-3}
O_2	2.1	5.2. 10^{24}	5.3 10^{24}
CO_2	$2\nu_2$ 1.2 ν_1 1.9	9.8 10^{22}	8.4 10^{21}
SO_2	FLUORESCENCE 120 at 2330 cm^{-1} 60 at 1300 cm^{-1}	1,3 10^{21}(50 ppm)	2.5 10^{18}
NO_2	75	1.8 10^{20}	2.5 10^{18}
O_3	2500	2.5 10^{18}	2.5 10^{18}
MIE SCATTERING	4 10^{20}	5.0 10^9	4.6 10^8
RAYLEIGH SCATTERING	1240	2.5 10^{25}	2.5 10^{25}

Table 5 - Used data for computation of the polluted atmospheric spectrum.

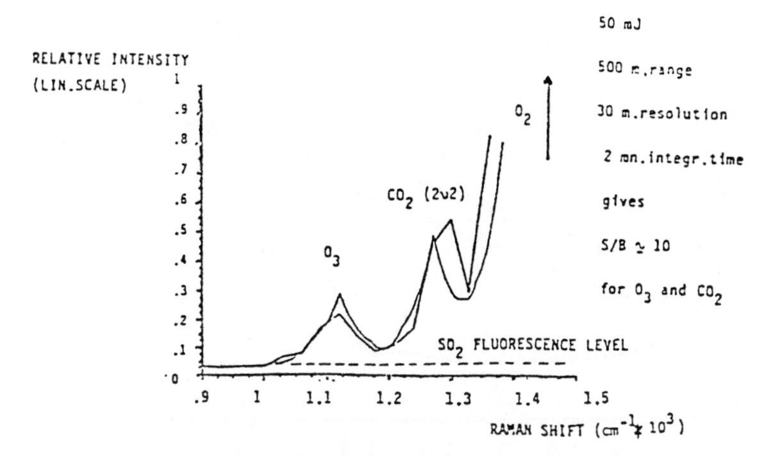

FIGURE 13 - COMPARISON BETWEEN THE RAMAN ATMOSPHERIC SPECTRUM AND THE COMPUTED SPECTRUM

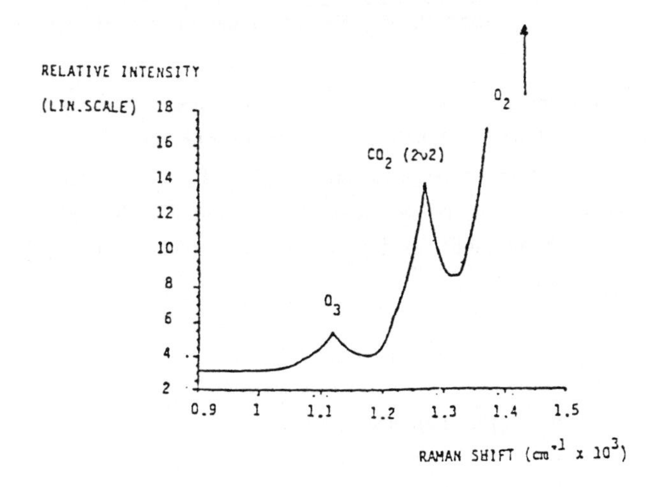

FIGURE 14- COMPUTED SPECTRUM : POLLUTED ATMOSPHERE

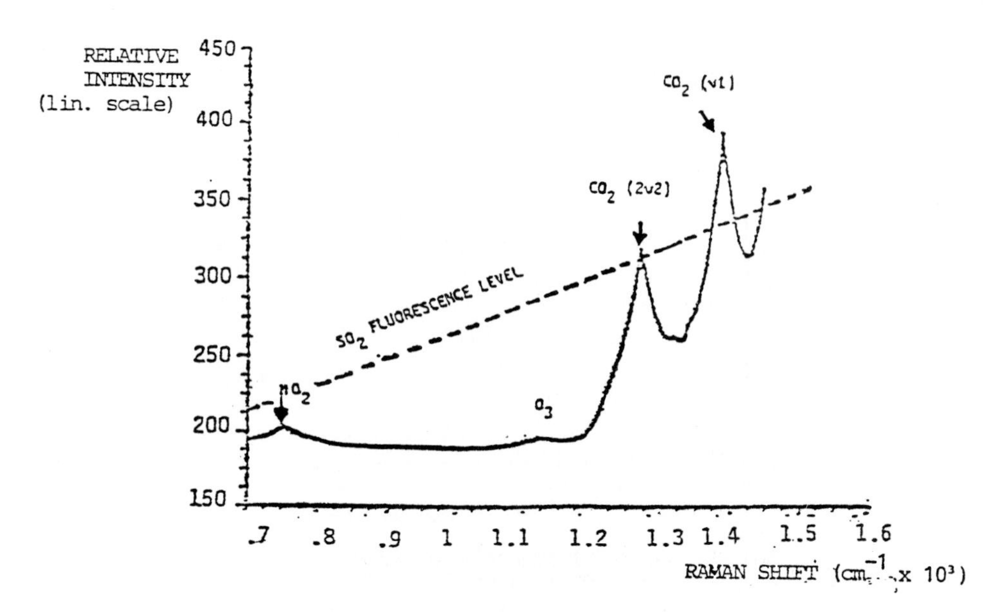

FIGURE 15 - Computed spectrum in plume taking into account SO_2 fluorescence effect at 266 nm.

plume center. The backscattering cross-sections of aerosols are calculated at 266 mn using Mie theory and Mc Cormick tables for a monodispersion of spherical particles. The computed spectrum in plume is shown Fig 15. SO_2 fluorescence is large enough to mask nearly all the minor gases Raman lines. These calculations indicate that detection of gases in a plume at 266 nm is not possible because of the SO_2 fluorescence.

5 CONCLUSION AND PERSPECTIVES

To summarize the presented results, it can be stated that the U.V. Lidar Raman is a very helpful system for meteorological studies and for the control of a polluted atmosphere.

In the meteorological field, U.V. Raman Lidar is able to measure boundary-layer water-vapor profiles with a good spatial resolution. Water vapor profiles are measured with a range of 3 km at day-time. Agreement with radio soundings is shown to be good.

As concerns minor species detection, first measurements of atmospheric ozone using resonance Raman scattering are presented. Identification and measurement of some pollutant gases is performed in a simulation tank. The detection limits of the system are investigated; they range from 10 ppb to 10 ppm depending upon the degree of resonance enhancement.

Calculations show that measurement of gases are possible in a polluted atmosphere but we must take care of some possible aerosol fluorescence effect.

For remote sensing in a plant plume, SO_2 fluorescence is thought to be large enough to mask nearly all the minor gases Raman lines so that detection at 266 nm is not possible.

Further technical modifications will improve the sensivity and the quality of day-time Raman Lidar measurements:

- use of a new laser generation having higher energy and repetition rate such as KrF or XeCl exciplex lasers,
- use of a multichannel detection system for measuring simultaneously several Raman lines,
- reduction of the aerosol fluorescence effect,
- automatisation of the Lidar system for routine monitoring.

(*) Commissariat Energie Atomique, C.E.N. Saclay, D.P.C., 91191 GIF, FRANCE
(**) Météorologie Nationale, E.E.R.M./C.R.P.A., 78470 MAGNY LES HAMEAUX, FRANCE.
(***) Electricité de France, D.E.R., E.A.A., 78400 CHATOU, FRANCE.

REFERENCES

1. G. MEGIE et al. (1978) Planet. Space Sci. 26, 27
2. G. MEGIE et al. (1981) IAMAP 3d Scientific Assembly, Lidar session, 17-28 August 1981, Hamburg, F.R.G.
3. J.L. LESNE (1979) Télédétection par Lidar, Mesucora, Paris.
4. J.L. GAUMET, D. RENAUT, J. BONNET, TRAN QUOC PHONG (1980) La Météorologie VIe série n°23.
5. R. CAPITINI, D. RENAUT, J.C. POURNY (1980) VIIe International Conference on Raman Spectroscopy, Ottawa, Canada.
6. D.A. LEONARD (1967) Nature 216, 142.
7. J. COONEY (1969) Laser Applications in the Geosciences, Huntington Beach, California.
8. S. MELFI et al. (1969) Appl. Phys. Letters 15, 29[illegible]
9. S. MELFI (1972) Appl. Optics 11, 1605.
10. J. COONEY (1971) J. Appl. Meteorol. 10, 301.
11. J.C. POURNY, D. RENAUT, A. ORSZAG (1979) Appl. Optics 18, 1141.
12. D. RENAUT, C. BRUN, J.L. GAUMET, J. BONNET (1981) Note technique EERM n° 94, Météorologie Nationale (France)

14. R.G. STRAUCH et al. (1971) Appl. Optics 10, 2665.
15. D. RENAUT (1977) NOte interne EERM n° 392, Météorologie Nationale.
16. J. COONEY (1972) J. Appl. Meteorol. 11,108.
17. J.A. SALZMAN et al. (1971) NASA, technical note, TN-D6336.
18. A. PETITPA (1977) Note interne EERM n° 387, Météor. Nat.
19. J.A. SALZMAN, T.A. CONEY (1973) Nasa, technical note, TM-X68250.
20. Y.F. ARSHINOV (1981) IAMAP 3[d] Scientific Assembly, Lidar session, 17-28 August 1981, Hamburg, F.R.G.
21. R. GILL (1979) J. Appl. Meteorol. 18, 225.
22. M.J. GARVEY, G.S. KENT (1974) Nature, 248, 124.
23. H. INABA, T. KOBAYASI (1969) Nature, 224, 170.
24. T. KOBAYASI, H. INABA (1970) Appl. Phys. Lett. 17, 139.
25. H. INABA, T. KOBAYASI (1972) Opto-Electron. 4,101.
26. S. NAKAHARA et al. (1972) Opto-Electron. 4, 169.
27. S. MELFI et al. (1973) Appl. Phys. Lett. 22,402.
28. T. HIRSCHFELD et al. (1973) Appl. Phys. Lett. 22, 38.
29. V.M. ZAKHAROV, V.A. TORGOVICHEV (1978) In remote Sensing of the Atmospher, A. Fynat, V. Zuev ed., Elsevier SPC, Amsterdam.
30. Y.G. VAINER et al. (1979) Sov. J. Quantum Electron. 9, 296.
31. S.K. POULTNEY et al. (1977) Appl. Optics 16, 3180.
32. D. RENAUT, R. CAPITINI, J.C. POURNY (1981) La Météorologie VI[e] série n° 25.
33. D. RENAUT, J.C. POURNY, R. CAPITINI (1980) Optics Letters, 5, 233.
34. D. RENAUT, C. BRUN, R. CAPITINI (1982) XI[e] International Laser Radar Conference, Madison, US.A. (Nasa Conference Publication 2228).
35. R. CAPITINI, E. JOOS, D. RENAUT (1981) IAMAP 3[d] Scientific Assembly Lidar session, Hamburg, F.R.G.
36. E. JOOS (1982) Thèse de Docteur-Ingénieur, Université des Sciences et Techniques de Lille, France.

REVIEW PAPERS

H. INABA Detection of Atoms and Molecules by Raman Scattering and Resonance Fluorescence in Laser Monitoring of the Atmosphere, Ed. E.D. Hinkley Topics in Appl. Phys. Vol. 14. Springer-Verlag, Berlin, 1976.

J.L. GAUMET, D. RENAUT, A. PETITPA Applications Météorologiques du Lidar Sciences et Techniques de l'Armement 4e fascicule, Paris 1980.

A.O. VAN GYSEGEM Meteorological Observations by laser indirect sensing techniques. Instruments and observing methods, report n° 12 WMO, Genève, 1982.

Optical Remote Sensing of Air Pollution,
Lectures of a course held at the Joint Research Centre, Ispra, Italy, 12–15 April 1983,
P. Camagni and S. Sandroni (Eds). 205–233

LIDAR APPLICATIONS TO AEROSOLS AND PARTICLES

P. CAMAGNI

1. INTRODUCTION

Remote sounding of atmospheric properties and atmospheric phenomena has received an enormous impact from the advent of optical interrogation systems, first introduced with the elastic-scattering Lidar (for general information, see Refs. 1-3). After the successful demonstration in mesospheric and tropospheric tests of the middle 1960s (4-7), the Lidar technique has been constantly improved, thanks to the general progress in laser and detection equipment and to the introduction of various refinements (automatic computing and control systems; multi-wavelength operation, polarization analysis). Therefore, the specific capacity of this instrument to probe scattering masses has been progressively extended to a variety of problems, such as atmospheric profiling, structure of mixing layers, visibility studies, cloud and fog formation, transport of aerosol and solid pollutants and many others. Meanwhile, other concepts were developed from the basic idea, leading to the introduction of differential-absorption Lidars and Raman Lidars (separately treated in this Course). Physical problems involved in these applications have generated new insight into the optics of particulate media, with positive fall-out on the operation of the classical Lidar. Nevertheless, a number of conceptual and practical limitations are still persisting on the way to better exploiting this instrument.

The aim of the present lecture is to outline some methodological aspects concerning the observation of particles and aerosols, via the Lidar experiment. The discussion of experimental facts is devoted to representative cases, which are best suited to illustrate the main problems and the techniques adopted for solving them (analysis of a Lidar profile; consistency of observations; identification of scattering components; multiwavelength operation). Concepts related to the optics of particulate media are only synthetically reviewed, as a support to this scheme. General scattering theory and the principles of Lidar instrumentation are assumed to be known to the audience, through the preceding lectures of Prof. Quenzel and

Dr. Varey (this Course). The treatment also avoids the specialized links of Lidar monitoring with atmospheric physics and meteorology, of which a pertinent review is afforded in Dr. Pettifer's lecture.

2. LIDAR MONITORING ON PARTICULATE MEDIA

2.1 The Lidar equation; basic concepts

After firing a pulsed laser beam through the atmosphere, the back-scattered light power is described by the well known Lidar equation, which after correcting for angular aperture (proportional to R^{-2}) can be written:

$$\left.\frac{P(R)}{P_0}\right]_{corr} = C \cdot \beta(R) \exp[-2\int_0^R \alpha(R')\,dR'] \qquad (1)$$

where:

R = ct/2 is the range (or depth) from which the light emitted at time zero is returned after time t,

$\beta(R)$ and $\alpha(R)$ are the volume back-scattering and extinction coefficients of the medium at R, including all the different contributions from molecular gases, particles and aerosols. Values of these quantities are expressed as reciprocal lengths (usually m^{-1} or km^{-1}); for β they are also referred to unit solid angle.

C = a constant embodying several instrumental effects such as pulse width of the laser beam, detector efficiency, effective receiver area and electronic gain. All these factors are independent of R.

P_o = the emitted peak power.

In the following it will be understood that the aperture factor and the emitted power are absorbed in the simplified notation P(R), indicating range-corrected and normalized intensity of the Lidar signal.

Equation (1) in its complete form has been developed and commented by various authors, quoted in Refs. 1, 8 and 9. Its validity implies that multiple scattering be neglected.

The Lidar equation is the natural tool for mapping extended volumes of scattering matter, via the correlation between the quantities β, α and the range R, which is established by observations of power P at different return times t. The profile of normalized P(R) is everywhere governed by the product of the local $\beta(R)$ with an integrated path-transmission T(R), each factor containing specific and background contributions which we separate for convenience as follows:

$$\beta(R) = \beta_s(R) + \beta_0(R)$$

$$T(R) = \exp\left(-2\int_0^R [\alpha_s(R') + \alpha_0(R')]\, dR'\right) \qquad (2)$$

where $\beta_s(R)$ and $\alpha_s(R)$ are the scattering and extinction coefficients of the medium under examination; β_o and α_o are the corresponding quantities for ambient atmosphere, which often are taken to be constant for simplicity. Note that this implies reference to the concept of standard air, whose optical properties are fixed by a natural content of gases plus water aerosols. Use of this concept requires: i) that a clear atmosphere be optically defined, as a function of temperature, pressure and humidity, and ii) that the defining conditions be constant over the range and time of the given experiment. Deviations of local atmosphere with respect to these conditions shall be embodied in the space-dependent contributions β_s and α_s.

Tables I and II give indications as to the magnitude of scattering and extinction effects in a real atmosphere, compared with those of a purely molecular atmosphere. These data are taken from various sources, quoted in Refs. 9 and 10. Notice that, even for the clearest conditions, β_o and α_o are substantially larger than predicted by Rayleigh scattering alone. This indicates the ubiquitous effects of trace dust and water aerosols, which eventually are enhanced when water-drop condensation occurs. In any case the values for clear standard air are smaller by two to three orders of magnitude with respect to those for moderate haze, fog and water clouds. The latter in turn may be much lower than the values observed inside a thick particulate, e.g. coal smoke. These observations justify the positions $\beta_s \gg \beta_0$, $\alpha_s \gg \alpha_0$, which often are made to simplify the monitoring of dense localized structures (such as vapour clouds, chimney plumes, etc.). With this assumption, one is assured that the overall ratio $\rho = \beta/\alpha$ is governed by scattering and extinction of the medium under consideration, regardless of background values and of their fluctuations. This, as we shall see, is one of the conditions for explicit inversion of the Lidar equation.

Appropriate guesses about the magnitude of scattering and extinction coefficients are also of great importance in the evaluation of experimental error. The chart of Fig. 1, reported from Ref. 9, helps to illustrate the mutual relationship of sky radiance (noise), background attenuation (visual range), observed range and Lidar signal. Interplay of these parameters decides the accuracy of measurements and the reliability of data which can be inferred from them. The chart describes the typical case of a Ruby Lidar with pulse

TABLE I
Backscattering and extinction properties of a real atmosphere at Ruby wavelength

	visual range	visual extinction coeff.	extinction coeff.	backscatter coeff.
	v/km	σ_{vis}/m^{-1}	$\alpha\,[m^{-1}]$	$4\pi\beta\,[m^{-1}]$
water clouds	0.2	2.0×10^{-2}	2.0×10^{-2}	1.0×10^{-2}
	0.5	7.8×10^{-3}	7.8×10^{-3}	3.9×10^{-3}
fog	1.0	3.9×10^{-3}	3.9×10^{-3}	2.0×10^{-3}
haze	2.0	2.0×10^{-3}	1.9×10^{-3}	1.0×10^{-3}
	5.0	7.8×10^{-4}	7.2×10^{-4}	3.6×10^{-4}
	10	3.9×10^{-4}	3.4×10^{-4}	1.7×10^{-4}
	20	2.0×10^{-4}	1.6×10^{-4}	8.0×10^{-5}
clear	50	7.8×10^{-5}	6.1×10^{-5}	4.9×10^{-5}
	100	3.9×10^{-5}	2.8×10^{-5}	3.1×10^{-5}
	200	2.0×10^{-5}	1.2×10^{-5}	1.6×10^{-5}
	(clear air, 1013.2 mbar, 0°C)	[Ref. 9]		

TABLE II
Volume coefficients of the Rayleigh component of atmospheric scattering for ruby lidar (λ = 0.6943 μ).*

Height (km)	Molec. density (m^{-3})	$4\pi\beta$ (180°) (m^{-1})	α (m^{-1})
0	2.55×10^{25}	6.55×10^{-6}	4.37×10^{-6}
5	1.52×10^{25}	3.93×10^{-6}	2.62×10^{-6}
10	8.60×10^{24}	2.21×10^{-6}	1.47×10^{-6}
15	4.06×10^{24}	1.04×10^{-6}	0.69×10^{-6}
20	1.85×10^{24}	4.75×10^{-7}	3.2×10^{-7}
	[Ref. 10]		

* Rayleigh scattering is proportional to λ^{-4}.

energy of 2 Joule and detector/optical characteristics of good quality. The inserted example shows that for diurnal, overcast atmosphere, with visual range of ~ 10 km, the maximum permissible Lidar range is 5 km, if we fix an acceptable signal-to-noise ratio $\geqslant\sqrt{10}$. With similar constructions, the diagram can give the actual values of S/N to be expected for any given Lidar range.

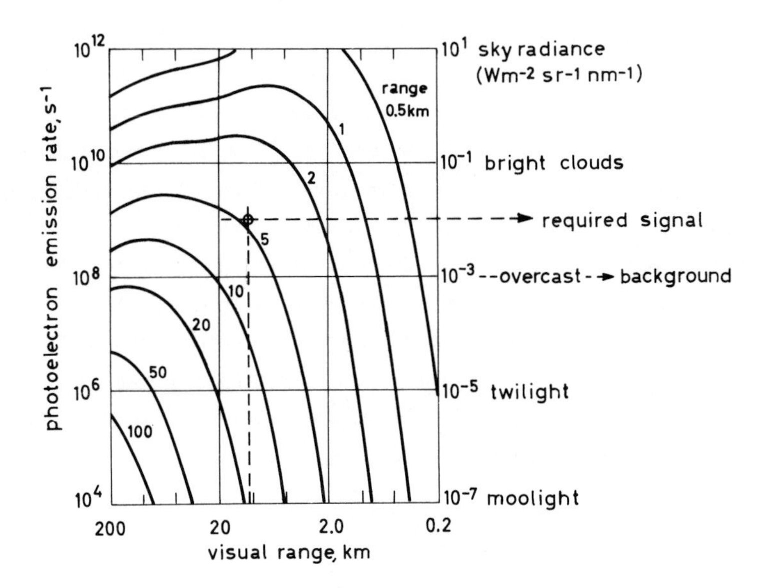

Fig. 1. Lidar signal from atmospheric backscatter as a function of the visual range, with target range as a parameter (Table I assumed for calculation; Ref. 9);

Knowing the signal-to-noise ratio, once can easily predict through eq. (1) the error to be expected in the derivation of background values β_o, α_o from a measure of P(R). This also establishes the error limit contained in a measure $P(R) + \delta P$, due to the presence of a composite medium $\beta_0 + \delta\beta$, $\alpha_0 + \delta\alpha$, hence the accuracy with which specific parameters can be separated from background parameters.

Experimental errors in the determination of P(R) may have a different impact, depending on whether the Lidar measurement is called to give relative or absolute determinations. The latter achievement is compulsory, if one wants to provide full characterization of particle size and concentration in complex media. On the other hand, an accurate profiling of appropriate ratios such as β_s/β_o or β_s/α_s may suffice to characterize simpler media. We shall review some of these aspects in later sections.

2.2 Definition of optical parameters

Notions concerning the scattering parameters of eq. (1) are recalled here in the simplest possible form, for the sake of future comments. For full discussion see general treatments (e.g. Refs. 11-12).

The problem of light scattering in a collection of identical particles moves from the calculation of two complex amplitudes $A_1(\theta;x;m)$ and $A_2(\theta;x;m)$ and their associated intensities $i_1 \equiv |A_1|^2$, $i_2 \equiv |A_2|^2$, which represent the flux of two orthogonal polarization components of the e.m. field, scattered into unit solid angle in a direction θ, as a result of interaction of an incident field of unit intensity with a particle of relative radius $x \equiv 2\pi a/\lambda$ (size parameter) and refractive index $\tilde{m} = m - ik$. All other parameters concerning light-scattering can be derived from these functions, starting with the definition of pertinent cross-sections σ for the various processes of interest (e.g. angular scattering, total scattering, total extinction, etc.). To each of these processes one can relate a dimensionless efficiency factor, obtained by reducing the corresponding σ to unit geometrical section of the target particle (radius a); general expressions are:

angular scattering efficiency: $F(\tilde{m};x;\theta) = \left[\frac{\lambda}{2\pi}\right]^2 \frac{i_1 + i_2}{2}$

total scattering efficiency: $Q_{sca}(\tilde{m};x) \equiv \int_{sphere} F d\Omega$ (3)

extinction efficiency: $Q_{ext}(\tilde{m};x) = [\frac{\lambda}{2\pi}] \, \mathrm{Re} \, \{\frac{A_1 + A_2}{2}\}_{\theta = 0}$

The last two definitions are proved to coincide in the case of non-absorbing particles, with real $\tilde{m}$.

In terms of efficiency factors, one can express the macroscopic volume coefficients for a collection of particles of the same type and of varying dimensions, making a weighted summation over a size distribution function dN(a)/da (number of particles per unit volume, per unit radius interval):

$$\beta_\theta = \int_0^\infty \pi a^2 \, F(\theta;\tilde{m};x) \frac{dN}{da} \, da$$

$$\alpha = \int_0^\infty \pi a^2 \, Q_{ext}(\tilde{m};x) \frac{dN}{da} \, da \qquad (4)$$

Owing to (3) and (4) the behaviour of α and β is expected to reflect the complex properties of the intensity function with respect to relative size, refractive index and scattering angle. Additional complications can be brought by the details of the weighing distribution: folding of these with intrinsic behaviour of F_θ and Q_{ext} may lead to difficult situations, impairing the prediction of experimental regularities with respect to θ and λ. Fortunately there exist some criteria of simplification. We comment here on qualitative trends, starting from a typical diagram of (i_1+i_2) vs x (Fig. 2).

Beyond the range of Rayleigh scattering, the shape of the intensity function is progressively dominated by marked stepping and sharp oscillations, which overcome monotonic behaviour as soon as $x \geqslant 1$ (i. e. for relative size parameters of practical concern in ambient aerosols). The strength and frequency of these oscillations increase on passing from low θ to back-scattering. In spite of these complexities, it is found that the envelope or smoothed behaviour always obeys some general rules. For instance, a running mean (abstracting from local oscillation) usually displays a series of broad, decaying oscillations superposed on a median value which is a direct, monotonic function of x. The same behaviour is reflected in the quantities F_θ, Q_{ext}, which are generated from (i_1+i_2). The following points are of particular interest for practical applications (see Refs. 13 to 15):

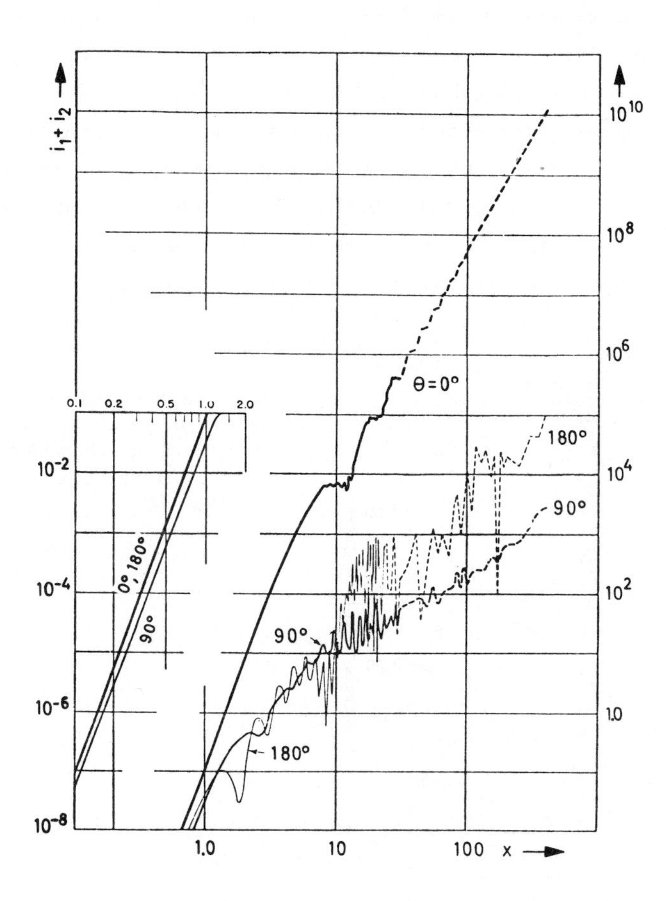

Fig. 2. Plot of intensity function $i_1 + i_2$ for spherical water particles as a function of relative size $x = 2\pi a/\lambda$, for scattering angles 0°, 90° and 180° [Ref. 13].

1) The period of the main oscillations is univocally related to the measure of relative size $x = 2\pi a/\lambda$. The first peak is also the absolute maximum; its position occurs at x values of the order of $1 \sim 10$ (for typical values of $\tilde{m}$). Successive peaks decrease in amplitude rather quickly.
2) The median value, in the region of the first peak, has weaker and weaker x dependence, on passing from forward scattering to finite angles. At $\theta = 180^o$, (i_1+i_2) assumes a nearly quadratic form; correspondingly the median value of F quickly becomes independent of x (compare with the well known dependence on λ^{-4}, valid in the Rayleigh limit). A useful compensation of x-dependence also occurs in the angular integral of this quantity, hence in the behaviour of Q_{sca} (see definitions (3)). Except for strongly absorbing matter, this will also reflect the behaviour of Q_{ext}.

A practical consequence of these concepts is the possibility of deriving the radius a from simple inspection of the main periodicity in $Q_{ext}(\lambda)$, irrespective of more complex details. This is illustrated by the diagram of Fig. 3 (Ref. 14), showing the remarkable conservation of the first peak in terms of a/λ. Application to measurements of the macroscopic extinction $\alpha(\lambda)$ is obvious. For instance, if two distinct radii such as a_1 and a_2 were prevailing in a hypothetical size-distribution, one would expect two dominant extinction peaks at the corresponding wavelengths λ_1 and λ_2.

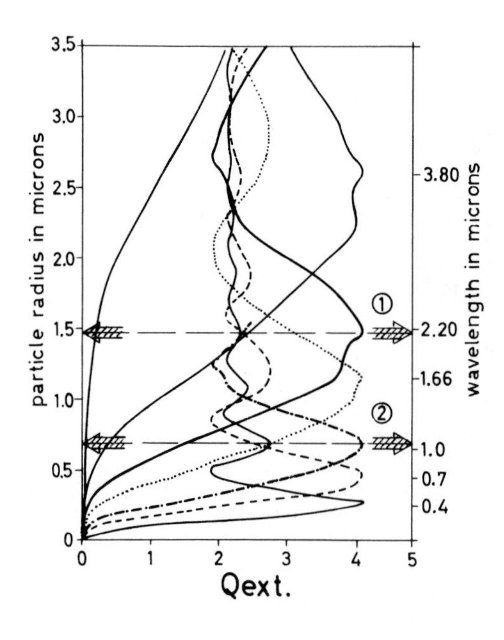

Fig. 3 - Extintion efficiency as a function of radius ($\tilde{m}$ = 1.33, water particles). The curve at the extreme left for 10.4 μm reaches a maximum at a radius of about 6 μm [Ref. 14].

2.3 Scattering/extinction relationships

The possibility to make explicit assumptions about the ratio β/α plays a key role in the inversion of the Lidar equation. Experimental and theoretical studies have shown that in many circumstances this relationship can be approximated by a simple power-law (Refs. 16 to 19):

$$\beta(\tilde{m}, \lambda) = C\,[\alpha(\tilde{m}, \lambda)]^{k} \qquad (5)$$

where C is constant for a given population of particles (fixed $\tilde{m}$) and the exponent k depends only weakly on size distribution.

One must be aware of the limits of eq. (5) before utilizing it for analytic work. In principle, there is nothing in the theory to predict that C and k should be unique for any given assembly, independently of its state of aggregation. Direct observation shows a large spread of values even for the same aerosol, in different surroundings. In this sense the above relationship is unreliable if one wants to predict visibility or extinction of an unknown medium from a knowledge of scattering alone (see also discussion in Dr. Pettifer's lecture). For Lidar analysis, however, the essential fact is that the exponent of eq. (5) may be taken as fixed in a given circumstance, regardless of absolute values. Of this there exists to-date sufficient justification (see Ref. 16-18 and quotations therein). Fig. 4 shows the β vs α relationship calculated for natural water aerosol in different conditions (Ref. 16), namely: i) for varying size-distribution, at constant particle number, without molecular scattering; ii) for varying number of particles and distinct distributions, in the presence of molecular scattering. In the first case the linearity of the log-log plot confirms that a specific k value can be assigned to describe very different size conditions, spanning two orders of magnitude of individual coefficients. The other results show the type of complications to be expected when competing amounts of a Rayleigh scatterer (atmospheric gas) are added to a specific Mie scatterer. This is also indicative of the situations that one encounters in complex, polydispersed media.

The numerical value of β/α varies from an ideal limit of ~ 0.12 (Rayleigh case) to $0.01 \sim 0.1$ for specific aerosols. For well characterized substances, (e.g. water aerosol above the micron range) smoothing effects make its dependence on average radius rather feeble, with overall variations not larger than 20% (Ref. 17). In view of this eq. (5) is often approximated by a simple proportionality, with $k \approx 1$.

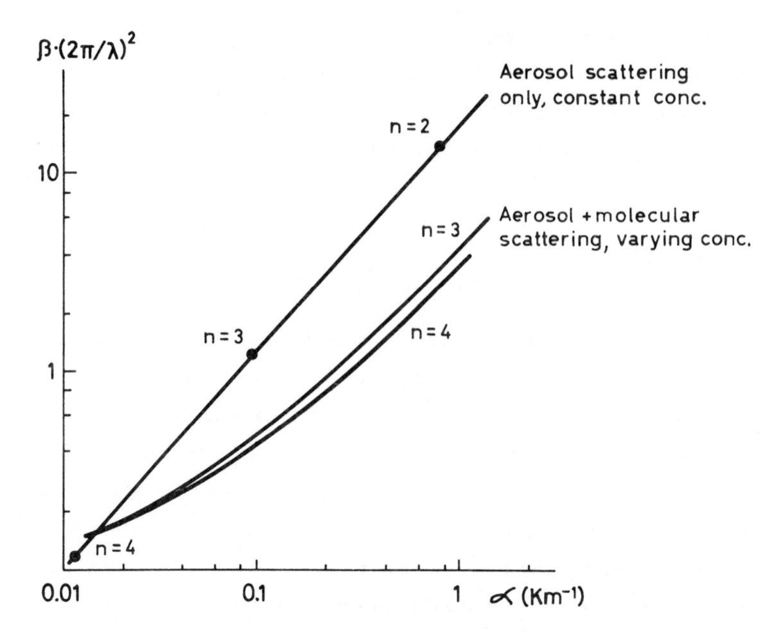

Fig. 4. Relation between back-scattering intensity and extinction coefficient of natural water aerosol, calculated for various size distributions of type $\sim a^{-n}$ (Ref. 16). Assumed $\lambda = 0.55\,\mu$m.

2.4 Inversion of the Lidar equation

Purpose of Lidar observations usually is to perform separate measurements of $\alpha(R)$ and $\beta(R)$, in relative or absolute units depending on later use. Problems which condition this programme can be commented starting from a simple Lidar experiment, referring to the case of a uniform scattering medium in a finite range R_1 to R_2, superposed to a fixed atmospheric background. Arbitrary values of α and β are chosen as indicated in Fig. 5).

The range corrected and normalized signal P(R) for this case is readily calculated from (1) and may be plotted in terms of a logarithmic signal

$$S(R) - S^* \equiv \ln(P(R)/P_0) - \ln(P^*/P_0) = \ln(\beta(R)/\beta^*) - 2\int_{R^*}^{R} \alpha(R')dR' \tag{6}$$

This has the convenient property of being independent of system origin and calibration constants, if one refers all points to a reference signal P^* coming from an arbitrary range R^*. The latter may be chosen so that the assumption $\beta^* \rightarrow \beta_0$ is justified; extrapolation of β_s/β_o, α_s and α_o then follows immediately from the combination of three independent measurements along the profile, e.g.: i) slope in the background region I or III; ii) slope in the interior of the

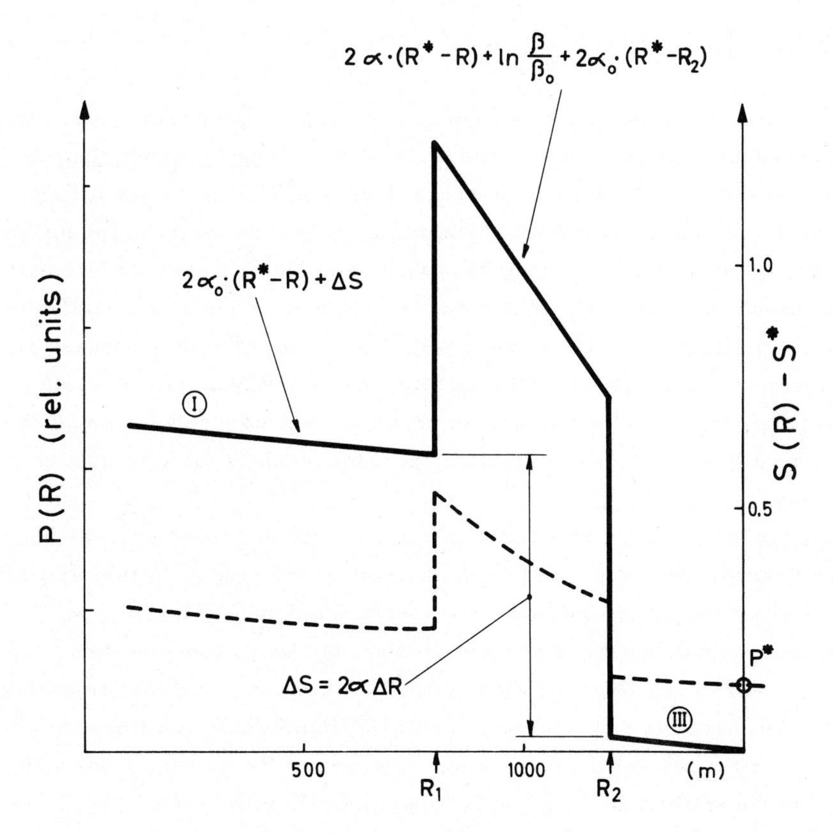

Fig. 5 - Elements of a Lidar profile from a uniform layer with $\beta_s \cong \beta_0$; $\alpha_s = 10\ \alpha_0 = 6 \cdot 10^{-4}$ m^{-1}

layer, and iii) attenuation ΔS across the layer (see inserts of Fig. 5).

The above scheme is subject to practical limitations even in simple cases. It is rather difficult to avoid horizontal or vertical structures of local atmosphere, due to diffuse pollution or natural haze gradients. This implies that the first part of the profile is hardly representative of a "clean", constant background. In the far range the signal P(R) is substantially attenuated and the determination of α_o may be inaccurate. Inside the layer itself the slope usually is not constant, owing to non-homogeneous conditions of the scattering medium. Therefore, direct inspection often reduces to measuring the attenuation step ΔS, i. e. the opacity or integrated optical thickness of the structure under examination.

An exact, point-by-point inversion is a difficult task. The differential equation corresponding to (6) is

$$\frac{dS}{dR} = \frac{1}{\beta(R)} \frac{d\beta}{dR} - 2\alpha(R) \tag{7}$$

which can be integrated only by assuming a specific relationship between the two coefficients. In the extreme case $d\beta/dR \ll 2\alpha\beta$ (slope approximation) the local derivative of S(R) gives a direct measure of α(R), but the scattering coefficient remains unaccessed. Furthermore, the consistency of the method is open to question: if β is small, the other coefficient is likely to be small and the assumption tends to break down, except in a very uniform scattering medium with $d\beta/dR \approx 0$. The above conditions are the opposite of those valid in a dense particulate, where both the scattering coefficient and its variations happen to be large. On the other hand, they may suffice for determining extinction profiles and visibility in diluted aerosol, such as found in light haze, mixing layers, etc.

Criticism of the slope approximation can be extended to several variants (ratio or slicing methods) based on subdivision of the range into thinner intervals, to which the same conditions are assumed to apply (ref. 19).

A general solution of the inversion problem is always possible if one assumes a power law relationship between β (R) and α(R), such as for instance eq. (5). Klett has shown how to integrate eq. (7) in this case, finding explicit solutions for the two coefficients, which are exempt from instabilities with respect to the trial values of k and to experimental error (ref. 19).

Full inversion is in any case a heavy analytical burden. Moreover, its validity rests on a proper choice of R^* and an accurate measurement of P^* in the far range, which is not easy to obtain. In view of this, a practical approximation is to take $k \approx 1$, i. e. $\beta/\alpha \cong$ const. , in keeping with the indications discussed in par. 2.3. Fig. 6 illustrates a simple scheme of application.

Suppose that two Lidar profiles $P_a(R)$, $P_b(R)$ were recorded in identical conditions, once firing through a given structure and next through clear air in a nearby direction. In a limited interval, curve b can be viewed as the local atmospheric signal, to be compared with an extrapolated background signal (heavy hatched line) which describes the unknown additional screening due to the medium between R_1 and R_2.

In order to obtain the corrected β(R) of the scattering structure, let us consider three independent measurements such as $P_a(R_1) \equiv P_b(R_1)$, $P_a(R_2)$ and $P_b(R_2)$, taken on the curves as indicated. From eqs. (1) and (2) the following ratios are immediately defined:

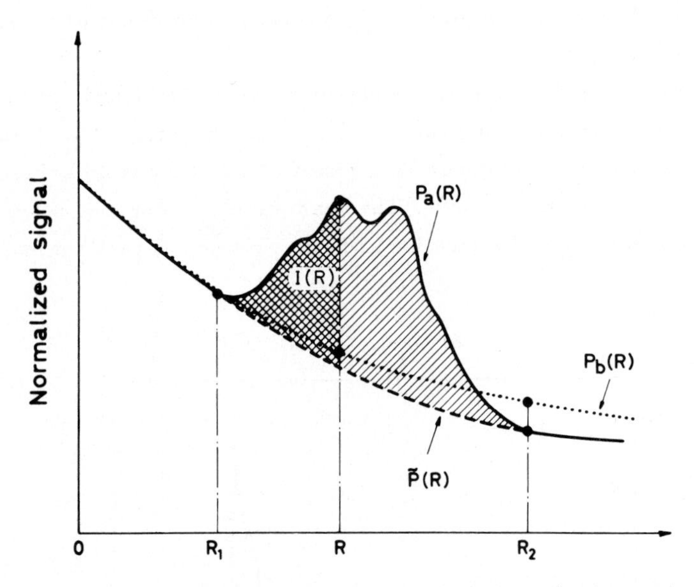

Fig. 6 - Scheme for simplified inversion of a Lidar profile (see text).

$$\frac{P_a(R)}{P_b(R)} = (1 + \beta_s/\beta_0)\, e^{-2\int_{R_1}^{R} \alpha_s(R')dR'}$$

$$\frac{P_a(R_1)}{P_a(R_2)} = e^{2\alpha_0 \Delta R} \cdot e^{2\int_{R_1}^{R_2} \alpha_s(R')dR'} \qquad \frac{P_b(R_1)}{P_b(R_2)} = e^{2\,\alpha_0\,\Delta R} \tag{8}$$

If one substitutes $\alpha_s(R) = \gamma\beta_s(R)$, assuming a fixed ratio of the two coefficients, the problem is readily solved at any R with good accuracy. It is sufficient for this to combine the last two expressions, work out the extinction integral

$$2\gamma \int_{R_1}^{R_2} \beta_s(R')dR'$$

(relative units) , and finally substitute for the similar integral of the first expression, in the same units; this requires a simple proportion between the partial area I and the total shaded area, which are limited by the curve $P_a(R)$ and by the extrapolated curve $\tilde{P}(R)$. The uncertainty with which the latter is defined between R_1 and R_2 has little effect on the procedure, owing to the integral argument used throughout. For the same reason, residual structure in α_o is also uninfluential as long as $\alpha_s \gg \alpha_0$.

With the above procedure self-absorption is effectively corrected and reliable measurements of $\beta(R)/\beta_o$ are obtained. Insertion of the results in eq. (7) then allows to calculate the total extinction coefficient, from which $\alpha_s(R)$ is easily separated in the assumption of a constant α_0 (independently available from the third of eqs. (8)).

The effects of correction on a complex profile can be quite important. Fig. 7 shows an example, illustrating the drastic alteration of peak ratios and relative areas that may occur as a result of recovery from screening effects in the far side of the range. The example has an obvious bearing for those experiments where distribution and conservation of particle masses is of concern.

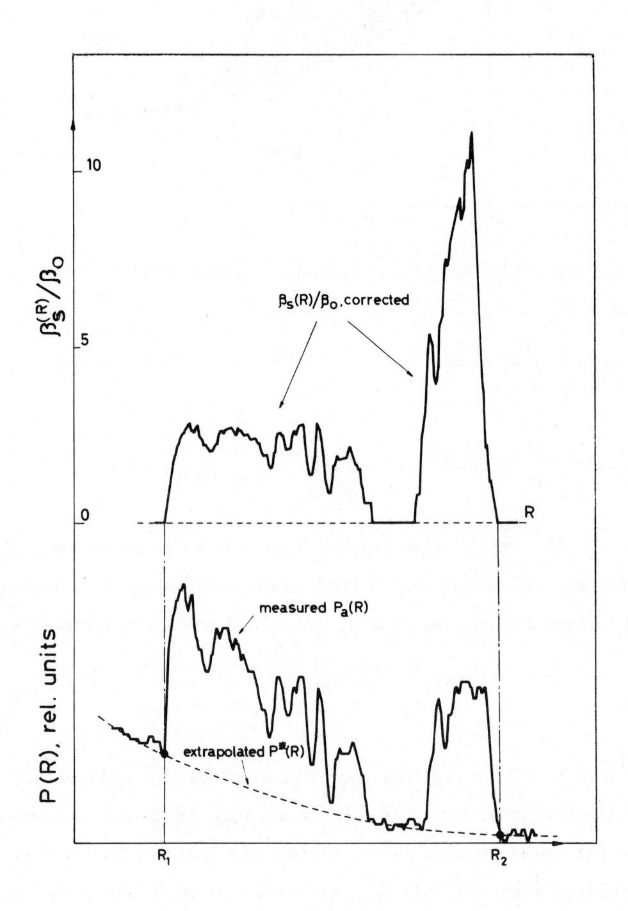

Fig. 7 - Effect of inversion on a Lidar profile through a multiple chimney plume; $\Delta R \cong 185$ m, $\alpha_0 \cong 1.10^{-3}$ m^{-1}

a) : back-scattered signal, direct record

b) : after correction for self-screening

3. LIDAR OBSERVATIONS ON PARTICLES AND AEROSOLS

3.1 Simple applications to emission and transport phenomena

The monitoring of thick plumes, or vapour clouds, often reduces to measurements of geometrical extension and a rough inversion of Lidar data can be tolerated. However, the difference between a crude Lidar profile and the corrected one is particularly critical when the evaluation of integral or statistical quantities is concerned.

A common case in Lidar experiments is the mapping of a localized particulate (e.g. a chimney plume) by means of repeated shots at close angular intervals. The corresponding intensity profiles are the starting data, from which position and distribution of the scattering mass are to be evaluated. Two useful quantities can be defined in this context for each trace, with reference to the scheme of Fig. 8:

Scattering power (particle burden)

$$M_i \equiv \int_{i\text{-th trace}} P(R)dR$$

Centre-of-mass

$$\vec{B}_i \equiv \frac{1}{M_i} \int_{i\text{-th trace}} \vec{R}\cdot P(R)dR \qquad (9)$$

where $\vec{R}$ is the range vector, expressed in terms of scalar range R and trace orientation.

With obvious extensions the above quantities can be combined to obtain centre-of-mass $\vec{B}$ and total burden M of any given section, by summing up the contributions of individual traces. As a final elaboration, a rms cross-section

$$\sigma^2 \equiv \sum_i \int_{i\text{-th trace}} |\vec{R} - \vec{B}|^2 \cdot P(R)dR / \sum_i M_i \qquad (10)$$

may also be defined, to compare with that predicted by statistical dispersion models.

It is clear that for meaningful evaluation of all these integrals the original data P(R) must be corrected, so as to provide a direct representation of $\beta(R)/\beta_o$. This implies an inversion procedure of the type illustrated in Figs. 6 and 7.

In recent experimental campaigns the Lidar and Cospec units of the JRC have made a series of combined experimental tests for the study of emissions

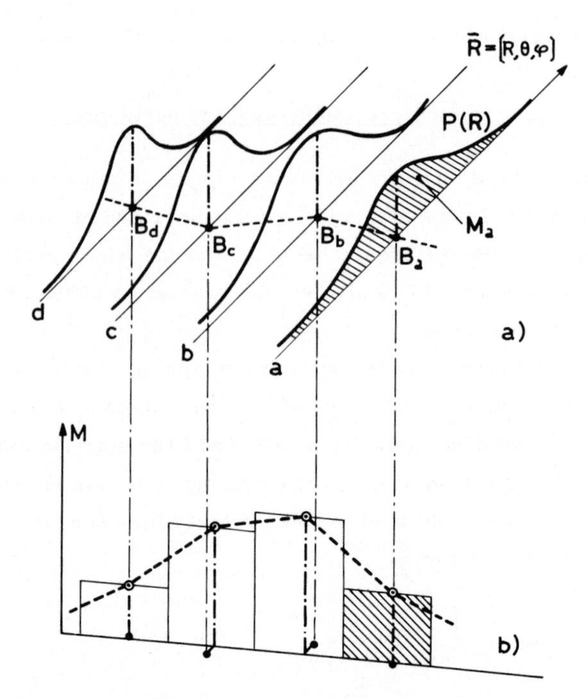

Fig. 8 - Diagram illustrating the definition of Lidar burden across a plume and its distribution in space.

from a fossil fuel plant (Turbigo, Italy). The main concern was comparative description of particulate vs gaseous (SO_2) emissions. A correct comparison in such cases implies that the Lidar should cope, at least on relative scales, with quantitative determinations typical of the COSPEC technique. In addition, Lidar must provide, through range-resolution, a consistent scheme of spatial mapping for both particle and gas distributions.

A typical example of these investigations is given in Fig. 9 (ref. 20), where the complex emission from a three-chimney power group is sampled during vertical rise (locally unstable conditions). The plot shows the distribution of isopleths in a vertical plane, for smoke particles and SO_2. To each point of a given isopleth there corresponds the same mass burden, as measured along a line of sight perpendicular to the plane of the figure. Burden data are obtained from the integration of extinction-corrected profiles, for the particle component; from direct Cospec measurements, for the gas component.

The role of quantitative analysis is further illustrated in Fig. 10, where downwind propagation of a plume is monitored by means of successive Lidar and Cospec sections along its trajectory (ref. 21). Again data for particle burden are worked out for each section from the corrected profiles of the

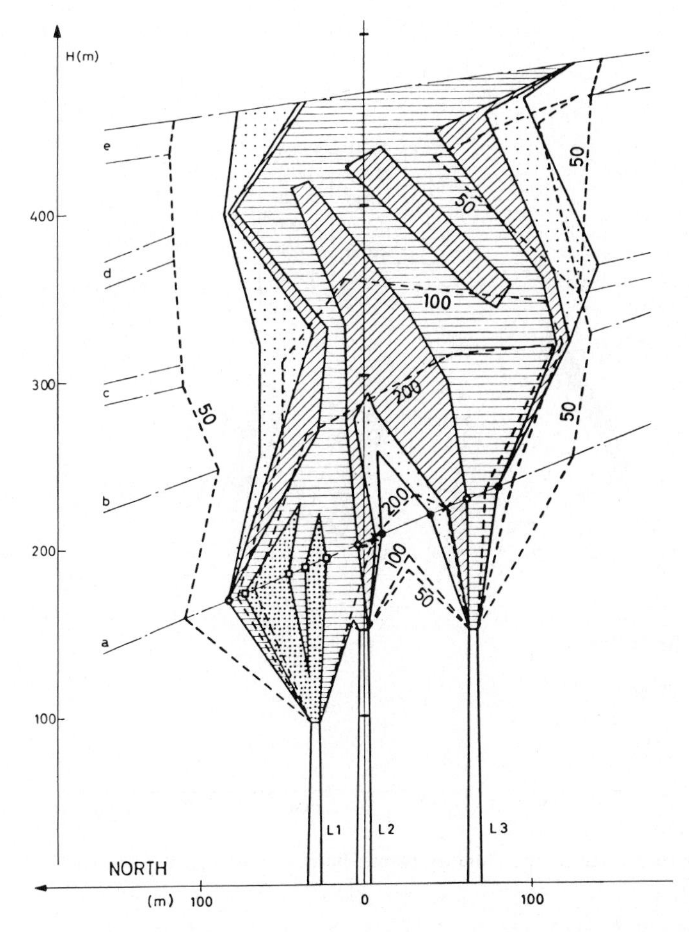

Fig. 9 - Comparison of isopleths for SO_2 (hatched lines) and particles (solid lines) during plume rise from the L1/L2/L3 chimney group of the Turbigo plant. a, b, c, etc. are the projections of gravity centres as defined by Lidar. Numerical inserts express the burden levels of SO_2 in ppm x metre. Symbols •, x, ○, □ indicate the relative particle burdens, in the succession 1 : 2 : 4 : 8.

constituent traces. The results are given plotting through the centre of each section a rectangular box (dark shade), the area of which is proportional to particle burden in units of (β_s/β_o) x m. Lateral dimensions of each box represent a rms cross-section Σ, also defined from the corrected Lidar profiles. Similar criteria were used in plotting SO_2 burden.

The remarkable constancy of Σ and of total burden, throughout the different sections, is a confortable test of the conservation of particle mass during propagation of the plume: this was physically expected for the case at hand, in view of stable conditions and negligible fall-out at short distances. Similar elaborations, performed with the use of raw Lidar profiles without inversion, had given for the same case a large scatter of data, which prevented any

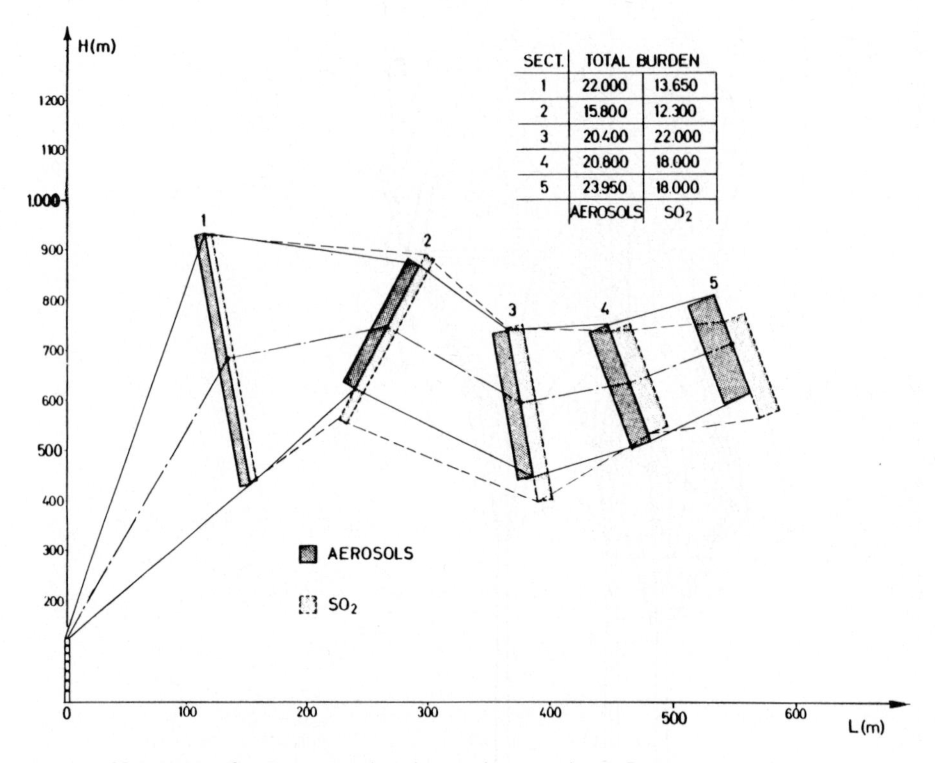

Fig. 10 - Monitoring of mass propagation along a chimney plume. Burden units: β/β_0 x m for particles, ppm x m for SO_2.

conclusion as to mass conservation. Incidentally, the data for SO_2 appear to be less regular with respect to burden propagation, as shown in the numerical insert: this is probably indicative of transversal escape, due to higher diffusivity of the gas phase.

Lidar studies of particulate emissions from industrial sites are quite numerous and we shall omit a systematic review (however, see refs. 22 and 23). Early work is often devoted to technical refinement, whereas particle valuation remains semi-qualitative. Good application to quantitative sampling and plume statistics can be found in recent work by Hoff and coworkers (refs. 24, 25) and by McElroy et al. (ref. 26), based on fixed- or airborne Lidar experiments.

3.2 Monitoring of particle properties

Other fields of intensive Lidar research are those concerning natural particulate, for which we list a few representative quotations:

- monitoring of tropospheric aerosol, natural haze, mixing layers and visibility (refs. 26-33);
- study of water aerosol, fog formation and water clouds (refs. 34-38).

These applications are often concerned with particle characterization, in relatively diluted aerosols. Therefore they are the natural testing ground for special concepts and new techniques, apt to improve the accuracy and capability of Lidar observations. This includes the definition of methods for separation of particle and molecular scattering or exact ratioing of β vs α (multi-angle or slant path techniques, refs. 33, 39); error analysis and calibration of Lidar experiments (resfs. 40, 41); use of auxiliary techniques in order to increase the information content of Lidar data. In this context polarization analysis plays a special role, to be discussed later. Another interesting development is the combination of distributed Lidar scattering with far-field target reflection, by which one obtains a separate measure of optical depth for a certain path, independently of its scattering profile. Uthe demonstrates in this way the utility of Lidar as a potential transmissiometer, for remote investigation of smoke or dust particles and clouds (ref. 42). Fig. 11 illustrates the principle of operation. Cloud formations may serve as a natural target.

Progress in the study of particles and aerosols has unfortunately a natural limitation, due to the fact that a single Lidar measurement is insufficient for mass and composition analysis of a polydispersed medium. This goal can be partially achieved only with the introduction of multi-wavelength and polarization techniques, to be commented in the following.

3.2.1 Multi-wavelength experiments. Can material properties of a population of particles be fully accessed by remote Lidar monitoring? This challenge is accompanying Lidar development in the course of the years; unfortunately, a complete answer to the problem is still controversial.

With current techniques and proper analysis, there are no difficulties in principle to an absolute determination of α and β. The problem then arises of using these data to get a good inversion of definitions (4), so as to unfold the contributing factors $F(\theta)$, Q_{ext} and $dN(a)/da$. This cannot be achieved with a single pair of coefficients measured at a fixed λ; in general, the use of Lidar measurements at several wavelengths will be required.

The information content of multi-wavelength experiments is discussed by

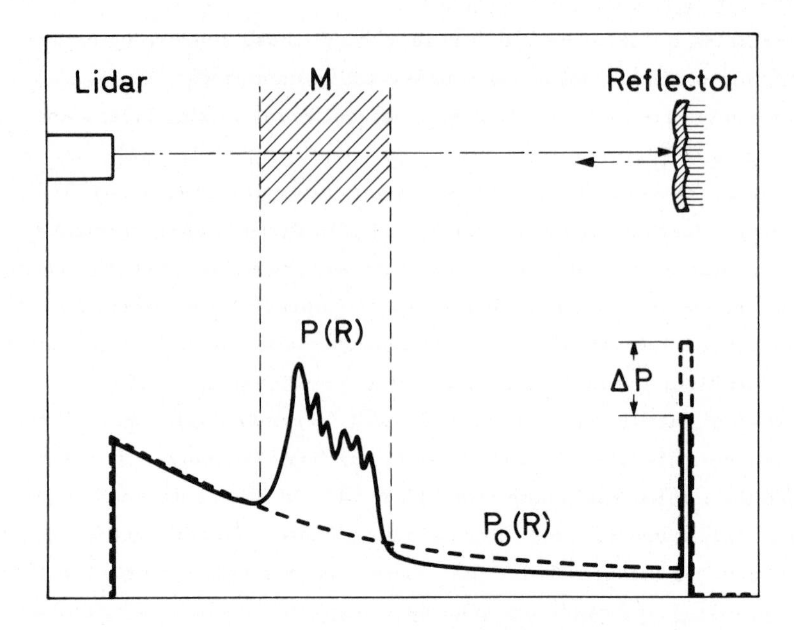

Fig. 11 - Extinction across a scattering medium M, observed by attenuation ΔP of Lidar signal from a reflecting target. P and P_0 are the profiles through M or clear air, respectively.

several authors (refs. 43-46). H. Quenzel and coworkers have recently summarized the general rules for the deconvolution of expressions (4). The main conclusions, concerning a population of identical particles, are as follows (ref. 44):

i) a trial model for dN/da and the assumption of a fixed index $\tilde{m}$ must be adopted anyhow;

ii) consistent retrieval of a model distribution, in terms of a multi-columnar histogram, can be obtained by fitting the unknown column strengths to measured values of $\alpha(\lambda)$ and $\beta(\lambda)$, provided a sufficient number of independent data are available. In general circumstances four data pairs at distinct wavelengths seems to be the necessary requirement. Alternatively, the data of only one coefficient, at 8 different wavelengths, might be used. For maximum consistency the choice of λ's should be made so as to scale the expected range of radii in the distribution. If fewer measurements are given, the information content of the inverted function deteriorates very rapidly;

iii) for a correct inversion, the rms error in the determination of absolute scattering coefficients must be kept well below 10%. On the other hand, the real and imaginary part of the index must be known (or guessed) to within 3% and 10% respectively.

These are severe conditions, even for well developed Lidars operating nowadays. The constraint on experimental accuracy can hardly be respected when operating multiple laser sources and polychromatic detection systems (for the applicative problems of multi-wavelength Lidars, see ref. 47). Also it is technologically very difficult to realize a power laser assembly capable to emit several distinct λ's, with a useful range from near u. v. to medium infrared, as would be required to match the scattering spectra to the dimensional spread of typical polydispersed aerosols (from 0.1 to ~10 μm). The most efficient and powerful sources in current use employ harmonic emissions from Ruby lasers and Neodymium lasers, or visible dye lasers, which satisfy this condition very partially.

The assumption on refractive index is also very restrictive. It is rarely satisfied in complex particle assemblies (such as continental aerosol), which are not characterized a-priori from the optical standpoint. Difficulties may arise even in well defined material (e. g. water aerosol; ash particles in a chimney plume) owing to unpredictable variations of optical properties, caused by irregular shape, surface reactions, heteronucleation, etc. Unfortunately, if the index $\tilde{m}$ is not a constant, the retrieval of size distribution from spectral data is often impossible.

For all these reasons a systematic approach to the problems of particles and aerosols, based on monostatic Lidar alone, is not yet defined and probably is not even recommended.

More practical ways to reach the same goal can be found, using appropriate combinations of Lidar with a complementary technique For instance, a spectral radiometer can profitably assist Lidar in the determination of atmospheric aerosol. The spectra of integrated optical depth, which are easily obtained with this instrument, have the advantage of being largely insensitive to the value of $\tilde{m}$, to the shape of the particles and to their density distribution in the atmospheric column. This permits a first trial of size-distribution models, which then are successively refined with the additional data on α and β, obtained by Lidar at a few wavelengths.

This line of approach has proved to be very fruitful in recent years, in particular for the observation of large scale phenomena such as aerosol layering in the troposphere and lower stratosphere, structure of continental aerosol, etc. For discussion and quotation of relevant work in this field, see for instance ref. 46.

3.2.2 Characterization of particles in simplified conditions. With two-wavelength or three-wavelength Lidars one can still have access to a variety of meaningful experiments. If a full analysis of size distribution is not requested, one can by-pass accuracy problems and the need of absolute calibrations: this is easily achieved by measuring some significant ratio such as $\alpha(\lambda)/\beta(\lambda)$, or $\alpha(\lambda_1)/\alpha(\lambda_2)$, instead of absolute values. Quantities such as these have a good information content in particular conditions. Werner and others (Ref. 48) were able to establish a correlation between the humidity content of continental clean-air and the back-scattering ratio $\beta(\lambda_1)/\beta(\lambda_2)$, taken at Ruby- and Neodymium laser wavelengths. Kent has analysed the data coming from various two-λ experiments (ref. 49) concluding that the above ratio is a significant parameter for qualitative sizing of aerosol types in continental air (transient nuclei $\sim$ accumulation mode $\sim$ coarse particle mode) and for the discrimination of urban pollution aerosols. Waggoner et al. (ref. 50) tested the ratio α/β for urban aerosol, combining Ruby Lidar and nephelometry. Values of this quantity were found to correlate directly with relative humidity above 75%.

Uthe (ref. 51) demonstrates that the quantity $\alpha(\lambda_1)/\alpha(\lambda_2)$, for proper choice of wavelengths, may be used for determining mean particle radii of single-mode distributions. Fig. 12 shows the correlation of mean diameter to the extinction ratio, for a variety of sub-micron assemblies. Fig. 13 shows the behaviour of extinction-to-mass, α/M, as a function of λ, for fly-ash aerosols of different size, as obtained from transmissiometry. This quantity is strongly dependent on particle size at short wavelengths, but is nearly independent in the infrared ($\lambda \geqslant 10\ \mu m$). Thanks to this, two simultaneous Lidar observations at, say 0.54 and 10.6 μm should be able to monitor average radius and total mass at any point of a given aerosol assembly, with the only help of a calibration on α/M. The extension of similar ratio criteria to both coefficients α and β, at three or more wavelengths, ought to improve quantitative applications even for plurimodal aerosols. Incidentally, this matter is another example of the fruitful connections existing between Lidar and transmissiometry.

3.2.3 Special methods - observations in polarized light. Exploitation of the above procedures may require special accuracy in the determination of total α, β and a good separation of the target aerosol. By definition

$$\alpha = \alpha_s(\text{Mie}) + \alpha_0(\text{Mie}) + \alpha_0(\text{Rayleigh})$$

$$\beta = \beta_s(\text{Mie}) + \beta_0(\text{Mie}) + \beta_0(\text{Rayleigh})$$

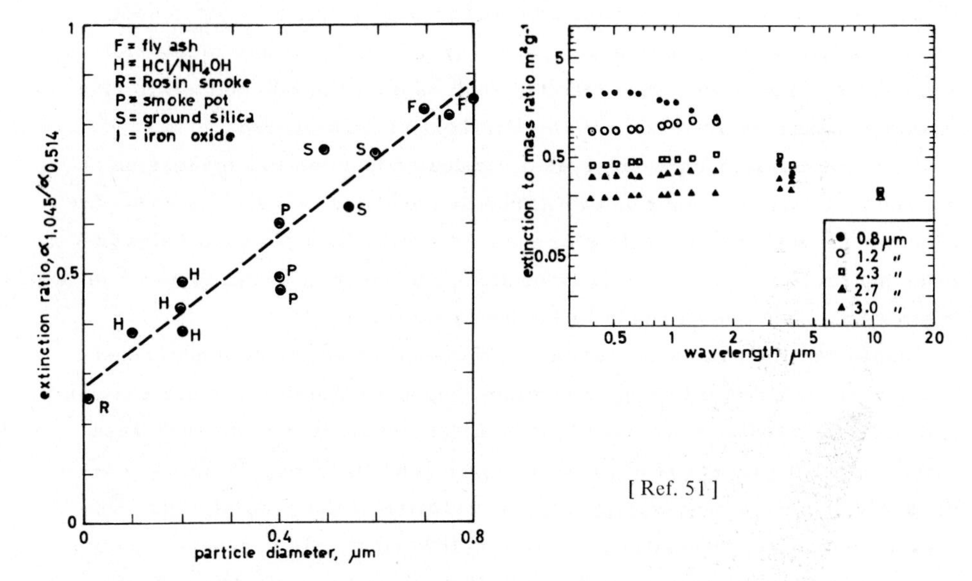

Fig. 12 - Ratio of extinction at 1.045 and 0.514 µm wavelengths as a function of mean (Sauter) diameter for submicron particles of different composition.

Fig. 13 - Extinction-to-mass concentration as a function of wavelength for various size fractions of fly ash (particle Sauter diameter determined from impactor measurements).

where the first term on the right describes the matter under examination; the others describe Mie-scattering and molecular scattering of background atmosphere. In very diffuse situations (e.g. continental aerosol or light pollution aerosol) the specific term might be of the same order as the sum of the other two; approximations such as $\beta_s \gg \beta_0$, $d\beta_0/dR \approx 0$, etc., would no longer be allowed. Independent monitoring of "clean air" background is essential in these cases. H. Shimizu et al. (ref. 52) report a method to calibrate β_0 (Rayleigh), based on the observation of Raman Lidar returns from atmospheric nitrogen. Following this, they derive the true ratio α/β_s (Mie) for natural water aerosol and find it linear with the average radius of water droplets, in a very large interval ($1 \leqslant \bar{a} \leqslant 30$ µm).

A powerful method for the separation of molecular effects is based on frequency discrimination between the Mie-scattering line and Doppler-broadened Rayleigh-scattering profile (ref. 53). A high resolution tunable Lidar working on this principle is able to obtain in one measurement accurate separation of aerosol and molecular coefficients (ref. 54).

On the way to better accuracy, one should also mention the development of instruments for the short-wavelength region. An ultraviolet or "solar-blind" Lidar operates in the region from 0.25 to $\sim 0.32\ \mu m$, where the solar flux at ground level is negligible, as a result of absorption by stratospheric ozone. Lidars working in this way have a greatly reduced sky-radiation noise and succeed in detecting weak effects, such as molecular Raman scattering. In this context they are dealt with by Dr. Capitini (this Course).

Work in polarized light deserves a particular mention in a discussion of methods. In fact, the addition of polarization analysis is a special bonus for Lidar measurements, for different reasons. In the first place, it helps in reducing problems of accuracy; secondly, it uncovers specific features of the scattering object, which would be hidden in natural light.

Figure 14a illustrates the scheme of a typical experiment. A polarized Lidar beam is emitted through a medium (supposedly uniform in our example) and the return beam is analysed by a polarization splitter S, so as to separately detect the fractions of back-scattered light that are polarized, respectively, *parallel* and *perpendicular* to the plane of the exciting light. One obtains in this way two distinct profiles $P_{/\!/}(R)$ and $P_{\perp}(R)$. In general, a consistent definition of these components will require the accessory use of a quarter-wave retarder Q, in order to subtract out a *randomly* polarized *residue*, due to elliptical depolarization in the scattering medium itself. Such a residue is negligible in the case of scattering from weakly absorbing, isotropic molecules or spherical particles (see refs. 11, 12).

From the Lidar equation, the expression of a so-called *depolarization ratio* (ref. 55) can be worked out:

$$\frac{P_{\perp}(R)}{P_{/\!/}(R)} = \frac{C_2}{C_1} \cdot \frac{\beta_{\perp}(R)}{\beta_{/\!/}(R)} \exp\left[-\int_0^R (\alpha_{\perp}(R'(- \alpha_{/\!/}(R'))\, dR'\right] \tag{11}$$

where each coefficient is related, with obvious notation, to a specific state of polarization of the back-scattered light.

The constant C_2/C_1, coming from intrinsic differences in the detectors of the two separate channels, is easily calibrated by a reference experiment: e.g. firing the Lidar beam through clear air, to a totally depolarizing target and taking the ratio of the reflected signals $T_{/\!/}$, $T_{\perp}$ as the normalizing factor. Essential simplification also occurs in the exponential factor of (11), due to the fact that the extinction coefficient is isotropic in the large majority of cases, $\alpha_{\perp} \approx \alpha_{/\!/}$. The reason for this is the randomizing effect of the atmosphere on the absorbing objects imbedded into it, resulting in the obliteration of orientational anisotropy (see ref. 55). It follows that in most cases the measured ratio $P_{\perp}/P_{/\!/} \equiv \delta$ can be taken to represent the quantity

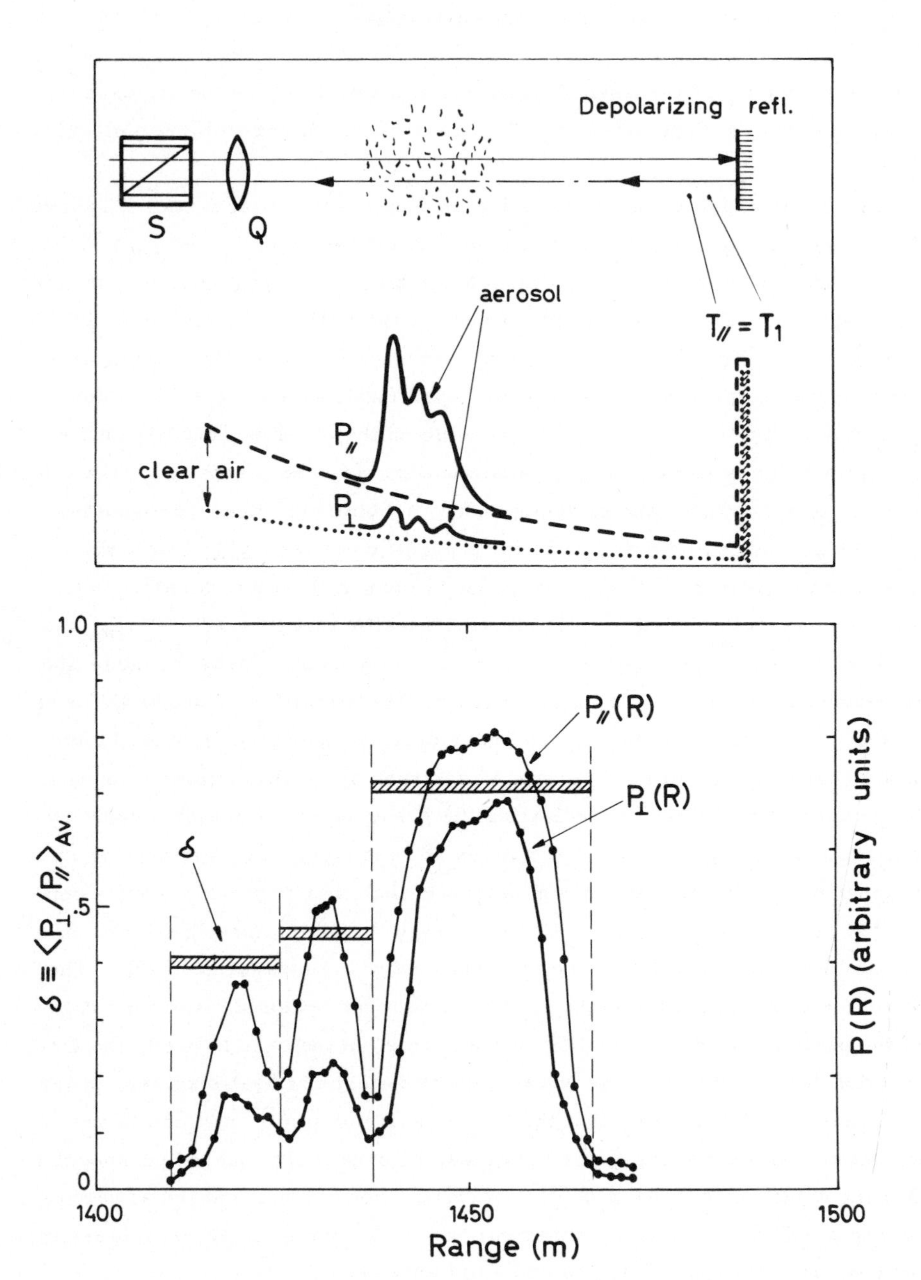

Fig. 14 - a) Scheme of Lidar observations in polarized light.
b) Polarized back-scattering profiles through a chimney plume. Horizontal bars indicate the average level of the depolarizing ratio in three different layers of the plume.

$$\frac{\beta_{\perp}}{\beta_{/\!/}} \equiv \frac{\beta_{\perp}(\text{Mie, aerosol}) + \beta_{\perp}(\text{Mie, background}) + \beta_{\perp}(\text{Rayleigh})}{\beta_{/\!/}(\text{Mie, aerosol}) + \beta_{/\!/}(\text{Mie, background}) + \beta_{/\!/}(\text{Rayleigh})} \tag{12}$$

where the notation expresses the general situation in which the atmospheric medium contains a specific aerosol, plus background aerosol and molecular gas.

Use of the depolarizing ratio is assisted by the following facts: i) in locally uniform atmosphere, the various background terms of $\beta_{\perp}$ and $\beta_{/\!/}$ are virtually independent of range; ii) the depolarizing ratio for Rayleigh scattering (atmospheric gas) is independently fixed for theoretical reasons (refs. 11, 56); iii) for background aerosol, with small spherical particles dominating, the corresponding ratio is very small (no depolarization, see refs. 11, 12 for generalities on Mie scattering). Depending on the problem, one can utilize this knowledge in various ways, in particular: 1) to help separating the effects of background aerosol and molecular gas, in nominally clean atmosphere, and 2) to obtain the depolarizing ratio for a specific aerosol, *via* comparison of the complete form (12) with the reduced form valid for clean air.

In the case of a dense particulate, the specific components of $\beta_{\perp}, \beta_{/\!/}$ are generally much larger than the background components. Measurements and interpretation of the depolarizing ratio are then simplified. Such is the case of Fig. 14b, showing two polarized Lidar profiles through a chimney plume (field observations of the JRC Lidar). A strong spatial structure is found in both states of polarization, indicating the presence of one major smoke layer and two subsidiary ones. As one notices at first sight, the depolarizing ratio is much larger in the former region than in the other two. This data is confirmed by plotting a histogram of the *average* δ for the three regions, obtained by smoothing point-by-point variations of the measured $P_{\perp}/P_{/\!/}$. The variation of δ across the depth of the plume must reflect a different nature of the particles contained in different streams. In the case at hand, the fact that depolarization increases where density is higher is satisfactory, under the reasonable assumption that the denser parts of smoke particulate are associated with large irregular particles. Alternatively, one might speculate that the structure of $\delta(R)$ is due to a varying proportion of spherical particles (water-sulphate aerosol) to aspherical particles, due to chemical interaction between water vapour, SO_2 gas and solid ashes.

In the case of dense, heterogeneous aerosols (as treated in our example) a full analysis of experiments in polarized light is still a difficult task, owing to theoretical complications introduced by multiple scattering and strong absorption. Therefore the applications remain in a qualitative stage, as an aid for the assessment of complex situations. For better defined media, polarized Lidar techniques have been positively developed and there are reasons to encourage further work.

Summarizing the previous discussion, there are two specific merits in this type of experiments. The first is the possibility of measuring $\beta_{\perp}/\beta_{//}$ directly, i.e. by-passing extinction problems and the inversion of Lidar profiles. The second is the great specificity of this quantity with respect to the scattering mechanism involved, as well as to shape and size of the particles. These aspects have been recently emphasized by the recognition of strong resonance patterns in the dependence of δ vs relative size, for various kinds of particles (Ref. 57). Such phenomena constitute a potential basis for direct particle fingerprinting.

Application of the above concepts has been actively pursued in recent years, though often restricted to semi-qualitative work for the complexities of particle optics. Example of particular interest are those concerning the study of dust and mixing layers, lower atmosphere and natural haze (Refs. 36, 58); structure of fogs and clouds (Refs. 36-38, 55); discrimination of ice precipitates in water aerosol (Refs. 55, 59).

REFERENCES

1 R. T. H. Collis, P. B. Russell: Laser Monitoring of the Atmosphere, Springer-Verlag N. Y. (1976), p. 71.
2 E. E. Uthe: Procs. Soc. Photo-Optical Instrum. Eng. 142, 67 (1978).
3 D. Deirmendjian: Rev. Geophys. and Space Phys. 18, 341 (1980).
4 G. Fiocco and D. Smullin: Nature, 199, 1275 (1966).
5 M. G. Ligda: Procs. 1st Int. Conf. Laser Technol., San Diego (USA) 1973, p. 63.
6 E. W. Barrett and O. Ben-Dov: Journ. Appl. Meteorol. 6, 500 (1967).
7 B. R. Clemesha, G. S. Kent and R. W. Wright: Journ. Appl. Meteorol. 6, 386 (1967).
8 R. T. H. Collis: Q. Journ. R. Met. Soc. 92, 220 (1966).
9 P. M. Hamilton: Phil. Trans. Roy. Soc. London, A. 265, 153 (1969).
10 R. T. H. Collis: Advances in Geophysics (Landsberg and Van Mieghen, Ed.), Academic Press, 1969, p. 113.
11 H. C. van de Hulst: Light Scattering by Small Particles, J. Wiley and Son, N. Y. 1957.
12 D. Deirmendjian: Electromagnetic Scattering on Spherical Polydispersions, Amer. Elsevier Publ. Co., N. Y., 1969.
13 R. Penndorf: Journ. Opt. Soc. Am. 52, 402 (1962).
14 H. Grassl: Appl. Optics, 10, 2534 (1971).
15 R. G. Pinnick, S. G. Jennings and P. Chylek: Journ. Geophys. Res., 85, 4059 (1980).
16 R. W. Fenn: Appl. Optics, 5, 293 (1966).
17 S. Twomey and H. B. Howell: Appl. Optics, 4, 501 (1965).
18 J. V. Mallow: Appl. Optics, 21, 1454 (1982).

19 J. D. Klett: Appl. Optics, 20, 211 (1981).
20 P. Camagni et al.: Nuovo Cimento 4 C, 359 (1981).
21 P. Camagni et al.: Procs. 2nd Eur. Symp. on Physico-Chemical Behaviour of Pollutants (Versino and Ott, Ed.); Reidel Publishing Co., Dordrecht, Holland (1981), p. 533.
22 W. B. Johnson and E. E. Uthe: Atmospheric Environment, 5, 703 (1971).
23 E. E. Uthe and W. E. Wilson: Atmospheric Environment, 13, 1395 (1979).
24 R. M. Hoff and F. A. Froude: Atmospheric Environment, 13, 35 (1979).
25 R. M. Hoff et al.: Atmospheric Environment, 16, 439 (1982).
26 J. L. McElroy, J. A. Eckert and C. J. Hager: Atmospheric Environment, 15, 2223 (1981).
27 B. G. Schuster: Journ. Geophys. Res., 75, 3123 (1970).
28 D. J. Gambling and K. Bartusek: Atmospheric Environment, 6, 181 (1972).
29 F. G. Fernald, B. M. Herman and J. A. Reagan: Journ. Appl. Meteor., 11, 482 (1972).
30 A. Cohen and M. Graber: Journ. Appl. Meteor., 14, 400 (1975).
31 C. Werner et al.: Rev. Sci. Instr., 49, 974 (1978).
32 R. M. Endlich, F. L. Ludwig and E. E. Uthe: Atmospheric Environment, 13, 1051 (1979).
33 J. D. Spinhirne, J. A. Reagan and B. M. Herman: Journ. Appl. Meteor., 19, 1980.
34 C. Werner: Opto-electronics, 4, 125 (1972).
35 S. T. Shipley, E. W. Eloranta and T. A. Weinman, J. Appl. Meteor., 13, 800 (1974).
36 J. D. Houston and A. I. Carswell: Appl. Optics, 17, 614 (1978).
37 A. Cohen: Appl. Optics, 14, 2873 (1975).
38 S. R. Pal and A. I. Carswell: Appl. Optics, 17, 2321 (1978).
39 J. J. De Luisi, B. G. Schuster and R. K. Sato: Appl. Optics, 14, 1917 (1975).
40 P. B. Russell, T. J. Swissler and M. P. McCormick: Appl. Optics, 18, 3783 (1979).
41 R. M. Hardesty, R. J. Keeler, M. J. Post and R. A. Richter: Appl. Optics, 20, 3763 (1981).
42 E. E. Uthe: Appl. Optics, 20, 1503 (1981).
43 S. Twomey and H. B. Howell: Appl. Optics, 6, 2125 (1967).
44 J. Heintzenberg, H. Müller, H. Quenzel and E. Thomalla: Appl. Optics, 20, 1308 (1981).
45 A. L. Fymat and K. D. Mease: Appl. Optics, 20, 194 (1981).
46 J. A. Reagan et al.: Journ. Geophys. Res., 85, 1591 (1980). See also J. A. Reagan and B. M. Herman: AIAA Journal, 10, 1401 (1972).
47 W. Carnuth, H. Jager, M. Littfass and R. Reiter: Laser 77 Opto-Electronics, München, 1977 (Scient. Techn. Press), p. 728; W. Carnuth: Laser und Elektro-Opt., 12, 36 (1980).
48 C. Werner, G. S. Kent and F. Köpp: Procs. 4th Int. Symp. on Meteor. Obs. and Instr. Denver, USA, 1978; J. Appl. Meteor., 18, 1649 (1979).
49 G. S. Kent: Appl. Optics, 17, 3763 (1978).
50 A. P. Waggoner, N. C. Ahlquist and R. J. Charlson: Appl. Optics, 11, 2886 (1972).
51 E. E. Uthe: Appl. Optics, 21, 454 (1982).
52 H. Shimizu, T. Kobayasi and H. Inaba: IEEE J. Quantum Electr., 13, 70D (1977) and Procs. 8th Int. Laser Radar Conf., Philadelphia (USA), 1977; Paper no. 57.
53 G. Fiocco, G. Benedetti-Michelangeli, K. Maischberger and E. Madonna: Nat. Phys. Science, 229, 78 (1971).
54 S. T. Shipley, E. W. Eloranta, D. H. Tracy, Procs. 9th Int. Laser-Radar Conf., München, 1979, p. 85. S. T. Shipley et al.: Final Report NASA-NSG 1057, University of Wisconsin, 1975.

55 S. R. Pal and A. I. Carswell: Appl. Optics, 12, 1530 (1973).
56 A. Cohen, J. Neumann and W. Low: Journ. Appl. Meteor., 8, 952 (1969).
57 T. R. Lettieri, W. D. Jenkins and D. A. Swyt: Appl. Optics, 20, 2799 (1981).
58 W. R. McNeil and A. I. Carswell: Appl. Optics, 14, 2158 (1975).
59 V. N. Smiley and B. M. Morley: Appl. Optics, 20, 2189 (1981).

Optical Remote Sensing of Air Pollution,
Lectures of a course held at the Joint Research Centre, Ispra, Italy, 12—15 April 1983,
P. Camagni and S. Sandroni (Eds). 235—258

LIDAR APPLICATIONS: PLUME TRACKING AND MODELING

D. Anfossi

1. INTRODUCTION

Transport and diffusion of effluents released from tall stack must be accurately evaluated in order to estimate ground level concentrations of the pollutants.

Accordingly many field experiments have been conducted over the last years at various power plants. Three techniques have been typically used: plume photography, airborne sampling and Lidar measurements. Plume photography, although inexpensive, has serious limitations, such as the limited downwind extension of measurements and the limit of resolution in defining plume boundaries and the distance of its centreline from the observer. Airborne sampling is, in principle, the best way to obtain information on plume geometry and composition but it is too expensive if an extensive measurement period is needed. Furthermore, it does not allow the reconstruction of plume geometry and trajectory in real time. The main advantage of Lidars is their ability to make remote aerosol measurements in the near field at a high resolution in space and time. It is assumed that the plume aerosols are small enough to be dispersed together with the stack gases and act as a tracer. This assumption allows us to study the gaseous part of the plume detecting its aerosol component with a Lidar.

Lidar technique is fundamental in plume rise and dispersion studies since it gives realiable measurements of the rise and three-dimensional dispersion of buoyant plumes. These data are the necessary input for developing formulas to predict plume trajectories and plume levelling heights. Much work has been done in this field but a certain number of problems still remain to be solved.

We will examine in the second chapter the most widely used plume rise models both for single and for multiple sources; the third chapter will deal with Lidar technique in studies of plume rise, trajectories and dispersion; the fourth chapter will present and discuss a few significant results illustrating the power of the method and its limitations.

2. PLUME RISE MODELS

2.1 Generalities

A detailed derivation of plume rise models is beyond the scope of this lecture but a qualitative description of plume behaviour and an outline of the general physical principle involved should be given (comprehensive reviews on the subject could be found in ref. 1,2,3,4 and 5. Qualitative descriptions can also be found in ref. 6 and 7).

In the treatment of plume rise it is assumed (ref. 3) that the exit conditions and location of the stack mouth are such that the plume may be considered to be rising through an atmospheric boundary layer where the mean properties, profiles of velocity, temperature and turbulence do not vary with time or with travel in the downstream or cross-stream directions.

In general such requirements will only be met by plmes from large industrial plants having stacks of the order of 100 m in height.The discussion which follows will be restricted to such emissions.

A jet of stack air moving through ambient air experiences a shear force at its perimeter giving rise to a boundary layer in which momentum from the jet is transferred to the stationary ambient air. This process produces a growth of the radius of the jet and a decrease of its average velocity. The ambient air, which initially had no velocity, at least in the direction of the jet motion, and which begins to move alongwith the jet, is said to be entrained by the jet.

A buoyant plume entrains air in the same manner as a pure jet, but at a different rate. Buoyance force helps to maintain the motion of the plume as it transfers momentum to the surrounding air. In fact a buoyant plume travels further than a pure jet. In general buoyancy becomes dominant over the initial vertical momentum,rather close to the source, generally after a few tenths of meters.

When the boundary layer is small compared to the diameter of the plume, the velocity , density and temperature within a circular plume cross section can be approximated by a uniform or "top hat" distribution: zero velocity and ambient temperature outside the plume and uniform plume velocity and temperature inside.

As the entrained air, which is at ambient temperature, mixes with the plume, the plume temperature decreases approaching the ambient temperature. As a consequence, the buoyancy force per unit volume decreases.

It is necessary to recall that in any plume rise model a closure assumption needs to be introduced. In fact, regardless of the conservation relationships used, they do not provide a complete set of equations because there is always at least one more unknown than there are conservation equations. This derives (ref. 2) from the impossibility so far to model the entire turbulent motion and therefore to quantitatively evaluate in details the entrainment process. To overcome this problem, a so called closure assumption is made (that is some "ad hoc" formula relating two unknowns already present in the equation set is written).

2.2 Single source

Two typical shapes of plume can be found: vertical or quasi-vertical in calm conditions and bent over in the presence of wind. Both have an envelope of conical shape which implies that plume radius increases linearly with rise $R = \beta z$ where β is the entrainment constant (this relationship is the closure assumption generally added to the conservation equation set). The main difference between the two configurations is in the different efficiency of en - trainement process, which is obviously greater in the bent over plume.

Two phases are distinguishable in the plume trajectory: a transitional phase and a final one. The transitional phase begins with a jet phase, near the chimney top (as stated above, the jet phase lasts only a few seconds or a few tenths of metres and, therefore is not considered in plume models). In the transitional phase, the growth of the plume is mainly due to the self-induced turbulence caused by the boyant plume. This first phase ends when the plume is nearly levelled off, that is when the buoyance becomes ineffective. In the final phase the growth of the plume is due to the atmospheric turbulence. Plume dimensions increase but the height of the plume centreline does not change (+) significantly any more.

Even if there is general agreement upon these qualitative features, the literature contains (following Briggs, ref. 1) more than 100 formulas predicting the plume trajectories or the final plume height. Here we will limit ourselves to two models, according to Moore and Briggs respectively, which are widely used and well supported by experiment (ref. 8). To derive their models these authors assume two different idealized geometrical forms of plumes:

(i) - Moore (ref. 9), A "lumpy" model in which the plume is assumed to consist of a series of more or less discrete lumps or puffs, which may or may not coalesce as they drift downwind. The model is therefore three-dimensional.

(ii) - Briggs (ref. 2). A two dimensional model, in which longitudinal mixing and breaking of plumes is assumed not to occur.

Additional assumptions common to both models are that the plume motion is stationary (the average values do not change in time), the wind velocity and direction are constant in time and in the vertical coordinate, the ambient potential temperature may vary only in the vertical direction and the Boussinesq approximation holds (i. e.: in the equations, the terms containing density fluctuations are disregarded unless they are multiplied by the acceleration of gravity).

(+) In neutral environmental conditions the plume centreline never stops growing since the difference between plume and air temperature never vanishes. In practice, however, the plume is considered to be levelled when the slope of the centreline is less than 5% or, better, when the plume has reached the height one would need, in the diffusion equation, to correctly calculate the maximum pollutant concentrations at ground level.

Taking into consideration these assumptions and the conservation of mass, momentum and buoyancy, the corresponding models are as follows:

Moore

- $u > 1$ m/s, all stabilities, both transitional and final phase

$$z(x) = 2.4\ Q^{1/4}\ u^{-1}\ x_*^{3/4} \qquad (1)$$

where $x_* = x$ for short distances or $= xt$ for large distances

and $xt = (4224\ .\ u)/\ (1239\ \Delta\vartheta'\ tu^2\ )^{1/2}$ (for $Hj > 120$ m)

being $\Delta\vartheta' = (0.08\ K)/100$ m.

The two asymptotic values are connected by the expression:

$$x_* = (xt\ .\ x)/(xt^2 + x^2\ )^{1/2}$$

Briggs

- neutral and unstable conditions, transitional phase

$$z(x) = 1.6\ F^{1/3}\ x^{2/3}\ u^{-1} \qquad (2)$$

- neutral and unstable conditions, final rise

$$z = 1.6\ F^{1/3}\ (3\ xf)^{2/3}\ u^{-1} \qquad (3)$$

where $xf = 2.16\ F^{2/5}\ Hj^{3/5}$ (for $Hj < 305$ m)

- stable windy conditions ($u > 1$ m/s), final phase

$$z = 2.6\ (F/u\ s)^{1/3} \qquad (4)$$

- stable calm conditions ($u < 1$ m/s), final phase

$$z = 5.0\ F^{1/4}\ s^{-3/8}\quad (+) \qquad (5)$$

For the transitional phase, under stable conditions, Briggs suggests to use eq.(2) with xf = const (u/sqrt (s)). There is not a general agreement about the value of the constant of proportionality. On ref.10 it is suggested the value 2; Briggs (ref.2) suggests 5, but reports other values from various authors, such as 1.55, 3, 4, 6.5. Kerman (ref.11) claims that 5 is the best value, even if from fig.1 of his paper it seems more precise to choose 3.

2.3 Multiple sources

If hot plumes emitted by two or more stacks close to each other combine, they rise higher than either isolated plumes and may lower

---

(+) Note that F (m** 4/sec** 3) = 3.8 10** (-5) Q (cal/sec) and F = g w r** 2 (Tp - Ta)/Tp; s = g/Ta. $\Delta\varphi a\ /\ \Delta z$

the maximum ground level concentrations. This is because, after merging, the buoyancy of the resultant plume is greater than that present in a single plume. In fact when the plumes meet, they entrain ambient air plus plume gas. This last is at a temperature greater than that of ambient air.

The interest for this topic in plume rise studies lies in the fact that, especially in the past, many power plants were constructed with lines of closely spaced stacks, each stack being connected to one or more generating units.

In general, two configurations are possible:

i) stacks having all the same height and emitting plumes with the same buoyancy flux;

ii) stacks of different heights and/or connected to units of different power (for a detailed analysis, see ref. 12).

Briggs (ref 13) proposed the following empirical formula for the enhancement factor $EN = \Delta HN / \Delta H1$ in configuration i);

$$EN = \left[(N+S)/(1+S)\right]^{1/3} \tag{6}$$

where: $S = 6 \left[(N-1)\ d/(\Delta H1 \cdot N^{1/3})\right]^{3/2}$. EN accounts for the enhenced rise of merged plumes over single plumes.

Anfossi et al. (ref. 14) proposed for the configuration ii):

$$He = Hi + C \left(\sum_{1}^{N}{}_{j} (Fi^{a} - (Hi - Hj)/C)^{1/a}\right)^{a} \tag{7}$$

where:

$$Hi = Hmax + \frac{\Delta Hmin - (Hmax - Hmin)}{1 + (\Delta Hmin - (Hmax - Hmin))/D} \tag{8}$$

and $C = \Delta Hmin / Fmin^{**} a$

$D = (N - 1)\ d.$

The values of a and $\Delta Hmin$ to be used in eq. (7) depend on which formula for the single plume rise $\Delta Hmin$ is used: a will be equal to 1/4 if eqs. (1) or (5) are used; a will be equal to 1/3 if eqs. (2), (3) or (4) are used. Eq. (7) is based on the assumption of a virtual stack of height Hi where the plumes are assumed to merge. The buoyancy flux for each source is then reduced so that the calculated rise for the single plume would be the same as before, but starting from this assumed virtual stack; these reduced buoyancies are then added and are treated as a single source emitted from the merging height.

Eq. (7) is more general than eq. (6). It can be also applied to the configuration i). In this case, setting all the heights and the buoyancy fluxes equal, the predicted enhancement is:

$$EN = (1 + N^{1/3} \cdot \Delta H1/D) / (1 + \Delta H1/D) \qquad (9)$$

In the previously quoted paper (ref. 14), it was shown that eqs (6) and (9) give essentially similar results. This fact is proba - bly fortuitous and however interesting in view of the different approaches involved in the derivation. Despite these differences, it can be seen that both models consider, as the main factors influencing the multiple plume rise, the number of stacks N and the dimensionless ratio $\Delta H1/D$ which takes into account all the meteorological and stack parameters influencing the rise and the distance between the chimneys.

3. LIDAR METHODOLOGY IN PLUME RISE STUDIES

3.1 Generalities

Two lectures of this course are devoted to the Lidar technique and its basic principles (see also ref. 15). It is only necessary to recall here that Lidar uses laser energy in a radar configuration. It directs a laser pulse towards the plume and measures the backscattered light as a function of time. The amplitude of the returned signal is proportional to the aerosol concentration in the plume, and the time of flight determines the range along the line of sight. The depth of plume intersection along the light path is obtained from the duration of the back-scattered signal. The vertical dimension is found by scanning the plume with a succession of pulses fired at different elevation angles. Returned signal are displayed on an oscilloscope triggered when the laser is fired.Lidars for plume tracking are generally installed on vans. The transceiver assembly is rotable through 360 degrees of azimuth and 180 degrees of elevation. Angular ranging may be performed manually or via computer programmed operations.

Data acquisition may be done by a Fast Transient Digitizer

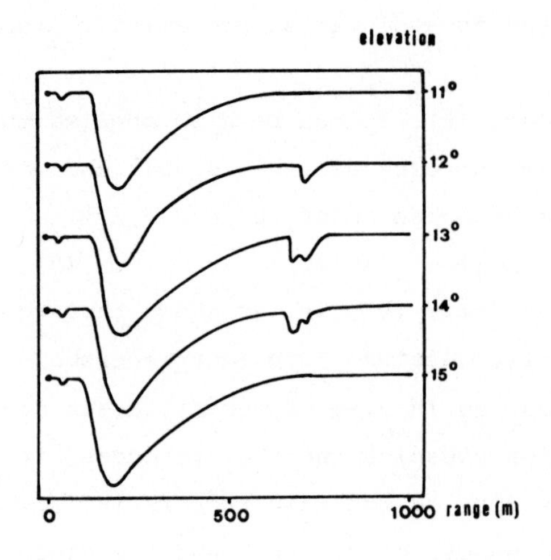

Fig.1 - Sketch of basic Lidar echoes.Peak appearing at 12,13 and 14 degrees are due to a smoke plume.

coupled to a minicomputer, or photographically. In this last case,the various echoes displayed on the oscilloscope are recorded by a Polarois camera (fig.1). Gross data as the depth of the plume can be derived directly from pictures, but detailed reconstruction requires a subsequent digitation.

3.2 Experimental procedure

First of all it is necessary to choose the positioning sites depending upon the operational facilities,the general trend of the wind direction in the area and the level of turbidity of the air. It is necessary to position the instrument so that the scanning plane will be approximately perpendicular to the plume trajectory and at a distance from the plume such as the plume echo (air plus plume)will be distinguishable from the background (air).

The typical location of Lidar with respect to the plume is shown in fig. 2.

Each vertical scanning of the plume is executed by setting a fixed azimuth angle and stepping the elevation angle between successive shots. Normally, many vertical scans are made at different azimuth directions, so that the plume cross sections are sampled at several downwind distances from the source.This allows the trajectory to be reconstructed and the results of measurements to be compared to the model predictions.

Vertical plumes are scanned by getting a fixed zenith angle

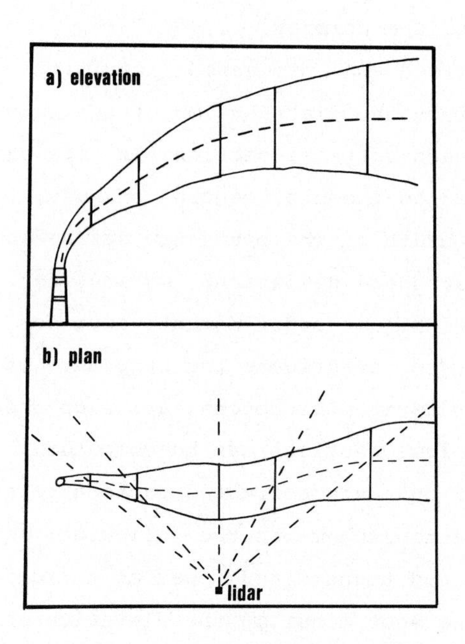

Fig.2 - Schematic view (elevation and plan) of typical Lidar operational technique for plume observations.

and varying the azimuth angle.

3.3 Analytical procedure

Depending on the way the echoes are read, i.e. manually or automatically, there are two methods of evaluating the significan parameters of a plume cross section.

In the first case (ref. 16) it is only possible to obtain, for each cross section, the direction, range and width. The minimum and maximum heights correspond to the bottom and the top of the plume. Their average gives the height ZM of the plume centreline at that distance from the stack. The contours of the plume are then projected to ground level. The two extreme values define the plume width along the scanning plan. The correct width of the plume is assumed to be that distance multiplied by the cosine of the angle between the scanning plane and the normal to the plume bearing. The direction of the plume is defined as the direction from the stack to the plume width mid-point. Analogously, the range is defined as the distance between the mid-point and the chimney.

If data are automatically recorded and processed (ref.10,17) and 18) it is possible to reconstruct the distribution of the aerosols and the centre of mass for each vertical section. In this case range and direction are defined as those of centres of mass, while the horizontal and vertical width of the plume are defined as twice the crosswind and vertical standard deviations σy and σz.

It is also possible to evaluate higher order moments such as the Skewness and Kurtosis. Finally to investigate the structure of plumes, integrated vertical and/or horizontal concentration of the tracer plume and Lidar-derivated plume sections can be computed.

All the computations are based on range corrected backscattered signals. An example of the computation scheme is the following: let Sij be the ratio between received and transmitted power as a function of the distance Lij (along the shot line), φj the elevation angle of the jth shot and i the subscript indicating the distances (on the jth shot) at which Sij are measured. We have:

$$YM = \frac{\sum_{1}^{N}{}_{j}\sum_{1}^{M}{}_{i} Sij\, Lij^{2}\, (Lij \cos\varphi j)}{\sum_{1}^{N}{}_{j}\sum_{1}^{M}{}_{i} Sij\, Lij^{2}} \qquad (10)$$

for the horizontal coordinate of the centre of gravity of received echoes in the cross-wind direction, and similarly:

$$ZM = \frac{\sum_{1}^{N}{}_{j}\sum_{1}^{M}{}_{i} Sij\, Lij^{2}\, (Lij \sin\varphi j)}{\sum_{1}^{N}{}_{j}\sum_{1}^{M}{}_{i} Sij\, Lij^{2}} \qquad (11)$$

for the vertical coordinate. The expression for the vertical dispersion coefficient (standard deviation) is:

$$\sigma z = \left[\frac{\sum_{1}^{N}{}_{j}\sum_{1}^{M}{}_{i} Sij\, Lij^{2}\, (Lij \cos\varphi j)}{\sum_{1}^{N}{}_{j}\sum_{1}^{M}{}_{i} Sij\, Lij^{2}} - (YM)^{2} \right]^{1/2} \qquad (12)$$

Similar expressions may be easily derived for the other quantities.

3.4 Complementary measurements

Lidar measurements of plume rise $(z(x))$ and dispersion (σ_y, σ_z) are generally part of a more complex experiment during which other instruments provide information on the meteorological conditions, on the SO2 vertical and g.l. concentration and on characteristics of the stack emission (see, for example, ref. 19 to 26).

The minimum information needed is:

i) plume rise

- wind speed at the stack mouth height
- temperature gradient in the lower layer of the atmosphere
- exit velocity and temperature of gases
- stack diameter and height

ii) plume dispersion

- SO2 g.l.c. at a certain number of stations

When possible it is useful to have also:

- wind profile in the first 1000 m
- temperature profile in the first 1000 m
- vertically integrated SO2 concentrations
- tracer experiments

4. RESULTS

4.1 Best fit curves and scatter of data

The first papers on the applications of Lidar to plume rise studies were by Hamilton (ref. 27 and 28) who demonstrated the power of this new method of quantitatively evaluating plume rise and dispersion. In his measurements of the plume of Northfleet Power Station (England), he reported the agreement between his Lidar observations and Lucas et al. (ref. 29) plume rise formula (+).

Hamilton discussed in details many problems related to such a topic. In particular, he examined the dependence of plume rise on wind speed. He found that the observations fitted ,on the average, the tested relationship, although there was a great deal of scatter. This latter is an important point. In fact the comparisons between plume rise theories and Lidar measurements reported in the

(+) This formula was then reconsidered and brought to its final version in 1974 by Moore (see eq. 1).

literature show, in general, large scatters. This is mainly due to the following reasons:

i) the measurement of a cross-section is not an instantaneous measurement. It takes, typically, about one minute.

ii) the actual meteorological conditions may vary during a cross-section scanning. This is particularly true during unstable and/or low wind speed situations.

iii) very often the meteorological parameters are available as half an hour averages while Lidar cross-sections are relative (see ii) to averaging times of the order of the minute.

iv) the lower layers of the atmosphere do not behave, in general, in the very simple and regular way assumed by models.

v) models refer to averaged or constant meteorological conditions. They are simple analytical tools to forecast the average behaviour of plumes, and give values to be introduced in diffusion models giving the average ground level concentrations.

In the following paragraphs we will present and discuss results obtained after the pioneering work of Hamilton.

4.2 Different Lidar estimates of ZM and plume dimensions

The height of the plume centreline at each cross section may be evaluated (see par. 3.3) either computing the average value between maximum and minimum recorded heights (ZM)1 or calculating the centre of gravity of received echoes (ZM)2. As these values are used to validate models, it is interesting to know the differences (if any) between the two evaluations. An example of such a comparison is shown in fig. 3. It refers to Lidar plume rise observations performed at the Turbigo Power Station (Northern Italy) by the author. Details of such measurements, almost all relative to merged plumes and the main results, are reported in ref. 10. This drawing is an unpublished part of such work. The comparison is relative to $\simeq$100 cross sections measured during neutral and stable conditions in a 100 - 1500 m range of downwind distances. Light and shared winds were prevailing.

It was found that the differences $\Delta = |(ZM)1 - (ZM)2|$

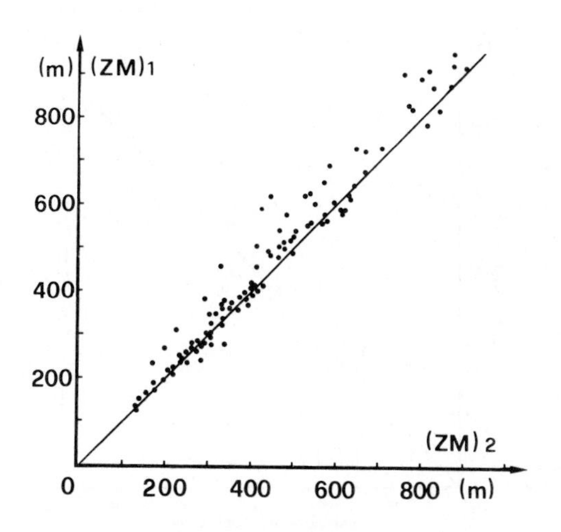

Fig.3 - Comparison between two different plume centreline height evaluations. (ZM)1=(Hmax+Hmin)/2 and (zm)2 from eq.(11).

having the following percentages were: $\Delta < 5$ m = 19%, $\Delta < 10$ m = 36%, $\Delta < 25$ m = 58%. The mean difference was 25 m. In a similar way the estimate of y and z were compared to the vertical (Az) and horizontal (Ay) plume semidimensions. The average results were the following:

$(Az)/\ \sigma z) = 2.0$ and

$(Ay)/\ \sigma y) = 2.4$

4.3 Difference in plume evaluations by different techniques

In the introduction it was said that three experimental methods can generally be used in the evaluation of Z(x). It is useful to verify how different the results obtained may be using different techniques.

Briggs (ref. 2) discusses the results of four simultaneous measurements at the same plume performed by Johnson et al. (ref. 19) and Proudfit (ref. 30) research teams. The first team used a Lidar and the second team used SO_2 sampling made by helicopter traverses through the plume. The ratio of the measured centrelines, ZM (Proudfit)/ZM (Johnson et al.) were: 1.2, 1.3, 1.4 and 0.7. These results illustrate and confirm very well the difficulty of defining plume rise even from the observational standpoint.

Figg. 4 and 5 show instead the results obtained by Edwards (ref 31) observing simultaneously the rise and the vertical radius of plumes by Lidar and photography. This comparison is based on 66 cases which are representative of a variety of meteorological conditions. Fig. 4 shows that the Lidar - photography observed plume

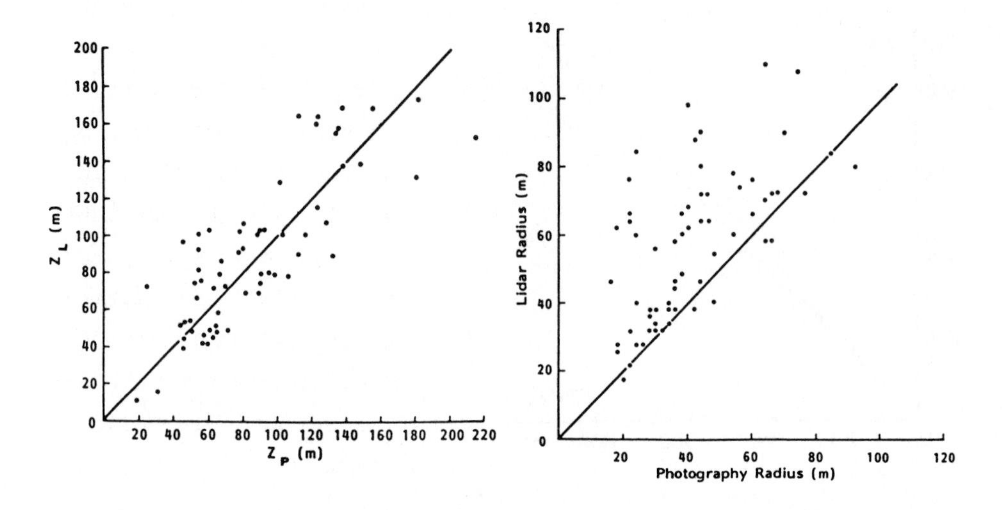

Fig.4 - Comparison of Lidar observed plume rise and photographycally observed plume rise (from ref.31)

Fig.5 - Comparison of Lidar observed and photographically observed plume radii (from ref.31)

rises are scattered symmetrically about the line of perfect agreement. This means that the two remote sensing techniques agreed quite well. There is a large scatter, attributed by Edwards to the different time averaging period of the two techniques. Fig. 5 compares the Lidar - photography observed vertical plume radii. It can be seen that Lidar results are larger than the photography (1.6 times). This fact is quite interesting since any particulate concentrations less than ten percent of the maximum had been neglected when calculating Lidar plume radii. This result is due to the different thresholds of the two techniques in estimating the plume boundaries, as photography may only detect the visible part of the plume. Therefore photographic observation, in general, will underestimate the plume spread.

4.4 Plume rise in complex sites

The shapes and trajectories of plumes rising in complex sites (hills, mountains, sea-side) are deeply modified by the topographic irregularities. These last induce, in fact, strong variations on the micrometeorological variables (such as wind and temperature profiles) which influence the motion of plumes.

The use of formulae to predict the fate of plumes in these cases becomes somewhat arbitrary. However, Lidar measurements, coupled to other meteorological observations, may help in understanding what is actually happening inside the plume and in contributing to correct the model forecastings.

Figg. 6 and 7 illustrate a peculiar example of what has been said. They refer (ref. 20) to a series of surveys carried out at La Spezia Gulf (Northern Italy).

Fig. 6 displays an example of horizontal view of a plume measured by Lidar. Fig. 7 shows the vertical projection of the same plume. In this last diagram, a simultaneous vertical wind profile is drawn. It suggests a possible explanation of the increased rise at about 500 m downwind in terms of a corresponding decrease of wind speed. At the same downwind distance (fig. 6), the plume also changes direction showing that the plume was rising through two vertical regiones occupied by two different air masses: sea breeze at the bottom and general circulation above.

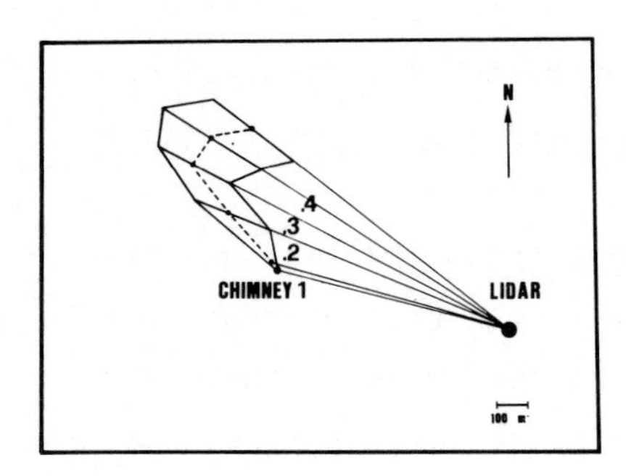

Fig.6 - Example of horizontal view of a plume measured by Lidar technique (from ref.20)

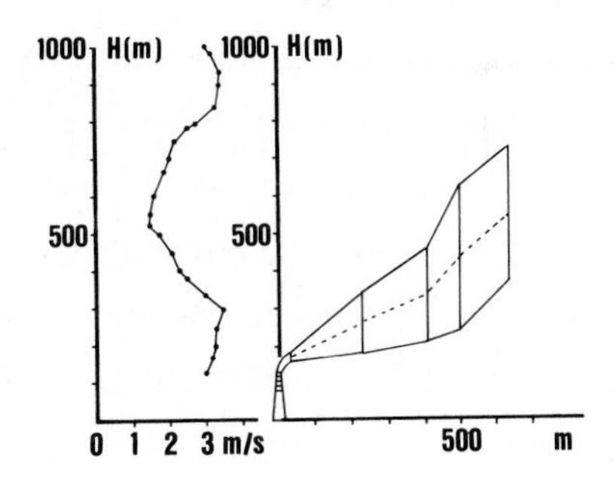

Fig.7 - Example of a vertical section of a plume measured by Lidar technique and of a contemporary wind profile (from ref.20)

4.5 Plume rise: single source

In spite of their analytical differences, both Briggs' and Moore's models (eqs. 1 and 2 to 5) proved to fit the experimental observations well. Some researchers compared their Lidar results only to Briggs' model (ref. 19, 21, 26 and 31). Others only to Moore's model (ref. 27, 32 and 33). Finally a few compared both models to their data (ref. 20, 34 and 35). After examining all these papers, a few conclusions may be derived:

i) both models describe well enough the plume trajectory and, in particular, the final height of the plume for windy conditions ($u > 1.0$ m/s)

ii) the differences between the two estimates are, in many cases, of the order of the scatters of data about the line of perfect agreement

iii) during calm conditions ($u < 1.0$ m/s) Briggs' models should be used (Moore's model does not consider such conditions, very rare in England but frequent in Italy, for example)

iv) the range of F values considered in the experiments is too little to clearly indicate a preference between exponent 1/3 or 1/4. The same also holds true for downwind distances.

v) the preference given by the various authors to one of the two models is not therefore based on experimental evidence but on subjective selection of which of the two formulae is considered better matching the physical bevaviour of plumes.

vi) as for plume dimension evaluations, Lidar observations evidentiated a substantial increase of σz in the kilometer nearest to the stack with respect to the classical curves (like those due to Pasquill - Gifford). This is because in the first part of its trajectory (transitional phase), the growth of the plume is principally due to self-induced turbulence caused by buoyant plume, while the atmospheric turbulence is relatively ineffective (ref. 17 and 36).

4.6 Plume rise: multiple sources

Only a few sets of data are available, at the moment, to test multiple source models. A first set of data (not obtained by Lidar) is quoted by Briggs (ref. 13). It constitutes the data set which served to test both Briggs and Anfossi et al. models. In addition to those data, three Lidar data sets are now available (ref. 10,20 and 26). A few conclusions may be stated:

i) all these papers confirm that multiple source rise of plume is significantly larger than that of the isolated plumes

ii) the average ratio $\sigma z/z$ (equivalent to the entrainment parameter β) found for merged plumes originating from two stacks (ref 26) is less than that for the single stack cases (0.47 and 0.62 respectively. This means that the internal dynamics for N combining plumes are not the same as those for a single plume with a buoyant flux equal to the sum of the N buoyant fluxes

iii) this is confirmed by Anfossi (ref. 10). From the study of the obtained series of cross-sections (see fig. 10), it was apparent that, after merging, plumes neither maintain their individuality nor mix completely in such a way to produce a new single plume. In other words the merging and entrainment processes produce a complicated distribution, with several minima and maxima of concentration inside the plume

iv) no influence of wind - stack line angle on the plume enhancement at the Turbigo Power Plant was found.

4.7 Plume structure

With the aid of Lidar, it is possible to observe the detailed internal structure and geometry of plumes (ref. 10, 17, 18 and 19).

This information is useful to the air pollution metereologists in that it should help them to develop more realistic theories of plume rise and dispersion. In fact a detailed examination of Lidar - derived plume cross-sections should reveal the changes in plume structure in response to changing meteorological conditions and yield further and better information on the basic nature of the entrainment process.

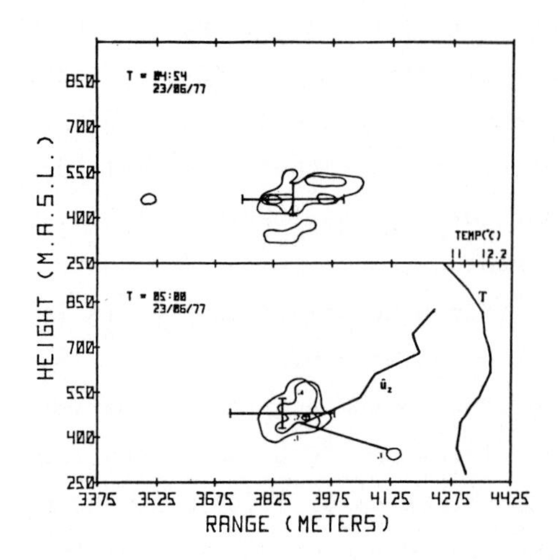

Fig.8 - Two cross-sections of a plume observed with time lag of 6 minutes in presence of a strong directional shear and a temperature inversion.Grid and cross hatches represent $2\sigma y$ and $2\sigma z$ in the scanning plane.Uz and T represent wind and temperature profiles obtained by minisondes (from ref 17).

Hoff and Froude ref. (7) reported interesting samples of the effects on the plume spreading of wind directional shears and of thermal inversions. Fig. 8 illustrates that the plume appears tied to the inversion base (notice that wind shear is 72 degrees).

Johnson and Uthe (ref. 19) provided other important examples of plume interaction with actual micrometeorological conditions. In fig. 9 (fig. 5 in their paper) we can see that the plume top is trapped by the temperature inversion at about 500 m, while the lowest 100 m is fumigating. The bubble one can see in cross-section 79 of fig. 9 is probably a convective thermal which has risen through the plume.

Uthe and Johnson (ref. 18) showed that the quasi-instantaneous profiles of concentration inside the plume are not Gaussian, but tend to resemble more closely a uniform, or "top hat" distribution.

Anfossi (ref. 10) evidentiated the complicated internal structure of merged plumes. Figg. 10 and 11 show an example of plume cross section and of the corresponding vertical and horizontal cross-plume integrated densities, respectively. These last indicate the profiles of vertical and horizontal aerosol concentration and represent what could be seen by two mobile integrating instruments viewing the plume from beneath or laterally respec -

tively.

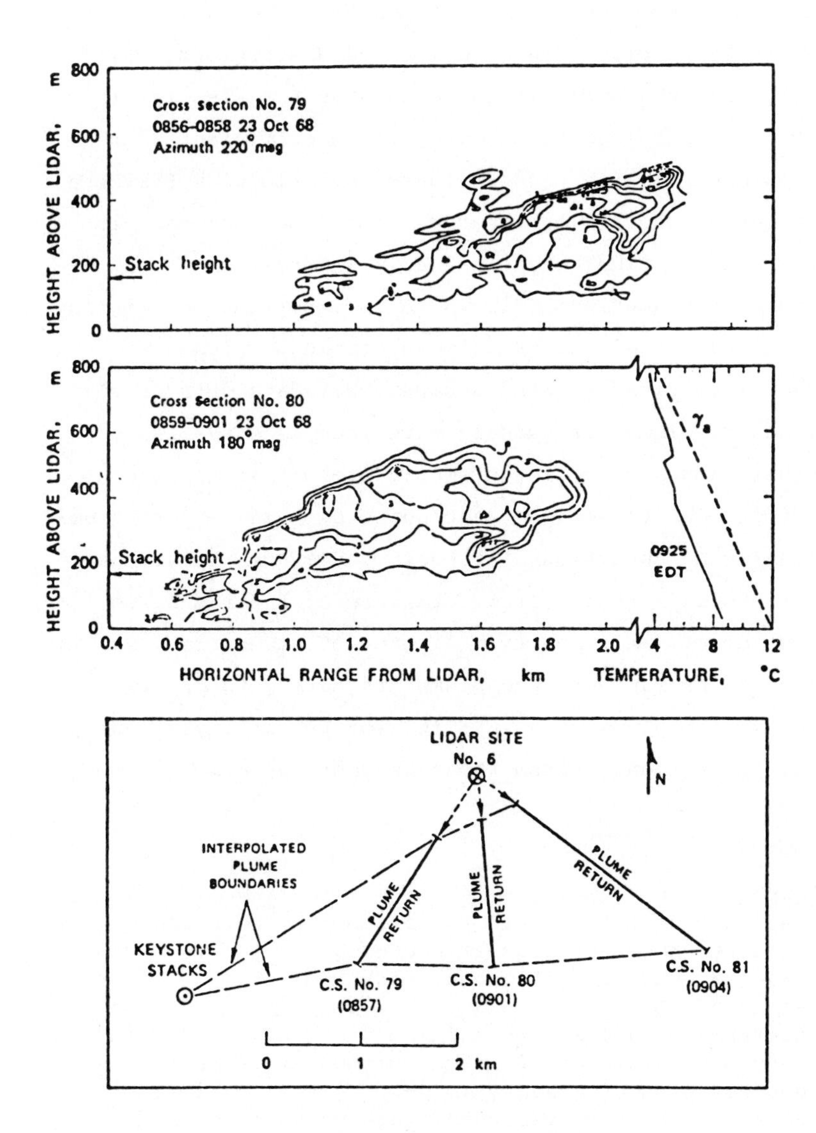

Fig. 9 - Plume vertical cross-sections and plan view of horizontal plume position (bottom) (from ref.19)

5. CONCLUSIONS

During the last two decades, many teams have been using the Lidar technique to study plume trajectories and dispersions. Their results clearly show the power of the apparatus but also its limits. In fact it is possible to measure with a sufficient degree of accuracy plume centreline and standard deviations as far as a few kilometers downwind from the source, depending upon the air turbidity of the site. Lidar gives valuable information on the internal structure of plumes and this is quite important in relation to the study of the interactions between combining plumes or among plumes and inversions or wind shears. Many data sets to test and possible modify plume rise models were thus made available.

On the other hand, as the measurements are not instantaneous, the cross-sections can be somewhat ambiguous especially during unstable or light wind conditions. If such conditions prevail,as is generally the case in Northern Italy, the use of Lidar must be limited to the study of the transitional phase of plume rise, and it cannot be used to extrapolate the plume trajectory to greater distances since, as the plume rises, it is kept by air layers moving, in some cases, indipendently from those present beneath.

6. NOMENCLATURE

a = exponent in eq. (7)
C = variable containing all the parameters not directly related to buoyancy flux at source
d = spacing between contiguous stacks (m)
EN = enhancement factor
F = buoyancy parameter (m4 s-3)
Fi = buoyancy parameter of merged plumes (m4 s-3)
g = acceleration of gravity (m s-2)
He = effective height of merged plume produced by N stacks (m)
Hi = merging point height (m)
Hj = height of the stacks (m)
Hmin,Hmax = heights of lowest, highest stacks (m)
Δ Hmin = maximum single plume rise from lowest stack Hmin (m)
Lij = distance, along the jth shot, at which Sij is measured (m)
N = number of stacks
Q = rate of emission of heat (MW)

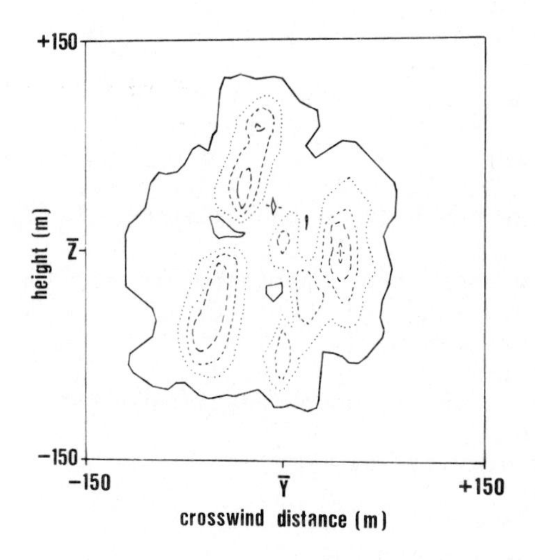

Fig. 10 - Lidar derived cross-section of two merged Turbigo-plumes. The concentration isoplets are at uniform interval (from ref.10)

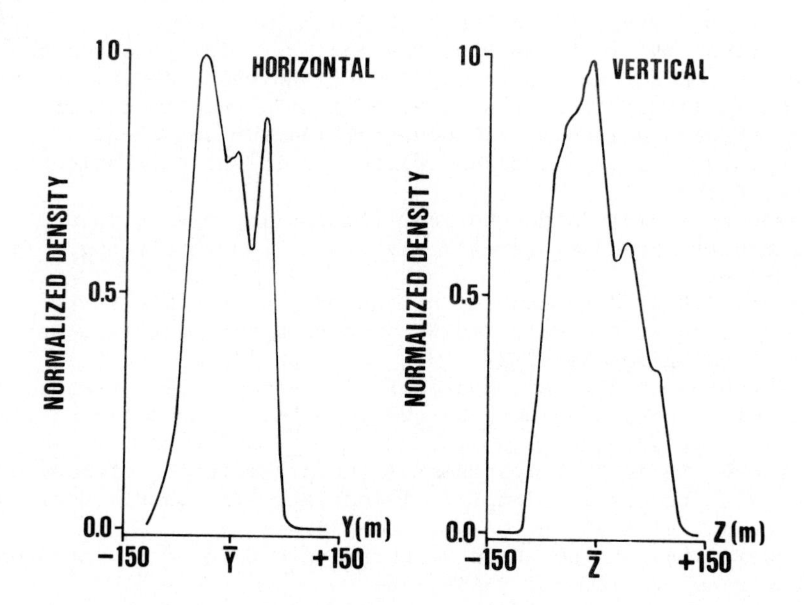

Fig.11 - Integrated densities corresponding to the cross-section drawn in fig.10 (from ref.10)

r	= internal stack radius (m)
R	= plume radius (m)
s	= stability parameter (m s-2)
Sij	= signal on the jth shot at a distance Lij
Ta	= ambient temperature (K)
Tp	= plume temperature (K)
u	= wind speed (m s-1)
x_f	= downwind distance at which levelling of the plume occurs (m)
x_*	= parameter which expresses the dependence of z(x) on downwind distance and stability (m)
YM	= horizontal center of gravity (m)
w	= efflux velocity (m s-1)
z(x)	= plume rise as a function of downwind distance (m)
z	= final plume rise (m)
ZM	= vertical center of gravity (m)
β	= entrainment constant
φ	= elevation angle of jth Lidar shot
σ	= crosswind standard deviation (m)
σ	= vertical standard deviation (m)

7. REFERENCES

1 G.A. Briggs (1969) "Plume Rise", U.S.A.E.C.

2 G.A. Briggs (1975) "Plume Rise Predictions", in: Lectures on Air Pollution and Environmental Impact Analyses, A.M.S.,Boston

3 D.J. Moore (1980) "Lectures on Plume Rise", in: Atmospheric Planetary Physics,edited by A.Longhetto,Elsevier,pp 327-354

4 G.T. Csanady (1973) "Turbulent Diffusion in the Environment", Reidel, Boston

5 A.Longhetto (1975) "Atmospheric Diffusion in Upper Levels of the Planetary Boundary Layer", Rivista del Nuovo Cimento, 5/4, pp 593-616

6 F.Pasquil (1974) "Atmospheric Diffusion", Wiley, New York

7 W.T. Kranz and D.P. Hoult (1973) "Design Manual for Tall Stacks" M.I.T. Publication No 73-3

8 G.P.N. Venter (1977) "A comparison of observed plume trajectories with those predicted by two models". Atmospheric Environment, 11, pp 421-426

9 D.J. Moore (1974) "A comparison of the trajectories of buoyant plumes with theoretical/empirical models". Atmospheric Environment, 8, pp 441-458

10 D.Anfossi (1982) "Plume Rise Measurements at Turbigo". Atmospheric Environment, 16, pp. 2565-2574

11 B.R. Kerman (1982) "A Similarity Model of Shoreline Fumigation" Atmospheric Environment, 16, pp. 467-477

12 G.A. Briggs (1981) "Plume Rise and Buoyancy Effects". To appear in: Atmospheric Sciences and Power Generation

13 G.A. Briggs (1974) "Plume Rise from Multiple Sources" in Cooling Tower Environment, Oak Ridge, N.C.

14 D.Anfossi, G. Bonino, F.Bossa and R.Richiardone (1978) "Plume rise from Multiple Sources: a New Model". Atmospheric Environment, 12, pp 1821-1826

15 R.T.H. Collis (1969) "Lidar". In: Advances in Geophysics", 13, pp 113-139

16 P.M. Foster; S. Sutton and R.H. Varey (1978) "Lidar Measurements on the Plume at Drax". C.E.R.L. Note No RD/L/N 53/78

17 R.M. Hoff and F.A. Froude (1979) "Lidar Observations of Plume Dispersion in Northern Alberta". Atmospheric Environment, 13, pp 35-43

18 E.E. Uthe and W.B. Johnson (1976) "Lidar Observations of Plume Diffusion at Rancho Seco Generating Section". EPRI Report NP 238

19 W.B. Johnson and E.E. Uthe (1971) "Lidar Study of the Keystone Stack Plume". Atmospheric Environment, 5, pp 703-724

20 D. Anfossi, P. Bacci, C.Giraud, A.Longhetto and A.Piano (1976) "Meteorological surveys at La Spezia site". In: Atmospheric Pollution, edited by M. Benarie, pp 531-540, Elsevier, Amsterdam

21 P. Bacci, G. Elisei and A. Longhetto (1974) "Plume Rise and Dispersion at Ostiglia P;wer Plant". Atmospheric Environment, 8, pp 1177-1186

22 B.E.A. Fisher, Y Gotaas, P.M. Hamilton, R.Houlgate, P.Maul and D.J. Moore (1977) "Observations and calculations of airborne sulphur from multiple sources out to 100 km". Atmospheric Environment, 11, pp. 1163-1170

23 A.Longhetto, P. Guillot, D.Anfossi, P.Bacci, G.Elisei,G.Frego, S.Sandroni and R. Varey (1982) "Atmospheric dispersion experiments in the near and medium field (fourth European Community Campaign, Turbigo Italy - September 1979". Il Nuovo Cimento, in press

24 R.A. Scriven (1979) "Measurements of Air Pollution at Drax during the 1976 CEC Remote Sensing Campaign". C.E.R.L. Note No RD/L/R 1984

25 P.R. Slawson, G.A. Davidson, W. McCormick, G. Roithby (1978) "A study of the Dispersion Characteristics of the G.C.O.S. Plume". Syncrude Canada Limited Report

26 J.C. Weil (1979) "Assesment of Plume Rise and Dispersion Models using Lidar Data". Martin Marietta Corporation Report PPSP - MP - 24

27 P.M. Hamilton, K.W.James and D.J. Moore (1966) "Observations of Power Station Plumes using a Pulsed Ruby Laser Rangefinder". Nature, 210, pp 723-724

28 P.M. Hamilton (1967) "Paper III: Plume height measurements at Northfleet and Tilbury Power Stations". Atmospheric Environment, 1, pp 379-387

29 D.H. Lucas, D.J. Moore and G.Spurr (1963) "The rise of hot plumes from chimneys. Air and Water Pollut.Int.J. ,7, pp.473-500

30 W.B.Proudfit (1970) "Final report on plume rise from Keystone Plant". Sign X Labs., Inc. Essex, NO 203, pp.767-1700

31 G.C. Edwards (1980) "Plume rise modelling". In: Second Joint Conference on Applications of Air Pollution Meteorology, edited by Amer. Met. Soc., pp. 140-145

32 P.M. Hamilton and D.J. Moore (1973) "Gas Turbine Plume Heights measured at Norwich Power Station - Preliminary Analysis".Atmospheric Environment, 7, pp. 991-996

33 A. Martin and F.R. Barber (1973) "Further Measurements around Modern Power Stations -I - III". Atmospheric Environment, 7, pp. 17-37

34 P. Bacci, A. Longhetto, D.Anfossi and C.Giraud (1975) "Discussion on Moore's three-dimensional approach to the plume rise problem". Rivista Italiana di Geofisica e Scienze Affini, II, No 1, pp. 27 - 31

35 P. Bacci and A.Longhetto (1980) "Plume Rise Observations in the Planetary Boundary Layer". In: Atmospheric Planetary Boundary Layer Physics, edited by A.Longhetto, Elsevier, pp.355 - 366

36 F. Pasquill (1976) "Atmospheric dispersion Parameters in Gaussian plume modelling. Part II", U.S.E.P.A. Report 600/4 - 76 - 030b, Research Triangle Park, USA.

Optical Remote Sensing of Air Pollution,
Lectures of a course held at the Joint Research Centre, Ispra, Italy, 12–15 April 1983,
P. Camagni and S. Sandroni (Eds). 259–277

MONITORING OF DISTRIBUTED EMISSIONS BY CORRELATION SPECTROMETRY

S. SANDRONI

1. INTRODUCTION

The air quality level and its trend in urban or industrial areas is usually controlled by a network of fixed monitoring stations connected to a central computerized unit. These stations, using physical, chemical or electrochemical methods, measure the local concentrations of pollutants as averages over 30 min or 1 h, as required by national laws. A classical network is expensive to equip and to operate; another limitation is the low spatial density and poor representativeness. For local complaints and controversies, surveys are performed by mobile laboratories equipped with meteorological and monitoring instrumentation.

Wider information can be obtained by remote sensors based on light absorption or scattering. Schematically, optical remote sensors may be employed in three arrangements (Fig. 1): (1) a double-ended active system such as the transmissometer, in which the detector measures the attenuation of a light beam emitted from a controlled remote source; (2) a double-ended passive system such as the Volz sunphotometer, for which direct sunlight or diffused daylight is the source and the pathlength is undefined; (3) a single-ended active system such as the Lidar, in which source and detector are mounted in a compartment and the detector measures the intensity of backscattered light. This last configuration, using a pulsed laser as source, allows the location of the pollutant cloud. At present two systems are routinely used, namely: (1) the visibility meter, in airports or along highways, and (2) the plume opacity monitor (this sensor is accepted by US-legislation but not yet in CEC Member States). In the first case the atmospheric transparency is measured as the attenuation of a light beam emitted from a source located at a known distance; in the second case, the plume opacity is given as the attenuation of daylight passing across the plume.

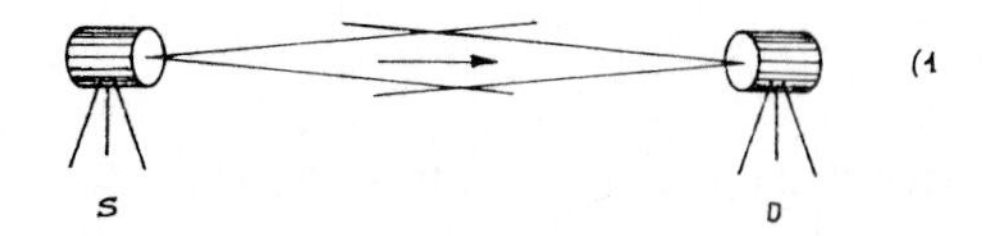

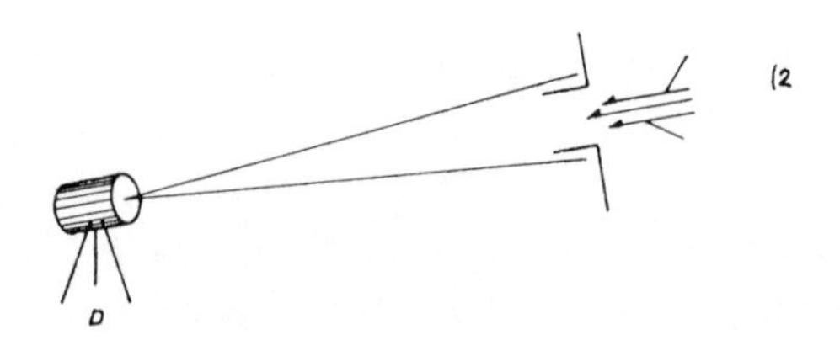

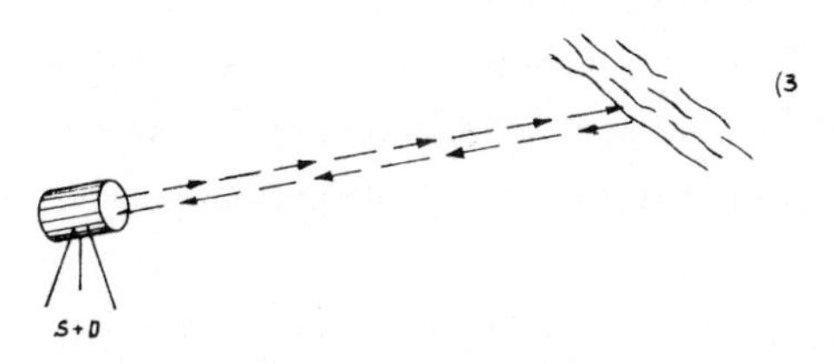

Fig. 1 - Arrangements for optical remote sensors:
S = light source;
D = detector.

The absorption correlation spectrometers (the principles of the technique are given in ref. 1) may be used in configurations (1 and (2; their applications are summarized in Table 1.

TABLE 1

Applications of correlation spectrometers to remote sensing of pollutants

Sensor	Light source	Applications
Ground-based	remote lamp	. monitoring of average concentration (active long-path, ALP)
Ground-based	daylight	. plume geometry and structure . source emission monitoring
Ground-mobile or airborne	daylight	. vertical burden measurement . pollutant mass flow . pollutant mapping over wide areas . emission inventory

The instrument displays the concentration-pathlength product (or burden) of the gas present in the line-of-sight; if the pathlength is defined, the average ambient concentration can be deduced.

In recent years commercial and research instruments have been developed; the classical one is COSPEC (I to V, Barringer Research), but TELETEC (CNR/Tecneco, particularly intended for ALP operation, refs. 2-3), RAMS (Onera) and the recent PLUMETRACKER (Moniteq) should also be cited.

This paper deals with some typical field applications of correlation spectrometers mentioned in Table 1 with reference to our direct experience.

2. ACTIVE LONG-PATH MONITORING

2.1 Features

As an optical system active long path (ALP) monitoring has several unique features which are of interest in establishing ambient air quality:

1) Area monitoring. Within a convenient radius it can be provided by periodically looking in several directions from a central point. In general, ALP can: a) establish to what degree a stationary monitoring site is representative of the area assigned to it in the monitoring network, and b) provide an experimental means to check the predictive accuracy of air quality simulation models (refs. 4, 5);
2) Sample integrity. In contrast to point monitors, ALP ensures sample integrity since no sampling is required;
3) Accessibility. ALP techniques make it possible to monitor line-of-sight paths which would otherwise be inaccessible (for instance, measurement across a valley). Such a feature lends itself to the characterization of pollution dispersion near roadways or airport runways, and to the measurement of trace gas averaged over elevated paths or cross-town paths in urban areas;
4) Modelling needs. For the validation of air quality models, air quality levels averaged over the considered area are at the same time accurate input data to be compared with computer outputs. The current pollution models are grid point models which assume that the input concentrations are averaged over some grid area. Therefore, there is a need to translate a point input/output into an area input/output and ALPs are suitable for this purpose. According to a recent model for laser-absorption computed tomography, an ambitious alternative to the DIAL system, spacings of hundreds of metres over ranges of tens of kms are sufficient for air pollution monitoring (ref. 6).

Experimentally, the simplest geometry one could adopt would consist of a series of several lamps located approximately 90° apart at the intersections of radial cardinal wind direction vectors from, and concentric circles around, the instrument site.

2.2 Active long path systems: COSPEC and TELETEC

At the JRC-Ispra the active mode of operation of correlation spectrometers has been particularly investigated. As several spectral frequencies (from 10 to 14) are involved, the light source should have a continuous spectrum at 2800 - 3100 Å (SO_2 spectral region) and at 4000 - 4400 Å (NO_2 spectral region). Of the lamps commercially available the best is the Xenon arc lamp which emits a continuous spectrum from UV to visible. The only limitation is the emission of lines in the region around 4000 Å; special care is therefore required in positioning the masks for NO_2.

Two instruments are suitable, the Cospec III and the Teletec, a research instrument developed by Tecneco and CNR-Bologna.

Cospec III, an instrument of the Cospec series, is able to operate with daylight as well as with an active source. The source used is a 75 W Xenon lamp installed in a Gregorian telescope and electronically modulated at 2.5kHz, a frequency which is distant from the frequency spectrum of atmospheric turbulence; the signal modulation also helps the optical alignment between lamp and instrument. As in the other Cospecs, a routine calibration is made by two internal cells. A special device which can be inserted into the light beam allows one to shift the incoming spectrum some Å away from the proper alignment; as the instrument shows no response to the target gas, the corresponding signal defines the zero baseline.

Teletec (Fig. 2) is a similar instrument based on a method developed at CNR Bologna and claimed to be different from Cospec (refs. 2, 3). The main features of Teletec as compared with Cospec are: a) a different signal processing; b) an automatic calibration obtained by a cyclic insertion of a calibration cell into the internal optical path of the instrument; and c) different collecting optics. The optical alignment of the instrument to the lamp is obtained by looking at the intensity of the signal received on an oscilloscope. As for Cospec, an accurate calibration is made by insertion in front of the receiving telescope of some fixed-length cells with known ppm-metre values.

The characteristics of the entrance optics of the two spectrometers are compared in Table 2. As Teletec has a smaller aperture than Cospec it is more suitable for active mode operation; while Cospec is more suitable for passive operation. At a distance of 1 km the surface seen from Teletec is 2.5 m^2, whereas for Cospec it is 20 m^2.

The performances of the two instruments for maximum distance and sensitivity are compared in Table 3.

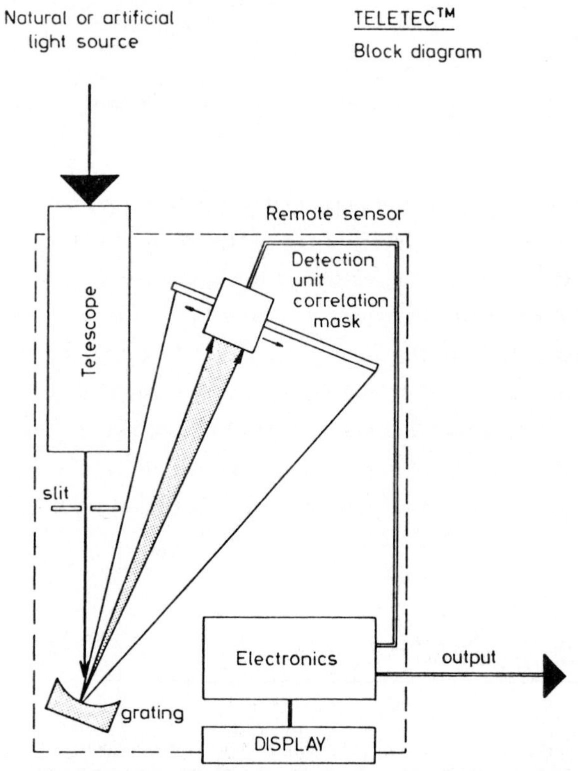

Fig. 2 - Block diagram of Teletec. In its latest version it can measure both SO_2 and NO_2 (ref. 7).

TABLE 2

The optics of Cospec and Teletec compared

	Cospec III	Teletec
Fore-optics f. o. v.	0.4 x 13.4 mrad	0.286 mrad
Telescope focal length	303 mm	700 mm
Telescope diameter	80 mm	110 mm
Entrance slit	0.25 x 8.13 mm	0.4 mm
Telescope spatial angle	$2 \cdot 10^{-5}$ sterad	$2.6 \cdot 10^{-6}$ sterad
Monochromator	0.25 mm Ebert-Fastie f/3.5	holographic f/6

TABLE 3 (refs. 7, 8)

Sensor	Source	Pollutant	Max. distance	Sensitivity
Cospec III	75 W Xe	SO_2	1.0 km	5 ppb. km
		NO_2	0.7 km	3 ppb. km
Teletec	150 W Xe	SO_2	8.5 km	10 ppb. km
			(12.5 km)	
		NO_2	2.0 km	3 ppb. km

The linearity range has been checked up to 700 ppm. metre for Cospec and 1500 ppm. metre for Teletec. By a mechanical shift of the detection unit towards UV (or visible), the linearity range can be extended (or decreased) while its sensitivity changes in the reverse direction. An important role is played by the contribution of diffused light to the received signal. For Teletec it is only 2% at 8.5 km, while it reaches 50% at 12.5 km on a clear sunny day (Table 3). Obviously, during the night, greater distances can be reached.

Although limited to two molecular species, this technique allows us to obtain remarkable results in comparison with other remote-sensing techniques (Table 4).

TABLE 4

Active long-path techniques of atmospheric SO_2 and NO_2

A) <u>for SO_2</u>

Technique	Sensitivity	Max. distance	Ref.
CO_2 laser abs.	3 ppm. km	1 km	9
Mask correl.	10 ppb. km	8.5 km	8
Diff. Lidar	20 ppb	1.2 km	10

B) <u>for NO_2</u>

Technique	Source	Sensitivity	Distance	Ref.
Diff. Lidar	dye laser	10 ppb	1-2 km	10
Diff. Lidar	dye laser	200 ppb	3.5 km	11
Diff. absorpt.	Ar+laser	100 ppb	1 km	12
Mask correl.	Xe lamp	3 ppb. km	2.0 km	-

In Table 4 costs are not considered since their evaluation is difficult for research apparatus, but they strongly favour the correlation spectrometer.

The possible extension of this technique to other molecular species has been investigated by Williams and Kolitz (ref. 13). Other species could be monitored by their absorption lines in the UV (e. g. Ozone) or in the IR (CO, NH_3, HCl, NO, hydrocarbons); a passive system for Ozone has been described by Tomasi et al. (ref. 14); in the IR interferences by water vapour and CO_2 are noticeable (ref. 15).

2. 3 Single-ended arrangement

The double-ended system previously described can be greatly simplified if the light source and the sensor are placed in one and the same compartment and a retroreflector is used. The optical path should be shorter than in the previous arrangement, but a reflector is much easier to install and to align than a lamp. This solution, apparently simple, has nevertheless met with some practical difficulties. The reflector should consist of a flat distribution of small corner cubes (cat's eyes), each unit having optically flat surfaces at 90° with an accuracy of 0.01°. The resulting beehive should be quite flat without dead space between the individual units; its dimensions should be proportional to the square of the distance at which the reflector will be placed, for instance 2.5 m^2 at 1 km. The estimated cost for a 1 m^2 reflector, built with Al-covered plexiglass corner cubes 10 cm long, made in a cast iron matrix, is higher than 10, 000\$, more expensive than a Xenon lamp and its associated optics. Because of these economic reasons, we have so far given preference to lamps.

2. 4 Atmospheric troubles in long-path measurements

The intensity of a light beam propagating in the atmosphere is altered by various phenomena such as:

. Scintillation. The distortion of a light beam caused by atmospheric turbulence produces a randomization of the intensity of the received signal coming from a collimated source. Generally, these phenomena are called "image dancing" or "scintillation". These phenomena are particularly important in horizontal light beam propagation near to the ground and are a complex function of the physical and dynamical characteristics of the local atmosphere. Some years ago we documented photographically the dancing of a light spot of an He-Ne laser during typical meteorological situations (ref. 16). The intensity fluctuations may even be of the order of $\pm$ 70% around a mean value along the propagation direction, increasing with the temperature gradient between air and ground and with the pathlength. The observed frequencies lie between a few Hz and some tens of Hz (ref. 3). This effect is partially reduced by lifting the entire system some metres above the ground. In the case of Cospec, which operates over a maximum distance of 1 km, the signal modulation is in its favour; for Teletec the integration time must be increased up to 1 min or more, although some short-term evolutions of dispersion phenomena are cancelled.

. Scattering losses. Rayleigh and Mie scattering occur both in active and passive mode giving an intensity attenuation and a modification of the spectral slope. In the passive mode the effect is complicated by the solar elevation angle (ref. 1).

. Interference by other molecular species. In the open air only Ozone at relatively high concentrations may interfere in SO_2 measurement (ref. 1).

. Diffused daylight contribution. Generally, the orientation of ALP is chosen in such a way as to avoid direct sunlight being received. However, the acceptance solid angle of the sensor allows some diffused daylight to contribute to the effective incoming radiant power. For Cospec, the use of a modulated source drastically reduces this contribution. In the case of Teletec the drastic reduction of the solid acceptance angle previously cited (Table 2) minimizes it. In field experiments an evaluation of this spurious contribution is made by comparing on an oscilloscope the signal intensities when the light source is switched on and off (see section 2. 2).

We emphasize once more that the active mode allows absolute measurements. On the other hand, the measurement performed in the passive mode depends on an indefinite path, on the light source distribution along the path and finally on the target gas distribution. Consequently the data are only relative.

3. FIELD EXPERIMENTS

3.1 Fall-out measurements from single sources

During the last 8 years we have taken part in several national and international campaigns. Since 1975, European-scale campaigns on the remote sensing of air pollution have taken place in different European countries under the sponsorship of CEC-DG XIIG. Four out of the six performed up to now have involved single sources, and their main objective has been to describe, and if possible to model, the dispersion and fall-out of pollutants released from the tall stacks of a plant, usually a coal-fired or oil-fired power plant. The investigated area has covered a 10-15 km radius around the source. The task assigned to ALP systems was to integrate the existing monitoring network in order to gain as detailed a picture as possible of pollutant fall-out.

For a correct location of ALPs as well as for the choice of the pathlength, some data are essential: (a) the site orography and climatology; (b) the stack heights and their emission; (c) the prevailing wind. From these data it is possible to deduce: (d) the sectors around the source on which the fall-out is expected; (e) the distance at which the maximum fall-out will occur; and (f) the corresponding plume width at that distance. For an evaluation of the last two points, the classical Pasquill-Gifford model can be used, the minimum distance being $2\,\sigma_y$.

In practice, considerable difficulties limit the location of these systems. As the two end-points of the optical path must be in direct view, we have always been obliged (except in the case of an airport) to lift the instrumentation to a considerable height above the ground. The most usual places are the tops of buildings and towers (if we can obtain the authorization!). An expensive solution is scaffolding. Another problem is the local visibility: in hazy sites such as the Po-valley, the pathlength must be shorter than desired if one wants to increase the daily hours of measurement.

Figure 3 shows the location of ALPs during the 2nd CEC-campaign held at Drax (England) in 1976. At Drax (a 2000 MW coal-fired power plant, one stack 270 m high, dominant wind direction from W-NW), fall-out was expected at 5 to 8 km. The optical paths defined by 2 Teletecs and their associated lamps were arranged from NE to SE in order to cover a large arc. A Cospec operating at a shorter range was placed close to some small sources.

During the 1979 CEC campaign at Turbigo (1360 MW oil-fired power plant, 5 stacks of different height going from 48 m to 150 m, dominant breeze from N in the morning and from S in the afternoon), the ALPs were located in the N- and in the S-sectors perpendicular to the N-S wind trajectories. According to the atmospheric stability class for that period (B or C, according to Pasquill's classification) and the wind speed, the maximum fall-out was expected from 3 to 7 km for the lower stacks and at a much greater distance for the tallest ones. In that campaign only the fall-out of the lower stacks was considered. The pathlengths were shorter than calculated because of the dominant haze during the day and of the fog during the night up to early morning (ref. 17).

In all cases ALPs have been integrated with point analysers.

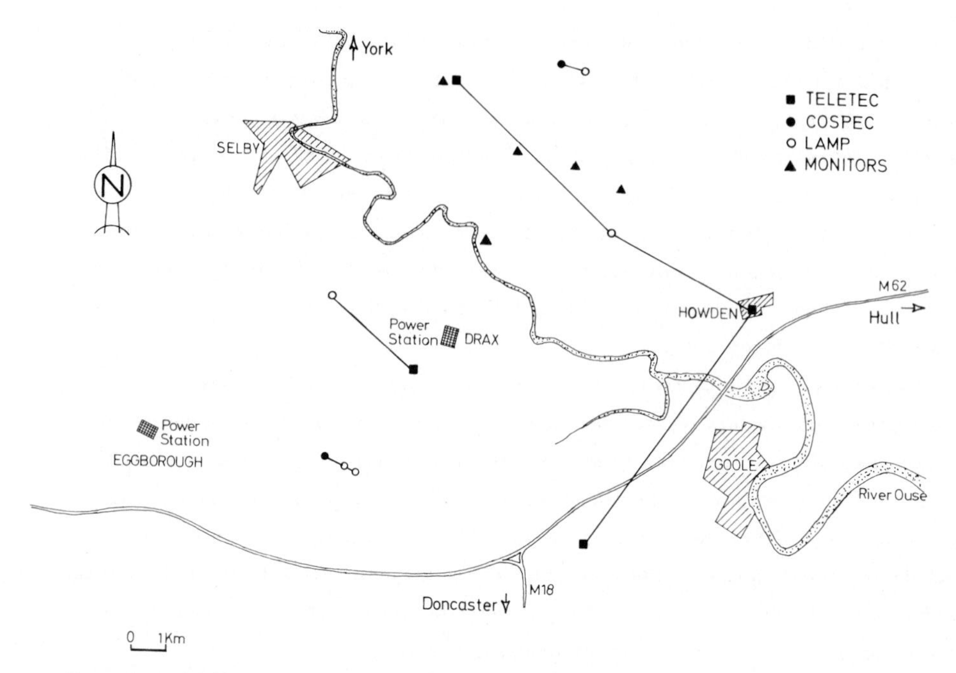

Fig. 3 - ALPs location at the CEC-campaign of Drax (1976).

3.2 Area survey - urban monitoring

Generally, the air mass circulation in urban areas is quite complex. The heat island generated by the city promotes a counterclockwise wind rotation which is superimposed upon the synoptic wind direction. The resulting complex wind field may not allow one to define any preferential wind direction over the city. The density of urban sources may be used as an index for the sectors which require particular care. The best geometry to be applied is a star configuration as sketched in Fig. 4. The sensor placed downtown is successively aligned with some lamps (or reflectors, in the case of a single-ended system) distributed along an arc around it.

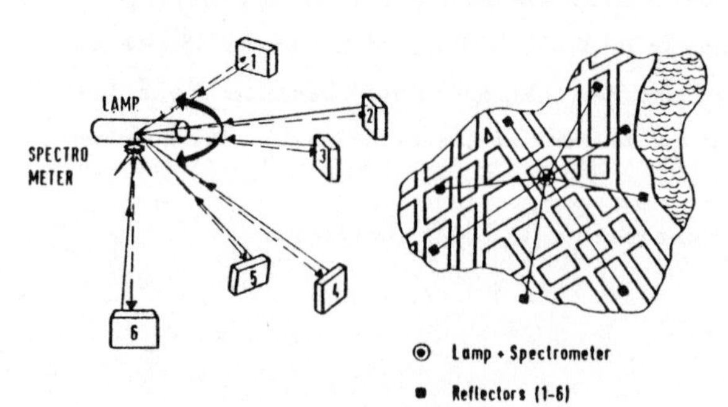

Fig. 4 - Sketch of ALP-distribution in an urban area (ref. 8).

An arrangement similar to that of Fig. 4 represents an alternative to classical urban monitoring networks. In less time, by successive optical alignments one could gather sufficiently complete information on the air quality level over the area under control. Furthermore, such surveys could be associated with other optical measurements, for instance visibility or turbidity measurement. One drawback is the fact that measurements are performed over the roofs and the data do not exactly represent the air effectively breathed by the population, but on the other hand a point monitor is a matter for wider debate. The best arrangement could consist of a combination of ALPs and point monitors located at crucial points. The pathlength should be compatible with the dimensions of the area under investigation, the local visibility and with the saturation level of the sensor. The maximum SO_2 concentration admitted by the recent EEC-Directive is 0.13 ppm.

As far as we know, only two short-term ALP experiments in urban areas have been performed, at Toronto(Canada) in 1969 by Barringer Research (930 m, ref. 18) and at Milano in 1978 by Tecneco (2500 m, ref. 7). Local and meteorological problems have limited these attempts. During the Milano experiment it was noticed that ALP detected the transport of a polluted mass over the city earlier than the monitoring network installed some metres below.

3.3 Comparison with point analysers

On the basis of experience gained from several campaigns, the combination of data obtained by ALPs and point analysers may allow us to deduce supplementary information. If the pollutant distribution is uniform along the optical path, the two sets of data coincide (ref. 19). In most cases this coherence is not found (Fig. 5) and other conclusions can be drawn.

The typical problem we faced was the fall-out distribution downwind of a stack. By a mathematical simulation based on the Pasquill-Gifford model, it is possible to calculate the ground concentration at the points where the monitors are located as well as the average ground concentration along the ALP trajectories.

Let us show some examples. Along a line perpendicular to the wind (plume) direction, 9.5 km downwind from a source, a spectrometer measures between 0 and 2 km while two point analysers are located at the extremes. Fig. 6 simulates the influence of wind speed variables from 1 to 4 $m. s^{-1}$ on the measurements at a constant atmospheric stability class (B). The simulation shows how critical is the wind speed below 2 $m. s^{-1}$ and the expected ground concentrations. Fig. 7 simulates the influence of wind direction (β) on sensors

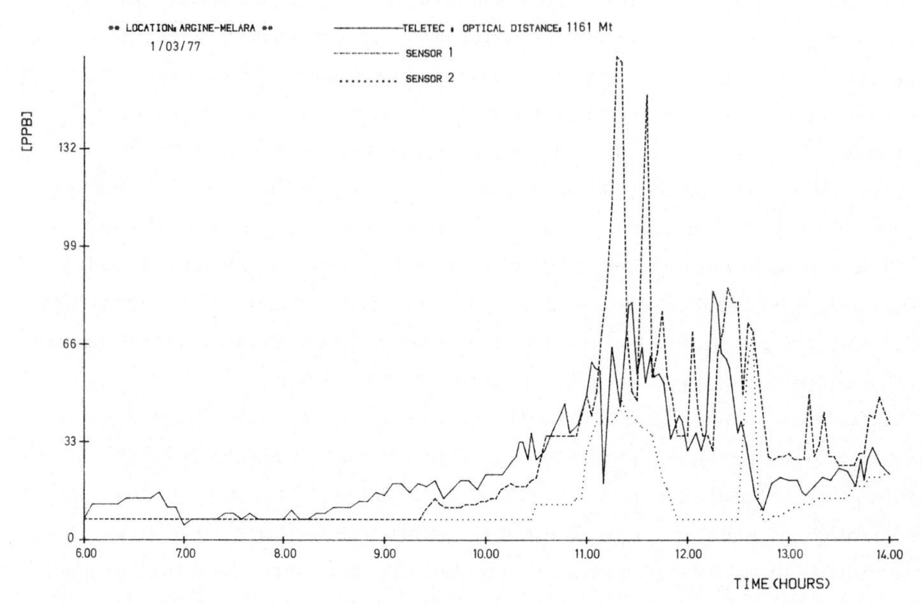

Fig. 5 - Comparison of a LP system and two monitors located at the extremes.

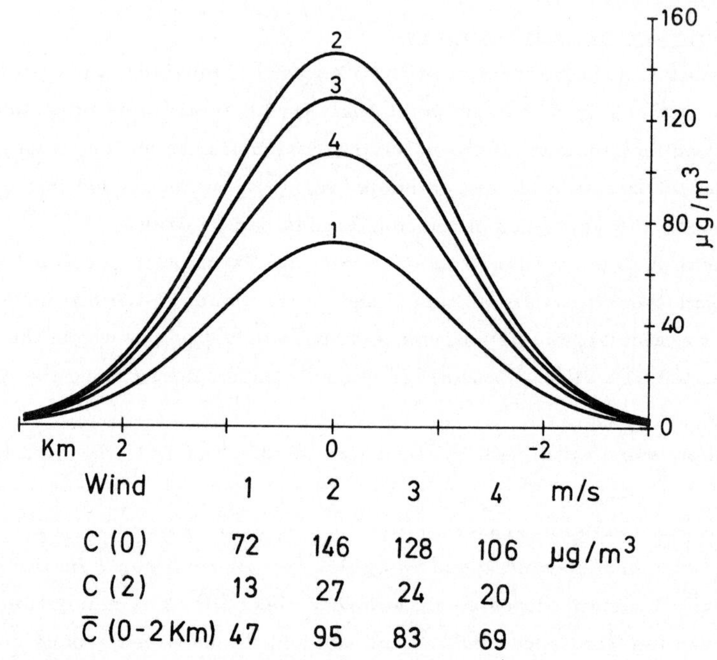

Fig. 6 - Simulation of a variation of wind speed from 1 to 4 m. s^{-1} for a B stability class; distribution of ground concentrations.

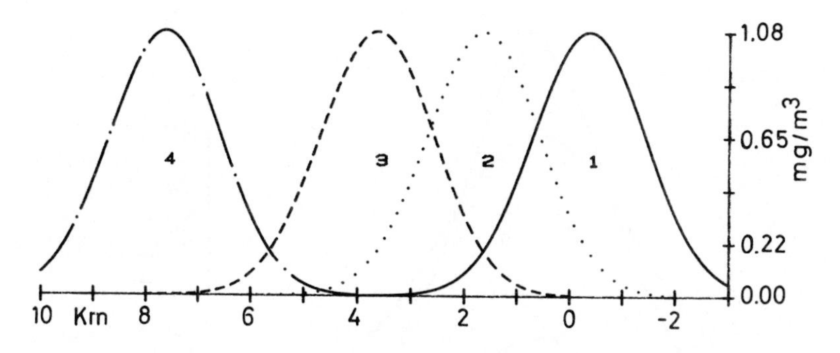

Wind direction (β^o)	$c(\mu g/m^3)$ at 0	at 5	at 7 km	$\bar{c}(\mu g/m^3)$ 0-5	0-7 km
0	1067	0	0	285	204
13	188	20	0	552	395
32	1	681	2	474	407
42	0	21	695	2	73

Fig. 7 - Simulated variation of wind direction at constant wind speed: (β = 0 (1); $\beta = 13^o$ (2); $\beta = 32^o$ (3); and $\beta = 42^o$ (4)).

distributed along a line. A spectrometer located at 0 is alternatively aligned to two lamps located at 5 km and at 7 km; the two optical paths are slightly shifted from each other. Three monitors are located at points 0, 5 and 7 km.

The influence of other parameters such as the stability class (Fig. 8), the emission rate, and the wind shear, can also be simulated. The degree of agreement between calculated and measured values represents a validity test for the assumptions previously made.

In general ALPs may be considered as a test for a regular distribution of the pollutant along the line-of-sight, if the path includes a large portion of the pollutant distribution at ground level. Furthermore, they smooth small-scale variations due to turbulent eddies, which are a handicap for a point analyser.

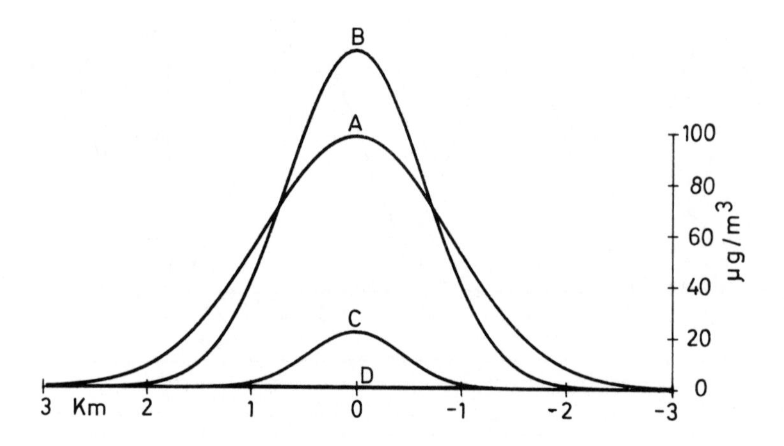

Fig. 8 - Simulation of a variable sky radiance (Pasquill stability class), at constant wind speed (2 m. s^{-1}) and direction.

4. COMBINATION OF LIDAR AND COSPEC DATA

As a further insight into the applications of correlation spectroscopy, we describe the results of its combination with another remote sensing technique, the Lidar. As mentioned above, a correlation spectrometer (Cospec, for simplicity) has a specific ability to detect a molecular component without any spatial resolution; on the other hand, a Lidar is able to detect and range particulate matter in the atmosphere. A combination of the two techniques may be a valid help for the study of localized pollutant masses such as stack plumes, in which gases and particles are present. At present the main application of Lidar is the measurement of plume rise and lateral spread from tall stacks. With some sophistication two-dimensional particle distributions can be obtained. This information can be combined with the gas density distribution within the plume, as given by a Cospec.

For this investigation the Cospec operates in the passive configuration (see Fig. 1). It is located close to a Lidar and scans the plume synchronously at several downwind points. Fig. 9 is a graphical representation of a plume sectioning and the spectral coordinates for a subsequent normalization.

When a Cospec is used for remote monitoring of a plume, diffused daylight is the source; the sampled volume is practically an unlimited cone subtended by the angular aperture of the collector (entrance iris). Hence, the measured volume is either undefined or too great to allow a spatial localization of the absorbing matter. A method for such a localization consists in obtaining from

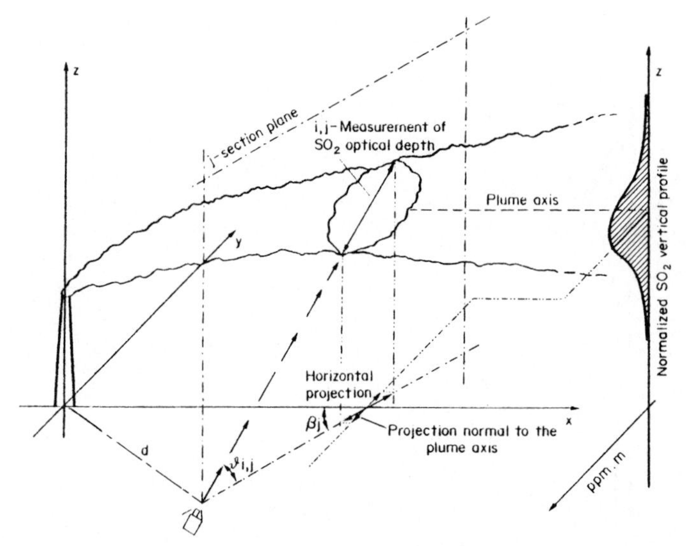

Fig. 9 - Graphical representation of spectral coordinates for one measurement plane (ref. 20).

congruent Lidar data, some knowledge of the spatial structure of the plume. According to the procedure developed by Camagni et al. (refs. 21, 22), the unknown centres of gravity of the absorbing matter are identified with the known centres of gravity of particles along the corresponding traces. This procedure allows us to compare the relative behaviour of gases and particles in a plume.

Two examples will illustrate the results which can be obtained. Fig. 10 shows a comparison of SO_2 isopleths (= equal burden data) with a qualitative view of particle density, schematically indicated by the shaded area which is defined by the projection of arbitrary isoturbidity contours. From the figure two observations can be inferred: (a) the main axes of the single plumes, as well as the axis of the composite plume are roughly the same for the two species; (b) there are remarkable differences in the lateral spread and local shearing, indicating perhaps a difference of diffusion regimes.

Further insight into the situation is obtained if one plots the total burden of the two species vs distance from the stack, making a weighted summation of the distinct constant-burden intervals so defined for each section. Fig. 11 shows the results for a vertically rising plume. For this case, a light but continuous decrease of SO_2 burden is observed in contrast to an increase in smoke particle concentration.

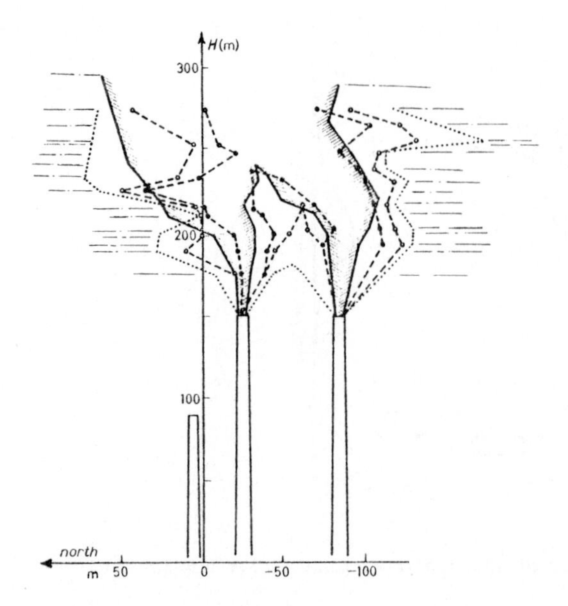

Fig. 10 - Comparison of SO_2 isopleths (hatched lines) with an arbitrary iso-turbidity profile (continuous line). Emission source: L2/L3 stacks of Turbigo power plant. Horizontal lines indicate the projections of Lidar and Cospec transverses. ... 0 ppm x m, o--o 100 ppm x m, •--• 400 ppm x m (ref. 21).

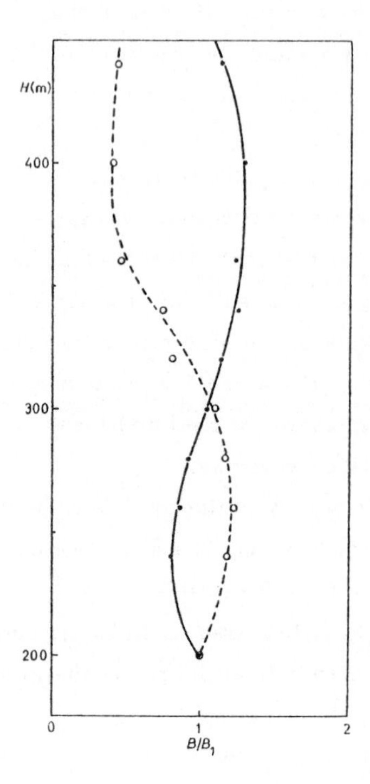

Fig. 11 - Integrated burdens of SO_2 (hatched line) and particles (solid line) over horizontal sections, as a function of section height above the ground. For each component, B/B_1 is the normalized ratio of the local burden to the burden measured in the lower section (ref. 21).

6. POLLUTANT TRANSPORT OVER AN EXTENDED AREA

A very recent use of research on correlation spectroscopy is the evaluation of pollution mass-flow over extended areas, urban or industrial, and in particular of pollution coming from surrounding regions as well as of pollution emitted from the area concerned.

The evaluation of pollutant emission from a single source is made by repeated surveys performed by a mobile Cospec (refs. 1, 23): the sensor is oriented vertically and measured the gas burden while the van in which it is installed moves perpendicular to the plume axis; the mass flow is given by the burden integrated over the plume section multiplied by the appropriate wind speed (ref. 23).

Very recent investigations are oriented towards the evaluation of the pollutant mass transported over extended areas, by performing surveys all around the area under investigation. Studies of this type concern the Milano urban area (ref. 24) and the Ghent industrial area (5th CEC campaign, ref. 25). On board data on time and van position are stored together with the measured data on a tape via a computer system; the processed data are later plotted on the map (Fig. 12). By introducing the synoptic wind field, input and output mass flows can be distinguished, while the area emission can be deduced by difference.

While avoiding a long discussion on experimental and data processing problems, it seems possible, from surveys performed in days typical from the meteorological point of view, to deduce information such as:

. input and output mass flow for different meteorological situations;

. the location and emission of some sources;

. the mixing height, from combined burden and ground concentration data.

In our study on Milano, the measured burden profile has been successfully compared with a profile simulated by an analytical model which made use of an emission inventory of urban sources as well as of industrial sources up to 50 km from the city (ref. 24). In the Ghent study (ref. 25), a multi-source area, accurate evaluations of emission from some single sources were obtained.

In our opinion, the applications here described (active long-path, combination Lidar/Cospec, pollutant transport) give ample evidence of the potentialities and of the wealth of new problems stemming from a remote sensor such as Cospec as well as from combined techniques, in the sense of contributing to a better understanding of atmospheric dispersion and transport mechanisms.

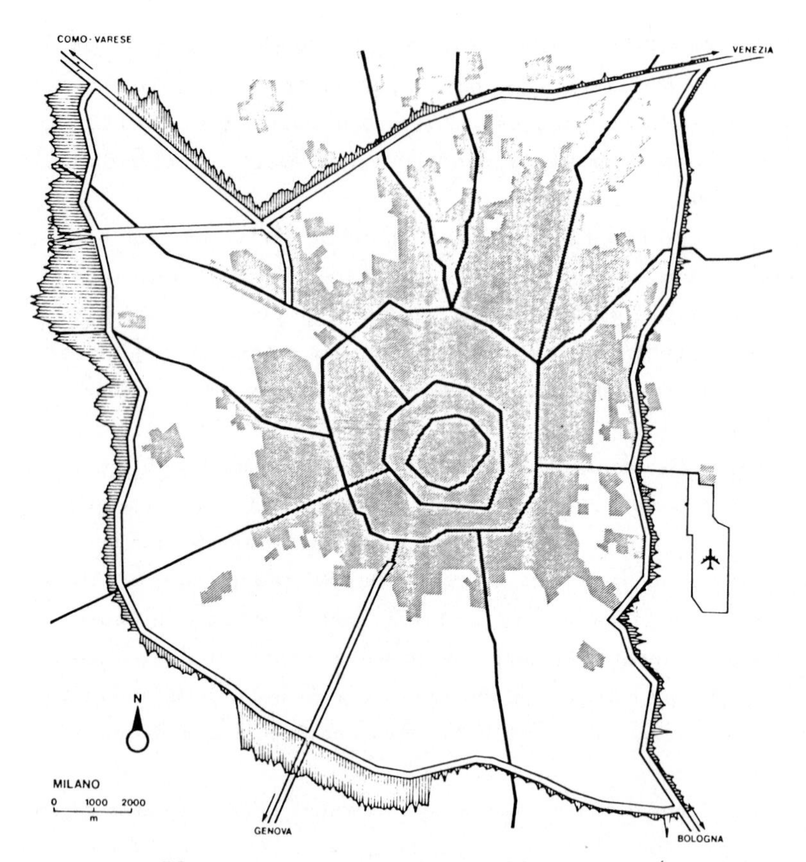

Fig. 12 - SO_2 burden measured around Milano area (18.02.81).

ACKNOWLEDGEMENTS

The author is indebted to "Atmospheric Environment" for the permission of reproducing some figures.

REFERENCES

1 M. M. Millan, lectures at this Course.
2 U. Bonafé, G. Cesari, G. Giovanelli, T. Tirabassi and O. Vittori, Atmos. Environ., 10 (1976) pp. 469-474.
3 F. Evangelisti, G. Giovanelli, G. Orsi, T. Tirabassi and O. Vittori, Atmos. Environ., 12 (1978) pp. 1125-1131.
4 A. C. Stern, Water, Air and Soil Pollution, 3 (1974) pp. 413-419.
5 W. A. McClenny, Int. Conf. Environ. Sensing and Assessment, 2 (1976) IEEE, N. Y.
6 D. C. Wolfe and R. L. Byer, Appl. Opt., 21 (1982) pp. 1165-1177.
7 G. Giovanelli, personal communication.
8 S. Sandroni in Atmospheric Planetary Boundary Layer Physics, A. Longhetto, Ed. (1980) Elsevier, Amsterdam, pp. 401-410.

9 A. Mayer, J. Comera, H. Charpantier and C. Jaussaud, Appl. Opt., 17 (1978) pp. 391-393.
10 D. J. Brassington, lecture at this Course.
11 K. W. Rothe, U. Brinkmann and H. Walther, Appl. Phys., 3 (1974) pp. 115-119.
12 H. Inaba, T. Kobayasi, M. Hirama and M. Harma, Electron. Lett., 15 (1979) pp. 749-751; I. Kobayasi, M. Hirama and H. Inaba, Appl. Opt., 20 (1981) pp. 3279-3280.
13 D. T. Williams and B. L. Kolitz, Appl. Opt., 7 (1968) pp. 607-616.
14 C. Tomasi, G. Cesari, F. Evangelisti, G. Giovanelli and T. Tirabassi, Atti Seminario Inquinamento Fotochimico, Roma, 1979, paper CNR-AC/3/146.
15 G. Restelli, lecture at this Course.
16 S. Sandroni and C. Cerutti, EUR 5427 (1976).
17 A. Longhetto, P. Guillot, D. Anfossi, P. Bacci, G. Elisei, G. Frego, S. Sandroni and R. Varey, Nuovo Cimento, 5C (1982) pp. 299-331.
18 Barringer Research Ltd., Contract DHEW-PH 2268-44 (1969) available from US Clearinghouse.
19 P. Guillot, G. Bonometti, H. Hasenjäger, A. van der Meulen, P. M. Hamilton, R. Haulet, J. Laurent, S. Sandroni, C. Cerutti, G. Giovanelli, T. Tirabassi, O. Vittori and P. Piccinini, Atmos. Environ., 13 (1979) pp. 895-917.
20 G. Giovanelli, T. Tirabassi and S. Sandroni, Atmos. Environ., 13 (1979) pp. 1311-1318.
21 P. Camagni, E. De Blust, C. Koechler, A. Pedrini, M. de Groot and S. Sandroni, Nuovo Cimento, 4C (1981) pp. 359-371.
22 P. Camagni, E. De Blust, C. Koechler, A. Pedrini, M. de Groot and S. Sandroni, 2nd Symp. on Physico-chemical Behaviour of Atmospheric Pollutants, Varese, Sept. 1981.
23 R. M. Hoff, M. M. Millan, APCA Journal 31 (1981) pp. 381-384.
24 S. Sandroni, M. De Groot, G. Clerici, S. Borghi, L. Santomauro, XIII NATO-ITM on Air Pollution Modelling, Ile des Ambiez (1982).
25 Final report of the 5th CEC Campaign at Ghent, June 1981 (in preparation).

Optical Remote Sensing of Air Pollution,
Lectures of a course held at the Joint Research Centre, Ispra, Italy, 12—15 April 1983,
P. Camagni and S. Sandroni (Eds). 279—299

LONG PATH MEASUREMENT OF ATMOSPHERIC TRACE GASES BY INFRARED ABSORPTION TECHNIQUES

G. RESTELLI

1. INTRODUCTION

For many years the use of infrared radiation has been proposed for the monitoring of trace gases in the atmosphere. Laser-based systems and Fourier Transform Spectrometers using a broad-band infrared source have been developed into mobile units to monitor atmospheric pollutants at ambient concentrations by long path absorption. Before entering into a detailed description of these monitors, it would be useful to list some points which should provide answers to the two questions:

. Why should one use infrared?
. How is infrared used?

Why should one use infrared?

The middle infrared, from 2 to 25 μm, is the region of the spectrum in which the fundamental vibration-rotation transitions of most gaseous molecules are located.

The spectral features, position and relative strengths of absorption lines (bands), act as a fingerprint for each molecule, ensuring specificity in atmospheric analysis and monitoring.

The line (band) intensities of many pollutants are sufficiently strong to provide adequate sensitivity to the absorption scheme.

At practical band-widths cooled photon-type infrared detectors are dark current limited and show very low values of minimum detectable power (NEP = 10^{-12} W $Hz^{-1/2}$). These detectors are then capable of revealing very small variations in absorption even where infrared radiation is of low power level; this in turn means that laser safety regulations can be met much more easily than with visible or near infrared radiation sources.

How is infrared used?

The technique is based on the observation of differential atmospheric absorption at wavelengths which correspond to significant spectral signatures of the pollutant; the infrared absorption can be used in the simple transmission scheme where a path averaged concentration of the pollutant is measured without any depth resolution. Bistatic systems are used where a radiation source and a detector share transmitting/receiving optics face one or more cooperative retroreflectors across an open-air path. Alternatively the differential absorption scattering scheme can be employed taking advantage of topographical reflectors or of atmospheric Mie scattering as a distributed reflector for range resolved measurements.

Coherent and incoherent infrared sources can be used to probe the atmosphere. The monochromaticity and the collimation of coherent sources make the use of infrared lasers the obvious choice. Infrared lasers can be divided into "fixed frequency" and "broadly tunable" lasers. According to ref. 1, a broadly tunable laser is one which is continuously tunable over 1 cm^{-1} or more and has a total tuning range of hundreds of cm^{-1}; a fixed frequency laser is one which cannot be tuned over more than a small fraction $(1 - 2 \cdot 10^{-3})$ of a cm^{-1}. Fixed frequency lasers such as molecular gas lasers may often be step tuned from one transition to another and are then indicated as line tunable lasers.

Additional considerations on laser linewidth, power levels, tunability methods, ease of use and reliability determine the selection of particular infrared lasers for the long path monitor as can be seen from the remainder of this chapter.

Photovoltaic or photoconductive infrared detectors made from different semiconductor materials cooled to the temperature of liquid nitrogen, are used. The selection of the semiconductor material is made according to the operating wavelength region to achieve best sensitivity, e.g. InSb for the 3-5 μm region, HgCdTe or PbSnTe for the 8-12 μm region. Direct detection and heterodyne detection techniques can be applied. The enhanced sensitivity of the second scheme permits in particular the application of differential absorption scattering.

The quantitative response follows from the Lambert-Beer law, using suitable calibration procedures. In order to increase the system sensitivity to levels adequate for ambient concentrations, the absorption signal is integrated over a beam path of the order of some hundred metres. At the same time, the minimum detectable pollutant concentration can be decreased by increasing the infrared beam intensity and by maximizing the absorption coefficient. As will be seen from the following, the increase in the IR source power cannot always be applied. It is more important to select the best spectral signature and to operate in resonance absorption which occurs when the laser wavelength coincides with one characteristic spectral feature. Since fortuitous coincidences between infrared laser wavelengths and sharp spectral features of pollutants are rare, the tunability of the infrared source is a parameter of extreme importance in terms of increased sensitivity, specificity and multipollutant capability.

From the above considerations it follows that, as far as detectors and spectral features are concerned, the infrared region is extremely valuable for analyses of the atmospheric spectrum aimed at the identification and monitoring of ambient pollutants. However, in spite of the many different systems proposed, or developed to a laboratory stage, use of long-path infrared absorption monitors in measuring campaigns has remained quite rare. A possible explanation for this limited success may be found in two types of problems arising on the one hand from the very nature of the atmosphere and on the other hand from the characteristics of infrared lasers.

The air, especially in the troposphere, contains huge amounts (with respect to pollutants) of water vapour, carbon dioxide, methane, etc., which give rise to strong absorptions in the middle infrared; and to a varying extent also in the far infrared. Large spectral regions are nearly blanked out except for the atmospheric windows extending approximately from 7 to 14 μm, from 4.5 to 5 μm and from 3 to 4 μm. It is thus necessary to work at wavelengths corresponding to an atmospheric window or in between the H_2O absorption lines, in those regions where sufficient separation exists.

An accurate analysis of the atmospheric transmission spectrum is then a necessary prerequisite for the identification of the suitable spectral regions and signatures for any pollutant of interest. This can best be performed using the AFGL magnetic tapes and computer programs for the generation of atmospheric transmission spectra (refs. 2-5). As a consequence of this analysis, there is considerable reduction in the number of spectral features available for the technique.

In Table 1 only the underlined signatures can be used for ambient air analysis at tropospheric level, since most of the features are, in fact, affected by severe interferences with the spectrum of atmospheric constituents. Moreover, the atmosphere gives rise to scattering and turbulence effects which primarily determine the intensity of a laser source returned to the detector from long-path transmission. Together with fluctuations in the laser emission which depend upon different instrumental effects, these problems have required sophisticated apparatus design and have considerably limited the performance observed with respect to that predicted on the basis of line intensities, available power of the laser source, beam path length and noise of the infrared detector. These calculations, in fact, point to values in the ppt range for most pollutants using μW laser sources and one km path lengths.

Table I - Location of absorption features of common atmospheric pollutants in the middle infrared spectrum. Those signatures free of spectral interferences at tropospheric level are underlined.

ABSORPTION FEATURES OF ATMOSPHERIC POLLUTANTS IN THE 2.3-15 MICRON SPECTRAL REGION

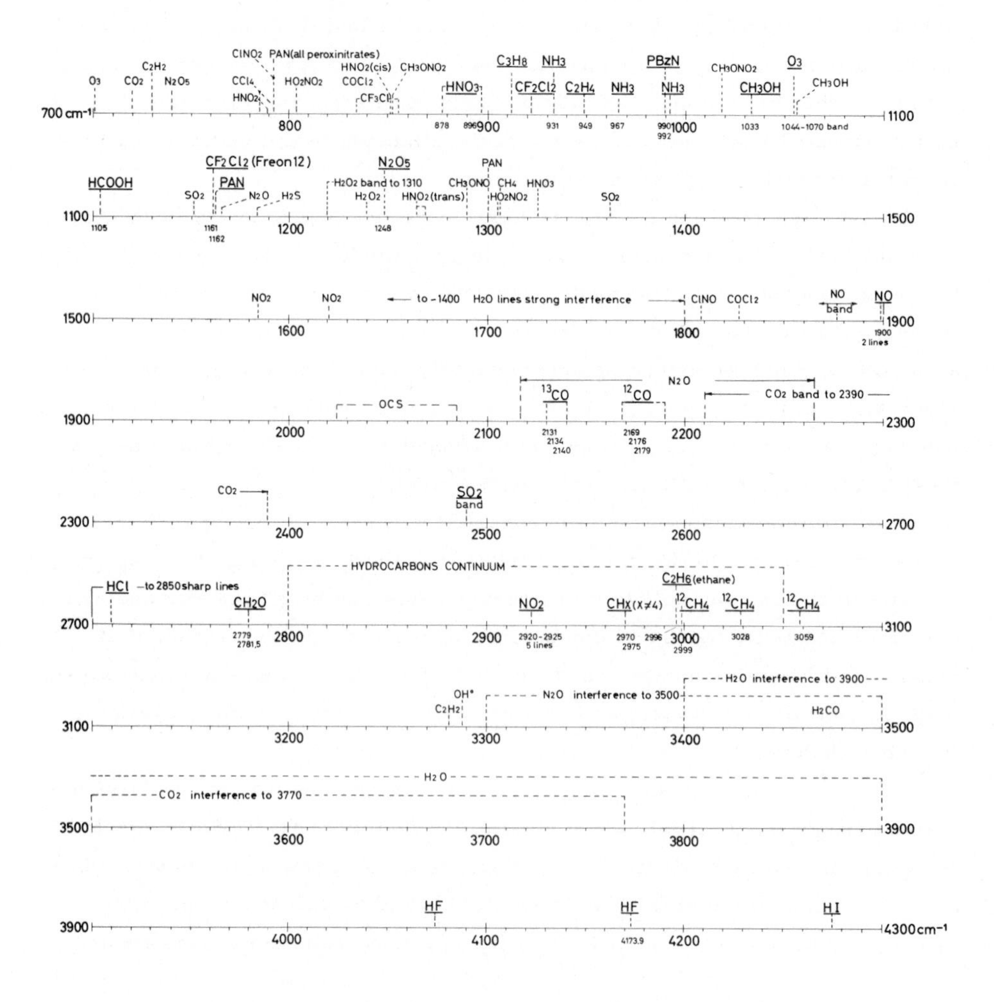

Long-path monitors (LPM) working on infrared absorption, under consideration for systematic application and practical use in measuring campaigns, are:
1) the tunable semiconductor diode laser (SDL) monitor;
2) the line tunable (molecular gas) laser monitor;
3) the long-path Fourier Transform Spectrometer (FTS).

This review will concentrate mainly on the SDL monitor. Monitoring based on infrared absorption is particularly favoured in the stratosphere by the progressively decreasing concentration of atmospheric constituents with altitude and by the modification of the spectral features due to narrowing of the absorption lines for the reduced effects of collisional broadening. These new conditions in terms of enhanced spectral features and reduced interference considerably increase the amount of information potentially available and have led to considerable success in the use of infrared absorption techniques using balloon or satellite borne FTS instruments and sunlight as infrared source.

2. TUNABLE DIODE LASER MONITOR

Lasers which can be tuned continuously over a certain wavelength range have the advantage of being operated at line centre, in the wing of a spectral line or at different absorption lines for detection of one or more pollutant gases. The characteristics of the infrared absorption scheme discussed in the preceding paragraph, together with considerations of size and portability, commercial availability and cost of the laser source, mean that the greatest attention has been paid to the semiconductor tunable diode laser (SDL) as opposed to alternative tunable infrared lasers (spinflip Raman laser, non-linear devices, high pressure gas lasers, difference frequency generators and photon mixers). The diode laser has been employed in open-path monitors and the characteristics of the systems have been investigated by different laboratories.

The heart of the monitor is a IV-VI semiconductor diode whose emission wavelength in the 2-30 μm range can be determined by an appropriate choice of alloy composition. Lasing is obtained by forward biasing (100-2000 mA) with the diode kept at a low (10-100 K) temperature. The laser emission wavelength is temperature tunable over tens of wavenumbers (up to 200 cm^{-1}) in segments and continuously over 0.3-4 cm^{-1} by changing the injection current. Discontinuities occur because of mode jumps; multimode emission is very frequently present, especially far from the threshold current and at high temperature. The emission power (cw) is generally below 1 mW. The characteristics and the advantages of SDL in infrared spectroscopy have been described in numerous articles (see e.g. refs. 6-7).

The design of a bistatic long-path carbon monoxide monitor using a tunable diode laser was pioneered at the MIT Lincoln Laboratory (ref. 8). The essential components of the laser and optical system are shown in Fig. 1. The laser, operated at cryogenic temperature, was mounted on the cold head of a closed cycle cooler (10-100 K) and the output beam collimated by an off-axis parabola, 12 cm in diameter, to a corner cube retroreflector located some hundreds of metres from the laser. The return beam was focused on a liquid nitrogen cooled InSb detector. Total beam path lengths in the range of up to 2 km could be used with less than 1 mW power in the laser output. The laser was initially designed for tunability over an interference-free absorption line of CO in the 5 μm region. The monitor was also tested for NO and for NH_3 by simply replacing the laser source. The entire system was incorporated into a mobile van and used in regional air pollution studies in 1974 and 1975 for monitoring of atmospheric carbon monoxide. In order to reach high sensitivity in open path monitoring, a fast derivative technique was developed.

Thermal fluctuations in the atmosphere give rise to fluctuations in the received signal level which destroy the capability of the system to detect small changes in the transmitted signal when the laser frequency is tuned across a spectral absorption feature. Beam defocusing and beam steering effects act in the same way. In a study (ref. 8) intended to overcome this problem, it was shown that the amplitude variations, due to atmospheric scintillation, of two

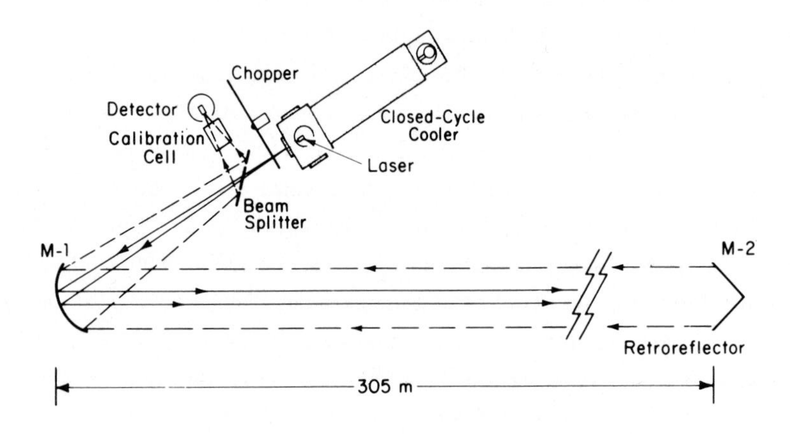

Fig. 1 - MIT SDL long-path monitor: essential components of the laser and optical system (after ref. 8).

pulsed signals transmitted over an open path, were reduced to a negligible level when an interval equal to 1 ms or less between the two pulses was used. The atmosphere appeared frozen and only long-term effects due to variations in laser emission intensity or to aerosols moving into the beam path were still present. The current tuning capabilities of the SDL made it possible to take advantage of this observation in the following operation scheme. The wavelength of the laser output was modulated at 10 kHz by superimposing a small sinusoidal current over the injection current (20 mA for a steady state current of 350-400 mA). In this way the laser frequency was tuned across the CO absorption line ($\sim$0.1 cm^{-1} full width at half maximum) and the absorption monitored by detecting the first derivative of the signal. The derivative was obtained by synchronously detecting the laser signal at the modulation frequency. The laser output was also amplitude modulated (170 Hz) by a mechanical chopper and the signal, detected by a second lock-in amplifier (direct transmission signal), was used to normalize the derivative signal. By taking the ratio of the first derivative and the direct transmission signal, the time-dependent turbulence and scattering effects were eliminated and the result was only proportional through constants to the average pollutant concentration along the beam path, as shown in the following mathematical treatment (ref. 8). The received laser signal I can be expressed as:

$$I = I_0 \exp[-K(\nu) \cdot C \cdot L] \exp[-\beta(t) \cdot L] \quad (1)$$

I_o is the transmitted laser intensity, $K(\nu)$, C and L have the usual meanings, absorption coefficient, concentration, beam pathlength, and $\beta(t)$ is an extinction coefficient which accounts for extraneous effects such as atmospheric turbulence, aerosol scattering, etc. The derivative signal $dI/d\nu$ can be obtained from Eq. (1)

$$\frac{dI}{d\nu} = -I_0 \, CL \exp[-K(\nu) \cdot C \cdot L] \exp[-\beta(t) \cdot L] \frac{d}{d\nu} K(\nu) \quad (2)$$

$\beta(t)$ can be considered wavelength-independent for the small increments used in the derivative. The ratio R between the first derivative and the direct transmission signal does not show any dependence on the time dependent turbulence and scattering parameter $\beta(t)$:

$$R = \frac{dI}{d\nu}/I = 2K(\nu_0) \cdot C \cdot L \cdot \nu'/\gamma^2 \, [1 + (\nu'/\gamma)^2]^2 \quad (3)$$

γ is the halfwidth at half maximum of the absorption line (of Lorentzian shape) and $K(\nu_o)$ is the peak absorption coefficient, ν' is the wavelength of maximum amplitude of the derivative signal and corresponds to the inflection point of the absorption line located at $\pm\ \gamma/\sqrt{3}$ from line centre. Eq. (2) assumes a point derivative. However, for atmospheric pressure broadened lines, a small finite variation in wavelength must be considered. A better approximation (ref. 9) to this situation is given in Eq. (4).

$$R = \frac{\Delta I}{\Delta \nu}/\bar{I} = [2/(\nu_2 - \nu_1)] \tanh \{C L [K(\nu_1) - K(\nu_2)]/2\} \quad (4)$$

ΔI is the difference in transmitted energy at wavelength ν_1 and ν_2 and $\bar{I} = (I_1 + I_2)/2$. Eq. (4) assumes a laser emission intensity independent of wavelength within the limits ν_1 and ν_2. It will be seen in the following that this assumption is a matter of major concern with regard to the ultimate sensitivity obtainable.

From Eq. (4) also follows the important observation that, if the ratio R must be a linear function of the concentration, this last parameter must fulfill the condition: $C \ll 2\sqrt{3}/L\ [K(\nu_1) - K(\nu_2)]$.

Calibration of the LPM for quantitative measurements was achieved by inserting gas cells and placing a retroreflector in front of the system for zeroing the open path. The cell contained a known concentration of CO in nitrogen at 760 Torr (3.8 Torr in a 10 cm cell are equivalent to 500 ppb in 1 km path). A minimum detectable concentration of 5 ppb of CO and an accuracy of $\pm$ 5% of the nominal reading were estimated for 600 m total beam path, using an integration time of 1 second (ref. 10).

Validation of the LPM by comparison with conventional point sampling instruments was performed during measurement campaigns. In the case of fairly uniform and static pollutant conditions, good agreement was obtained for CO between path integrated results, point readings and a gas filter correlation instrument.

The possible sources of error in the response of the system are discussed in detail in refs. 9-10. They are:

1) multimode emission of the laser;
2) shifts in the laser emission wavelength;
3) interference by nearby absorbing lines due to other species;
4) ambient temperature effects on the absorption line;
5) inaccuracies in the calibration (non-linearities);
6) variation of the laser emission intensity over the tuning range.

In order to overcome the problems generated by the occurrence of multi-mode emission in commercial diode lasers, a monochromator was inserted as mode selector between the laser source and the collimating optics in the test system of ref. 11. This modification, together with the use of large diameter optics, resulted in a system which was difficult to transport and had to be critically aligned.

The optical system consisted of Cassegrain and Newtonian telescopes (7 cm and 25 cm in diameter, respectively) in a Coudé mount permitting a wide mobility of the telescope without moving the laser source. The monitor was operated over a total beam path of 400 m, using a cooperative retroreflector (30 cm side corner-cube). The laser wavelength was tuned on the P(4) line of CO. Fig. 2 shows the complete system layout. Instabilities attributed to wavelength fluctuations of the laser emission originated by fluctuations in the diode heat sink temperature were observed to limit the accuracy and the sensitivity of the system ($\pm$ 40 ppb with 0.3 s integration time).

The problem of establishing the zero system and the linearity of the system response as a function of the pollutant concentration are extensively discussed in ref. 9. Also in the case of single mode emission of the diode laser (as in the case of ref. 8), any output power variation of the diode over the tuning range appears as a zero offset.

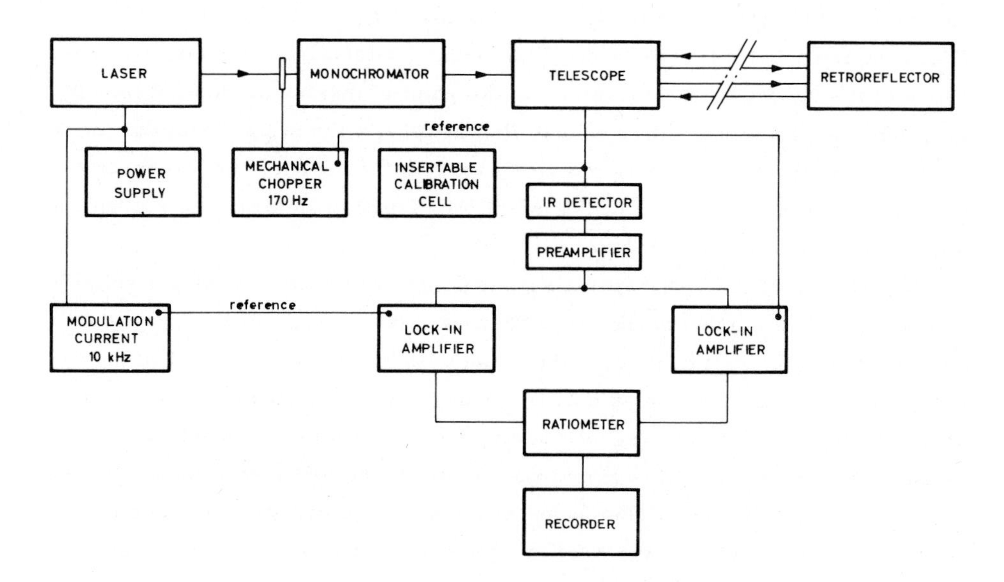

Fig. 2 - JRC-Ispra SDL monitor; system layout (after ref. 11).

As previously indicated, multimode emission from the laser is the most common situation, and if no mode selection is provided, the linearity of the system is maintained only for $(K(\nu) \cdot CL)$ values close to zero. The problem has been considered using computer simulation for a typical situation of two simultaneous laser modes. The overall analysis shows that the situation is a complex one with the following causes of error:

a) the ratio of the intensities of the two modes during a derivative scan is not constant;
b) the two modes are not co-linear (no equivalence between in-house calibration and long-path measurement);
c) the intensity of one mode may be attenuated to a level comparable to the detector noise (saturation effect) so that only the remaining mode contributes to the signal.

As previously indicated, even in the case of single mode emission, other effective sources of error derive from laser intensity fluctuations and from drifts in the laser emission wavelength. If the output power of the laser shows a significant variation during the derivative scan, the effect appears as a zero offset in the calibration. On the other hand, increased uncertainty and reduced sensitivity result from drifts and fluctuations in the laser emission wavelength. A shift in the laser wavelength (ν' in the mathematical formulation of Eq. (3)) may occur during operation due to a change in diode temperature. The exchange of radiation between the diode and the vacuum of the cryocompressor, in turn affected by ambient temperature changes, also modifies the diode operating temperature. Even a small frequency detuning is very effective because of the operation of the system on the point of maximum slope of the absorption line, and results in spurious fluctuations in the signal level observed.

An improved long-path SDL monitor based on the design of ref. 3 but intended to overcome the main problems of wavelength instability and multimode emission, is described in ref. 12.

Stabilization of the laser operation wavelength was obtained using a suitable feedback loop. A portion of the laser beam was taken from the beam splitter and sent through a cell containing the same pollutant to be measured in the atmosphere. The signal modulated at frequency f was detected by a lock-in amplifier at 2f to approach the second derivative: the output was used as the feedback error control. This signal was fed to the laser current supply which was driven to nullify the 2f component of the signal (a variation in diode temperature can be counterbalanced by an opposite variation in the injection current). This control loop made it possible to change the ambient temperature from +18° to +37°C without appreciable variation in the system response (<1% loss in system sensitivity). The modulation frequency was 1 kHz, and the amplitude of the modulation current corresponded to approximately one half-width at half maximum of the absorption line ($\sim 0.06\ cm^{-1}$); an optimum time constant of 10 s was empirically determined.

Mode selection was achieved using a Fabry-Perot étalon. This solution has some advantages over the monochromator of ref. 11. The characteristics of the étalon, in terms of requirements and performance, are described in ref. 12. In addition to the effective mode rejection, the FP étalon can replace the mechanical chopper for slow amplitude modulation since the driver of the étalon can be used to switch the transmission band on and off the laser frequency.

The system was tested at EPA-RTP over a 400 m path for carbon monoxide monitoring. The full scale sensitivity was set at 4.6 ppm and the system operated for 6 h runs. The drift in sensitivity over the whole period was 1.4% of full scale and the average peak to peak noise equal to 60 ppb. Fig. 3 shows a schematic layout of the system which, for the collimating/receiving optics, does not differ appreciably from that of Fig. 1. The increase in complexity of the apparatus with respect to previous SDL monitors is obvious.

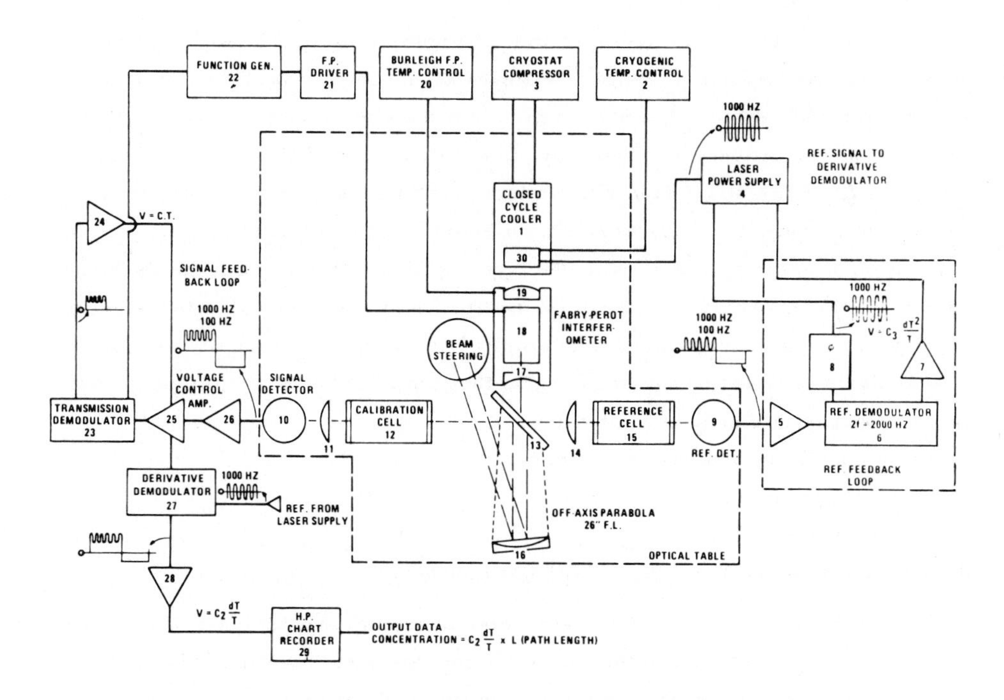

Fig. 3 - EPA long-path laser monitor block diagram (after ref. 12).

A diode laser system for "in situ" automatic monitoring of SO_2 along a 60 m path is described in refs. 13-14. The SDL was located in a dewar cooled to liquid nitrogen temperature and operated in pulse mode. The laser wavelength was set between 1127.45 and 1128.05 cm^{-1}. In this spectral region the SO_2 still preserves characteristic absorption signatures in spite of the fact that the single absorption lines are merged into one another by pressure broadening. The laser beam was collimated to a corner cube retroreflector through a beam splitter and returned to a PbSnTe detector. A scanning mirror regularly directed the laser beam to a calibration cell containing a known amount of SO_2 and to a short path to measure the laser output intensity. A minicomputer controlled the scanning mirror which directed a certain number of laser pulses through each of the three optical paths. The average concentration of SO_2 along the laser beam path was calculated using correlation procedures. A pulse repetition rate of 100 Hz was used to complete the measurement including computation, in 20 seconds. The detector noise and the non-perfect reproducibility of the laser pulse would set an ultimate limit of 10 ppb on the minimum detectable concentration of SO_2 (120 m total beam path length). However, for "in field" operation, laser beam wandering weakened the system performance by an order of magnitude, giving a lowest detectable limit for the SO_2 concentration of 100 ppb (ref. 14).

In the previously described systems, the sensitivity of the long-path monitors appears to be limited by their inability to measure a variation of 0.1% in the transmitted signal. This value is many orders of magnitude greater than the limit set by the infrared detector, also in the case of the limited power available from the SDL ($\leqslant$1 mW). An increase in sensitivity leads to an increase in the number of atmospheric pollutants that can be monitored. The optimization of some parameters that might improve this situation, is discussed in ref. 15; a sensitivity of about 0.1% has proved to be achievable by selecting a suitable modulation technique and employing a new optical configuration. The analysis is based on the assumption that the sensitivity is limited by four major sources of signal fluctuations:

a) noise due to atmospheric turbulence;
b) laser source noise;
c) optical feedback noise;
d) changes in the transmittance of the atmosphere.

Extended investigation of the laser noise spectrum and analysis of the signal transmission have led to the following conclusions. The laser should be operated near to the threshold current and the modulation frequency set between 4 and 10 kHz. A further decrease of the laser noise can be achieved by a suitable design of the diode configuration to completely eliminate the vibrational noise observed at 10 kHz. The reflecting-receiving optics must use large diameter retroreflector-telescope systems. The use of a cats' eye retroreflector with a beam/divergence of 0.15 m rad and a field of view of 0.5 degrees together with a 30 cm diameter primary mirror is suggested for the receiving telescope. In general, a large retroreflector allows, by aperture averaging (ref. 16), a greater reduction of the noise originated by atmospheric turbulence. A reduction of the effect of atmospheric turbulence is also obtainable using procedures based on the correlation existing between two beams of the laser emission having the same wavelengths and travelling over the same volume of air (ref. 17). This was demonstrated by splitting the retroreflected beam into two beams, detected individually and ratioed. The ratioed signal showed a definitely better signal to noise ratio because the noise in the two channels was correlated and thus partially cancelled out in the ratio.

Conventional analog techniques can be replaced by digital ratioing techniques with a significant improvement in amplitude resolution of the ratioing system and then in the signal-to-noise ratio. A signal-to-noise ratio of 1700 was achieved for a beam-path length of 40 m in the laboratory, using a 12 bit A/D converter. Fluctuations in the diode laser emission wavelength could be solved by locking it to the narrow absorption line of the gas kept at low pressure in a cell using the second derivative signal, as previously described.

The use of the second derivative as monitoring signal is suggested in ref. 18. This operation mode should be preferred to the use of the first derivative which is more sensitive to wing absorptions and thus more affected by errors in the case where spectral interference with atmospheric constituents or other pollutants occurs. The sensitivity of the technique expressed as the ratio of the second derivative signal to the direct transmission signal increases with the amplitude of the wavelength modulation signal. However, at the same time, the deviation from linearity of the calibration curve above a given value of absorbance ($K(\nu)\cdot CL$) also increases. Fortunately, by decreasing the modulation amplitude, the effect of reducing the linearity deviation is more pronounced than the decrease in the signal level. By using a modulation of one half the halfwidth of the line, a small deviation ($<10\%$ from linearity) occurs up to a ($K(\nu)\cdot CL$) value equal to 3 (which means, for CO lines and 1000 m total beam path length, an average concentration of up to about 6 ppm). The first derivative of the signal is used for locking the laser emission wavelength. The monitor made use of a monochromator for mode selection, a mechanical chopper and for the optics, of a Newtonian telescope together with a corner cube retroreflector. The system was operated in the field for carbon monoxide monitoring with excellent long term stability and system linearity.

A substantial improvement in sensitivity resulting from a detailed analysis of the limiting noise mechanisms followed by a novel technique of harmonic signal detection has been presented recently in ref. 19. A sensitivity of 0.01% has been shown to be obtainable as compared to the value of 0.1% of ref. 15 and of 0.7% of ref. 8. The basic observation is the fact that the wavelength modulation used to scan atmospheric pressure broadened lines, also produces an intensity modulation which cannot be considered reproducible, as was previously assumed (refs. 19-20). In the derivative technique, the signal is the sum of the harmonic contribution from the wavelength and from the laser intensity modulation. A mandatory requisite for accurate/sensitive measurements is the reproducibility of the harmonic of the intensity modulation ("background signal"). Air turbulence and optical feedback into the laser are the main causes of a time-varying background signal. To minimize the background signal with respect to the signal resulting from laser wavelength modulation due to absorption, the 6th or the 8th harmonic was used (ref. 19) instead of the 1st or 2nd derivative previously employed. The signal decreased (from 1 to 0.24 and to 0.18 going from direct absorption to the 6th and 8th harmonic) but the background signal was reduced correspondingly from 1% of the 2nd harmonic to less than 10^{-2}%. Normalisation to the average transmitted power was obtained by ratioing in an analog circuit, the demodulated harmonic signal (6th harmonic at 800 Hz) with the demodulated 1st harmonic background signal which, if obtained using an aperture to select a uniform portion of the laser beam, is linearly proportional to the power. The system was thus practically insensitive to turbulence, misalignment and other sources of fluctuation in the received signal. The SDL monitor described in ref. 19 contained coaxial collimating and receiving optics. However, it was observed that the feedback into the laser cavity of a small fraction of the laser intensity considerably increased the background signal. To avoid this effect, the use of non-coaxial transmitting/receiving optics is suggested.

The system with a single mode diode laser was tested at low $(K(\nu)\cdot CL)$ values using the N_2O line at 1142 cm^{-1} detected at the 6th harmonic. With a total beam path of 240 m the noise level was found to be equivalent to 0.01-0.02%. The system appears capable of detecting isolated absorption lines in the atmosphere with a sensitivity of 10^{-6} m^{-1} for a measurement time of a few seconds.

Table II (ref. 19) shows a list of molecules of atmospheric interest with their respective limits of detection calculated according to the above result.

TABLE II - Detection limits of gases of atmospheric interest calculated for an SDL monitor of sensitivity 10^{-6} m^{-1} (ref. 19)

Molecule	Spectral region (cm^{-1})	Spectral feature	Estimated sensitivity (ppb)
N_2O	1145	line	20
	2240	line	0.1*
CH_4	1304	line	2
CO	2111	line	0.25
NH_3	1103	5 overlapped lines	0.5
	1074	line	0.75
HNO_3	1334	bandstructure	0.35*
NO	1890	doublet	0.5
NO_2	1610	bandstructure	0.5*
SO_2	1360	bandstructure	3
	1140	bandstructure	12

* In these spectral regions a sensitivity of 10^{-6} m^{-1} may not be attainable due to severe attenuation of the laser beam by the atmosphere.

As concluding remarks the following points can be highlighted. Resonance absorption using a tunable laser source demonstrates positive advantages in terms of specificity and sensitivity. The semiconductor diode laser (SDL) which offers the least cost and complexity of operation of the various infrared tunable lasers, has been shown capable of functioning in a long path atmospheric monitor. A mobile system has been used extensively for ambient monitoring of CO by EPA in regional air pollution studies with a full-scale sensitivity of 4.6 ppm, 1.4% of full-scale drift over 6-hour runs and 60 ppm average peak-to-peak noise.

Because of the limited wavelength tuning range, the SDL monitor is usually dedicated to a single pollutant. However, other molecules could be measured simply by changing the laser source: NO, N_2O, HCl, NH_3, HNO_3 are the most interesting candidates for this long path monitor but still wait for "in field" validation experiments. In recent years, several laboratories have analysed in detail the reasons why, with the ultimate limits set by the infrared detector, the sensitivities obtained remain relatively low. As a consequence, significant improvements have been introduced to the original design and a sensitivity increased by two orders of magnitude with respect to field operated monitors, has been achieved in laboratory studies.

Some problems, such as SDL frequency instabilities and atmospheric turbulence effects, have been solved. On the other hand, the need for a single mode emission laser and the response of the instrument over a limited range of pollutant concentrations are difficulties which have been only partially overcome. With regard to this second point, the monitor is best suited to the detection of very low pollutant concentrations such as those found in ambient air studies.

3. LINE TUNABLE LASER MONITOR

Molecular gas lasers can be used as the source of a large number of discretely tunable monochromatic wavelengths located in different regions of the middle infrared spectrum. The use of these systems as monitors is based on the occurrence of fortuitous overlaps between emission wavelengths of a molecular gas laser and the absorption features of the gaseous species of interest (ref. 21). The pollutant is measured by transmitting a laser beam with two wavelengths, one fitting as close as possible an absorption feature of the pollutant, the other in a region of comparatively little or no attenuation. Many coincidences between the emission wavelength of a molecular gas laser and the absorption feature of pollutants have been found especially using CO_2 lasers (see e.g. refs. 22-24). CO, DF and isotopic CO_2 lasers have also been suggested and used to extend the range of the technique. In refs. 23, 25-38 the reader can find descriptions of the systems, discussions on the selection of lasers and of the best pair of wavelengths for a wide range of pollutants, together with results of field measurements.

The two wavelengths are emitted by using two different lasers, by coupling two laser cavities with the same medium, or by an alternative change of the laser cavity. Pulsed or cw operated lasers are used. In the case of sequential emission of the two wavelengths, the problem of atmospheric scintillation can be partially overcome by averaging a certain number of laser pulse pairs.

The path averaged pollutant concentration C is related to the differential absorption by the equation:

$$C = \frac{1}{[K(\nu_1) - K(\nu_2)] \cdot L} \ln \frac{\epsilon_1 P_{t,1} P_{r,2}}{\epsilon_2 P_{t,2} P_{r,1}}$$

$K(\nu_1)$ and $K(\nu_2)$ are the absorption coefficients at ν_1 (in coincidence with the pollutant absorption feature) and at ν_2 (off coincidence); $P_{t,1}$ and $P_{r,1}$ are the received and transmitted intensities at ν_1, and $P_{t,2}$ and $P_{r,2}$ at ν_2; ϵ_1 and ϵ_2 are the total efficiencies of the optical system at wavelengths ν_1, ν_2.

The considerable power available from molecular gas lasers (~1-3 Watt cw for a single line) together with direct detection or heterodyne detection have been successfully employed in the development of bistatic and monostatic ground-based and monostatic airborne monitors, using cooperative retro-reflectors or topographic diffusely scattering targets such as foliage, wood or the earth's surface.

Heterodyne detection is performed using a small portion of the laser output with a suitable wavelength offset as local oscillator. Beam path lengths up to some km have been used to detect pollutants such as O_3, C_2H_4, NO, N_2O, CH_4, HCl, etc., with sensitivities in the ppb range. Various studies have demonstrated that many less common pollutants such as those present in industrial environments (e.g. vinyl chloride, acrolein, styrene, hydrazine, etc.) could be monitored with path-averaged detection sensitivities in the 10-100 ppb range (refs. 23, 37).

An analysis of the problems encountered when using this monitor, zeroing and calibration, is given in ref. 9 for a bistatic sequential wavelength CO_2 laser used to monitor ambient ozone. Atmospheric turbulence effects, signal emission normalization, spatial distribution of the two wavelengths and interferences with the atmospheric spectrum are the main sources of system uncertainty and loss in sensitivity.

An airborne laser absorption spectrometer for remote measurement of tropospheric ozone is described in refs. 27, 34.

The instrument uses two CO_2 grating tunable lasers and heterodyne detection of the radiation backscattered off the earth's surface. Ozone concentrations in the range 30-60 ppb were measured from 0 to ~1700 m altitude with an uncertitude of ± 20 ppb primarily determined (ref. 34) by variation in surface material differential albedo spectra (± 15 ppb) and by calibration uncertainties (± 10 ppb). The state of the art and perspectives for the development of an infrared differential absorption LIDAR are reviewed in chapter VII of the book.

4. LONG PATH FOURIER TRANSFORM SPECTROMETER

The Fourier Transform Spectrometer (FTS) was shown by Hanst (ref. 39) to be valuable for multiple pollutant monitoring of air point samples. Since then this instrument has been extensively used for studies of atmospheric chemical reactions in smog chambers and for "in situ" detection of stratospheric species in long path absorption (the sun is used as infrared source) or emission measurements.

Recently an FTS-long path system for monitoring of ambient pollutants has been set up and used at EPA (refs. 40, 41). The system consists of a commercial infrared Fourier Transform Spectrometer and auxiliary equipment installed in a van and used as a mobile instrument to obtain the spectral signatures and from them the concentrations of gaseous pollutants by long-path absorption spectroscopy. The high throughput, the resolution up to 0.06 cm^{-1} FWHM, the multiplexing and the speed of response of the FTS in comparison with other infrared spectrometers, make this instrument by far the most suitable for low level infrared signal analysis. The use of the FTS for multi-pollutant monitoring is conventional: spectra are recorded and subsequently the pollutant concentrations derived from suitable spectral signatures using calibration spectra. Calibration spectra of known optical depths are obtained for the various gases in conditions approaching the "in field" measurement to minimize instrumental effects. For many pollutants, concentrations in the 1-10 ppb range can, in principle, be detected over a 1 km path length. A 1 kw quartz iodine lamp was used for long-path measurements together with Dall-Dirkham telescopes (f/5) as light source collimating and receiving optics. The receiver telescope focused the transmitted energy on the interferometer aperture, adjustable for compatibility with the desired spectral resolution. Using a 1.44 km path propylene, C_2H_4 and SO_2 were measured at 0.5 cm^{-1} instrumental resolution. Hydrofluoric acid was also detected using the 4173.9 cm^{-1} line at the resolution of 0.125 cm^{-1}. The accuracy of long-path concentration measurements was estimated to be generally better than $\pm$ 10% (ref. 41).

5. CONCLUSION

Pollutant gases in the troposphere can be monitored by analysing long-path atmospheric transmission in the infrared.

Discrete wavelength or continously tunable IR lasers have been employed in conjunction with cooperative, natural or artificial retroreflectors, as well as broadband infrared sources with a Fourier Transform Spectrometer.

Long-path monitoring yields gas concentrations averaged over the beam-path with a sensitivity adequate for the analysis of various pollutants at ambient level. It represents a viable alternative to point monitors and is particularly useful for reactive gas monitoring and monitoring across otherwise inaccessible regions for global and regional studies of air quality.

The tunable diode laser (SDL) monitor has been the object of extensive study and has been developed into "in field" systems employed in air pollution studies

Table III

Feature	Point monitor	Long path (IR) monitor (LPM)	Comments on LPM
Type of measurement	A localized air volume is sampled; the problem of ensuring representativity can be severe	Large area monitoring can be achieved e.g. by placing many retroreflectors around a central location of the LPM.	LPM is preferable in the case of regional studies and for monitoring areas of difficult access.
Intrusivity and sample integrity	A sampling technique is required and problems exist with sample integrity in the case of reactive pollutants.	Ensures non-intrusivity and sample integrity.	LPM is preferable for reactive gases and in cases where non-intrusivity is required.
Control of sample; sample conditioning;simulation	A control exists on the sample to be measured so that synthetic samples can be used to establish reference levels (zero and multipoint calibration). The control of the sample also permits in some cases sample conditioning such as sample conversion, etc. Simulation of monitoring conditions in the laboratory is possible.	Control of the sample is impossible, reference levels are indirectly established. No sample conditioning is possible and it is difficult to simulate atmospheric conditions in the laboratory so that field testing of the monitor is a necessity.	Use of LPM is less indicated due to difficulties in the procedures for zero setting and for calibration of the monitor.
Detection techniques	Choice available amongst many analytical techniques. Problems arise in the choice if real-time response is desired.	The technique is based only on the observation of differential absorption at one or at different wavelengths. Offers real time response.	LPM is restricted in the pollutant range by the spectroscopic characteristics of the atmosphere - advantages exist in the stratosphere.
Cost	Monitors are generally low cost, fairly easy to operate and require little maintenance.	LPM monitors (laser systems) are expensive and require especially trained personnel for operation and maintenance.	LPM is much more expensive.

principally on CO, but also on SO_2 and NO. In recent years, laboratory studies have shown the great potential of the SDL monitor, whose use may be envisaged for ambient level monitoring of other very important pollutants such as NH_3 and HNO_3.

The molecular gas laser system would appear to be the best suited for monitoring uncommon pollutants. The same consideration holds for the FTS. Ambient levels of these compounds are expected to be extremely low, however, with the single exception of high concentrations existing at fairly localized sites (e.g. industrial sites).

The development of an airborne IR Differential Absorption Lidar using molecular gas lasers as described in ref. 34 looks very attractive, but the high cost of such an instrument severely limits the development of this monitor for routine application (ref. 42).

Generally speaking, judgement on the cost effectiveness of the infrared monitors described above is favourable only in a few circumstances, as can be seen from Table III, drawn up mainly on the results of the analysis performed in ref. 9.

REFERENCES

1 E. D. Hinkley, P. T. Ku and P. L. Kelley, Techniques for detection of molecular pollutants by absorption of laser radiation, in E. D. Hinkley (Ed.) Laser Monitoring of the Atmosphere, Springer-Verlag, Berlin, 1976, pp. 237-295.

2 L. S. Rothman, Appl. Optics 20 (1981) 791-5.

3 L. S. Rothman, A. Goldman, J. R. Gillis, P. H. Tipping, L. R. Brown, G. S. Margolis, A. G. Maki and L. D. G. Young, Appl. Opt. 20 (1981) 1323-8.

4 A. Chedin, N. Husson, N. Scott, I. Gobard, I. Cohen-Hallaheh and A. Berroir, La Banque de Données GEISA, Lab. de Météorologie Dynamique du CNRS, Note interne no. 108 (1980).

5 C. Pagny, F. Cappellani and G. Restelli, Spectrogaz: Fichier de Données Spéciales et Elaboration des Spectres de Transmission Infrarouge à Partir de Rubans Magnétiques AFGL, JRC report EUR8259FR (1982).

6 E. D. Hinkley, K. W. Nill and F. A. Blum, Infrared spectroscopy with tunable lasers, in H. Walther (Ed.) Laser Spectroscopy of Atoms and Molecules, Springer-Verlag, Berlin, 1976, pp. 126-196.

7 J. Hesse and H. Preier, Lead salt laser diodes, in H. J. Queisser (Ed.) Advances in Solid State Physics, Pergamon XV, 1975, pp. 229-251.

8 R. T. Ku, E. D. Hinkley and J. O. Sample, Appl. Optics 14 (1975) 854-861.

9 W. A. McClenny and G. M. Russwurm, Atmospheric Environment, 12 (1978) 1443-1453.

10 R. T. Ku and E. D. Hinkley, Long path monitoring of atmospheric carbon monoxide, Interim report to the NSF and EPA, MIT Lincoln Laboratory, April 1976; and
R. T. Ku, Final report, MIT Lincoln Laboratory, November 1977.

11 F. Cappellani, G. Melandrone and G. Restelli, Long path infrared absorption system for gaseous pollutants monitoring of the atmosphere, in M. West (Ed.) Lasers in Chemistry. Elsevier Publ. Co., Amsterdam, 1977, pp. 61-69.

12 L. W. Chaney, D. G. Rickel, G. M. Russwurm and W. A. McClenny, Appl. Optics 18 (1979) 3004-3009.
13 E. Max and S. T. Eng, Opt. Quant. Elect. 9 (1977) 411-418.
14 E. Max and S. T. Eng, Opt. Quant. Elect. 11 (1979) 97-101.
15 R. S. Eng, A. W. Mantz and T. R. Todd, Appl. Optics 18 (1979) 3438-3442.
16 D. L. Fried, J. Opt. Soc. Am. 57 (1967) 169-171.
17 A. G. Kjelass, P. E. Nordal and A. Bjerkestrand, Appl. Optics 17 (1978) 277.
18 W. Wiesemann and W. Diehl, Appl. Optics 20 (1981) 2230-2232.
19 D. T. Cassidy and J. Reid, Appl. Optics 21 (1982) 1185-1190.
20 P. Pokrowsky and W. Hermann, GKSS Forschungszentrum Geesthacht GmbH, Report GKSS 81/E/23 (1981).
21 G. B. Jacobs and L. R. Snowman, IEEE J. Quantum Electron. Q. E-3 (1967) 603-611.
22 W. Schnell and G. Fisher, Appl. Optics 14 (1975) 2058.
23 A. Mayer, J. Comera, H. Charpentier and C. Gaussaud, Appl. Optics 17 (1978) 391-393.
24 G. L. Loper, G. R. Sasaki and M. A. Stamps, Appl. Optics 21 (1982) 1648-1653.
25 L. B. Kreuzer, N. D. Kenyon, C. K. N. Patel, Science, Vol. 177 (1972) 347-349.
26 T. Kobayasi, H. Inaba, Optical and Quantum Electronics 7 (1975) 319-327.
27 R. T. Menzies and M. S. Shumate, Appl. Optics 15 (1976) 2080-2084.
28 E. R. Murray, J. E. van der Laan and J. G. Hawley, Appl. Optics 15 (1976) 3140-3148.
29 E. R. Murray and J. E. van der Laan, Appl. Optics 17 (1978) 814-817.
30 B. Marthinsson, J. Johansson, S. T. Eng, Optical and Quantum Electronics 12 (1980) 327-334.
31 N. Menyuk, D. K. Killinger and W. E. DeFeo, Appl. Optics 19 (1980) 3282-3286.
32 K. W. Rothe, Radio Electron. Eng. 50 (1980) 567-569.
33 M. Hamza, T. Kobayasi, H. Inaba, Optical and Quantum Electronics 13 (1981) 187-192.
34 M. S. Shumate, R. T. Menzies, W. B. Grant and D. S. McDougal, Appl. Optics 20 (1981) 544-552.
35 S. Lundqvist, C. Fält, U. Persson, B. Marthinsson and S. T. Eng, Appl. Optics 20 (1981) 2534-2538.
36 C. Weitkamp, GKSS81/E/57 report (1981).
37 N. Menynk, D. K. Killinger and W. E. DeFeo, Appl. Optics 21 (1982) 2275-2286.
38 U. Persson, J. Johansson, B. Marthinsson and S. T. Eng, Appl. Optics 21 (1982) 4417-4420.
39 P. L. Haust in Advances in Environmental Science and Technoloty, J. N. Pitts and R. L. Metcalf (Eds.) J. Wiley, New York, 1971, Vol. 2, Chap. 4.
40 W. F. Herget and D. Brasher, Appl. Optics 18 (1979) 3403-3420.
41 W. Herget, Appl. Optics 21 (1982) 635-641.
42 G. E. Schweitzer, Environ. Sci. Technol. 16 (1982) 338A-246A.

Optical Remote Sensing of Air Pollution,
Lectures of a course held at the Joint Research Centre, Ispra, Italy, 12—15 April 1983,
P. Camagni and S. Sandroni (Eds). 301—327

VERTICAL MASS LOADING OF AEROSOL PARTICLES BY SUN-PHOTOMETRIC MEASUREMENTS

C. TOMASI

1 INTRODUCTION

The sun-photometer at multiple wavelengths is a reliable instrument used to take ground level measurements of direct solar irradiance at the main "window" wavelengths in the visible and near infrared range of the atmospheric attenuation spectrum (refs. 1 and 2). Following the sun-photometer technique based on the comparison between ground-level measurements and calibration values of solar irradiance at the top of the atmosphere, precise estimates of vertical atmospheric attenuation can be obtained at the various wavelengths. By separation of these estimates into the different contributions made by the atmospheric constituents, one can evaluate the particulate matter optical thickness and the turbidity parameters. From these, numerical evaluations of the vertical mass loading of aerosol particles can be made by using appropriate and realistic models of particulate matter extinction.

Generally, the sun-photometer technique is affected by several error terms which, if not properly accounted for, can cause considerable errors in the evaluation of the particulate matter optical thickness. As pointed out by Shaw (ref. 3), these errors are of three kinds: (i) instrumental errors, (ii) calibration errors, and (iii) atmospheric errors. The main instrumental characteristics and the procedures suggested by sun-photometry to minimize errors are examined in the next sections.

2 THE SUN-PHOTOMETER AND INSTRUMENTAL ERRORS

A drawing of a multiple wavelength sun-photometer is shown in Fig. 1. As can be seen, a sun-photometer substantially consists of a tube containing the following elements:

(1) a thin quartz window at the entrance of the tube to prevent dust and humidity from entering the instrument and damaging the interfercence bandpass filters or the lens.

(2) An entrance diaphragm which acts as a field stop to reduce the aber-

ration effects of the lens.

(3) A rotating mechanism which supports a series of bandpass filters and sequentially places them in the sunlight path.

(4) An achromatic quartz lens of rather large diameter and of focal length L of ten centimeter or so.

(5) A diaphragm of small diameter d located at distance L from the lens to define the instrumental field of view, whose angular diameter is equal to 2Φ with $\mathrm{tg}\,\Phi = d/2L$.

(6) A thin frosted quartz disk located behind, and glued to the focal diaphragm. This disk is needed to diffuse the monochromatic sunlight so that the whole sensible surface of the detector is uniformly illuminated.

(7) A detector which is usually a silicon solar cell or a silicon diffuse PIN photodiode characterized by good spectral sensitivity and linear responsivity in the visible and near infrared spectral range.

(8) A thermostating box containing the detector to avoid the offset voltage drift of the sensing element with temperature.

The bandpass filters must be chosen with peak wavelengths that correspond to the main atmospheric "windows", in which the absorption by water vapour and other atmospheric gases is weak. The most commonly used window wavelengths are listed in Table 1. Moreover, the transmission curves of the filters must have a halfwidth of no more than 150-200 Å, for which the measurements of direct solar irradiance at ground level can be correctly examined in terms of the Bouguer-Lambert-Beer law for monochromatic extinction.

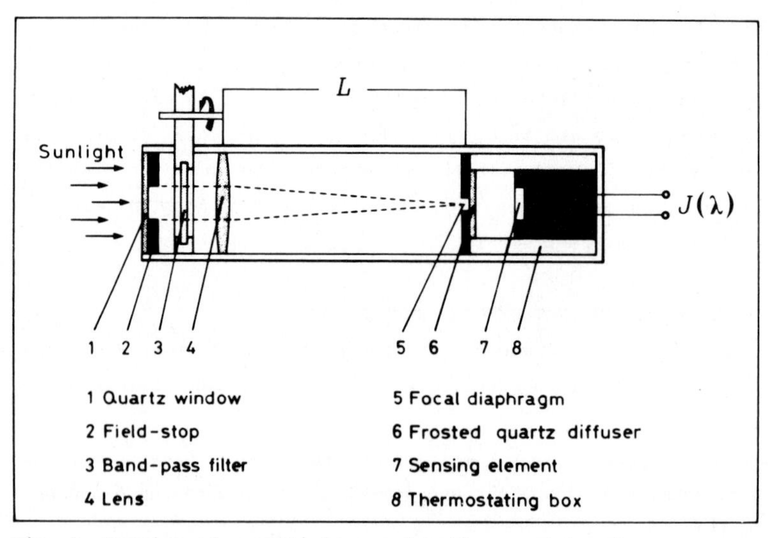

Fig. 1. Drawing of a multiple wavelength sun-photometer.

In addition to the window bandpass filters, use is commonly made of a bandpass filter with peak wavelength at about 0.94 μm. The output voltage $J(0.940\ \mu m)$ of the sun-photometer turns out to be strongly attenuated by water vapour absorption (σ band). Therefore, the ratio between $J(0.940\ \mu m)$ and the voltage $J(\lambda_r)$ taken "simultaneously" with a bandpass filter centred at the near "window" wavelength λ_r varies as a function of the water vapour content along the sun-path. If the form of this relationship curve is determined by means of calibration measurements, the precipitable water can be easily evaluated from the measurements of this hygrometric ratio, as shown by Volz (ref. 2), Tomasi and Guzzi (ref. 4), Pitts et al. (ref. 5), and Bird and Hulstrom (ref. 6).

One of the most important specifications of the sun-photometer is the angular field of view. Since the sun-photometer has the purpose of measuring the direct solar radiation flux and the angular diameter of the sun is of about 30', optimal values of 2Φ should be used of no more than 1°. In fact, when too wide an angular field of view is adopted, a considerable amount of diffuse sky light enters the instrument and is the cause of large errors. Shaw (ref. 3) calculated that an error of 1.6% is made in evaluating the particulate matter optical thickness at visible wavelengths, for a 5° diameter field of view, a solar elevation angle $h = 12°$, and normally occurring atmospheric turbidity conditions and surface albedo. From Deirmendjian's (ref. 7) calculations, we found that the errors caused by diffuse light are, for practical purposes, negligible in relatively clear atmospheres for $2\Phi < 1°\ 30'$, provided that h is larger than 10°.

Important instrumental errors are also due to detector nonlinearity, long-term changes in sensitivity, and detector dark current. These kinds of error can be considerably reduced by using high-performance detectors. Moreover, a set of standard lamps can be very useful for checking the instrumental sensitivity of the detector. The errors made in reading output voltages are generally of little importance.

3 THE CALIBRATION PROCEDURE

The output voltage $J(\lambda)$ of a sun-photometer pointed at the sun is proportional to the radiation flux entering the instrument. This quantity is composed of both attenuated direct solar radiation and diffuse sky radiation in the instrument's field of view, so that the output voltage $J(\lambda)$ may be written in the analytical form

$$J(\lambda) = R\, J_o(\lambda)\, \exp\left[-m(h)\ \tau(\lambda)\right] + \int_{\Omega} I_d\left[\lambda, \vartheta, \zeta, A(\lambda), \tau(\lambda)\right]\, d\omega \qquad (1)$$

In eq.(1), R is the correction term for the earth-sun distance at observation time (ref. 8), $J_o(\lambda)$ is the solar radiation measurement at the top of the atmosphere (m = 0) and at mean earth-sun distance (R = 1), m(h) is the atmospheric air mass relative to the vertical direction, as traversed by a solar ray at elevation angle h, Ω is the solid angle of view of the instrument, and I_d is the diffuse sky light intensity, which depends on wavelength λ, on zenith angle ϑ and azimuth angle ζ of the sun, on ground albedo $A(\lambda)$, and on the atmospheric optical thickness $\tau(\lambda)$.

As mentioned above, when the measurements are taken in relatively clear atmospheres by using sun-photometers having a small angular diameter of view and the sun is high enough above the horizon, the term of the diffuse light can be neglected in eq.(1), so that the Bouguer-Lambert-Beer law in the simple form

$$J(\lambda) = R\ J_o(\lambda)\ \exp[-\ m(h)\quad \tau(\lambda)] \tag{2}$$

can be correctly used to examine the sun-photometer measurements. The term R is inversely proportional to the square of the earth-sun distance so that it can be easily calculated on each measurement day: it varies from a maximum of 1.0335 (January, 1) to a minimum of 0.9666 (July, 1). The relative air mass m(h) can be determined from the apparent solar elevation angle h at measurement time, by making use of the following formula proposed by Kasten (ref. 9)

$$m(h) = [\sin h + 0.15\ (h + 3.885)^{-1.253}]^{-1} \tag{3}$$

or of tables such as those given by Kondratyev (ref. 10).

The determination of the quantity $J_o(\lambda)$ at each window wavelength can be made through the following different procedures: (1) application of the so-called Langley plot method, (2) use of standard lamps of known spectral irradiance, and (3) intercomparison with other calibrated sun-photometers.

3.1 Langley plot method

This method is based on the assumption that both atmospheric optical thickness $\tau(\lambda)$ and sun-photometer sensitivity remain constant during the period in which measurements of $J(\lambda)$ are taken at different solar elevation angles. If this assumption is true, the data-points giving the logarithms of $J(\lambda)$ plotted as a function of the relative air mass are found to form a line having intercept $J_o(\lambda)$ at m = 0 and a slope equal to $-\tau(\lambda)$. But generally, optical thickness $\tau(\lambda)$ considerably changes in time as a consequence of diurnal fluctuations of the aerosol particle population features. Thus, underestimated values of both $J_o(\lambda)$ and $\tau(\lambda)$ are obtained when the Langley method is applied to measurements taken in atmospheres in which $\tau(\lambda)$ increases as the relative air mass decreases. An example is shown in Fig. 2, in which the "true" zero-air-mass value $J_o(0.50\ \mu m)$ is equal to 8.5 mV and the atmospheric optical thickness $\tau(0.50\ \mu m)$ increases linearly in time from 0.35 (at 07:30 local time) to 0.59 (at 11:30 local time), while air mass $m(h)$ decreases from 9.03 to 1.87. For these time-variations of $\tau(0.50\ \mu m)$, we calculated $J(0.50\ \mu m)$ from eq.(2) at the various solar elevation angles. The logarithms of $J(0.50\ \mu m)$ turn out to be best-fitted by a Langley line having the wrong value $J_o(0.50\ \mu m)$ = 5.3 mV as intercept, a slope of -0.294, and a very high correlation coefficient of -0.999. This example clearly shows that the incorrect use of the Langley method gives considerably underestimated values of $\tau(\lambda)$. Therefore, the Langley plot method cannot be used indiscriminately.

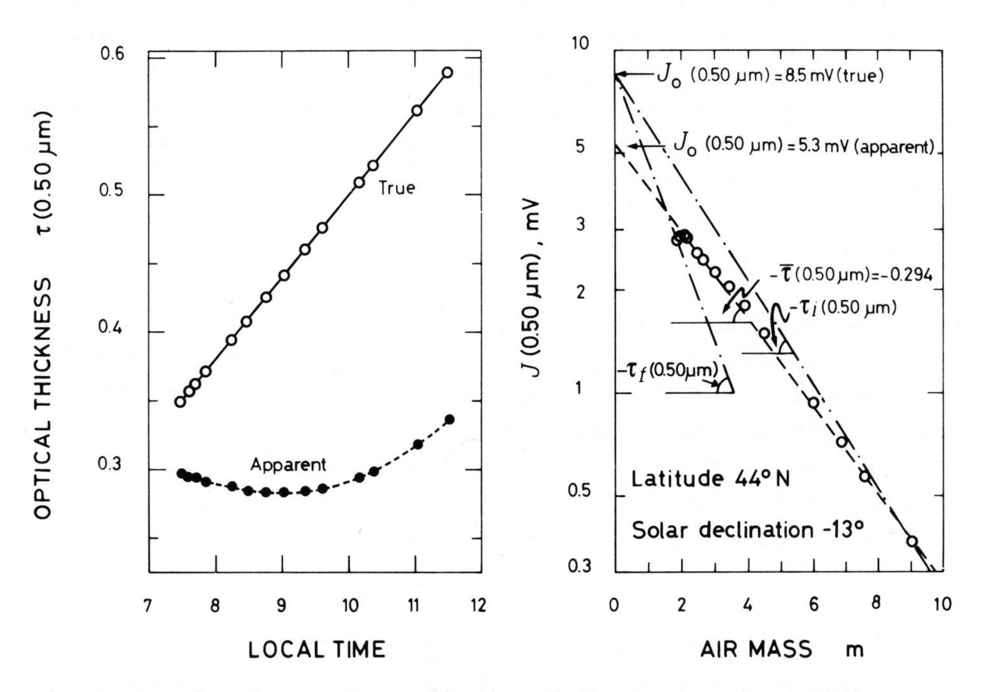

Fig. 2. Example of erroneous application of the Langley plot method to a data-set taken during a period in which the atmospheric optical thickness varies in time. On the left: open circles give the linear variation of the true optical thickness $\tau(0.50\ \mu m)$ as a function of time; solid circles give the time-patterns of $\tau(0.50\ \mu m)$ as found by using the "wrong" value of $J_o(0.50\ \mu m)$. On the right: logarithms of $J(0.50\ \mu m)$ as a function of air mass m and Langley best-fit line giving the apparent zero-air-mass value of $J_o(0.50\ \mu m)$, which is underestimated by 38% with respect to the "true" value. Coefficients $\tau_i(0.50\ \mu m)$ and $\tau_f(0.50\ \mu m)$ give the "true" values of the optical thickness at the initial and final measurement time, respectively.

Since the fluctuations in atmospheric optical thickness are closely related to the evolving features of the boundary layer, it appears wise to take calibration measurements suitable for use in the Langley plot method only with the following conditions: (i) at high mountain stations with relatively low surface albedo (no snow nor glaciers), (ii) on days during anticyclonic periods, (iii) on days without intense ground heating by the sun (no summer days), and (iv) for clear sky conditions, in the absence of thin cirrus layers at angular distances of less than 15° from the sun. Moreover, it is important to take daily calibration measurements over a wide range of the relative air mass with a variation of m of at least 4. Calibration measurements taken on several days give different estimates of $J_o(\lambda)$, from which reliable average values can be obtained.

3.2 Standard lamps

Standard lamps of known irradiance can be used to calibrate a sun-photometer. If $F_o(\lambda)$ is the extraterrestrial solar irradiance at mean earth-sun distance (ref. 11), $F_s(\lambda)$ is the irradiance of the standard lamp at wavelength λ, and $J_s(\lambda)$ is the output voltage of the sun-photometer, when the lamp is viewed from the working distance suggested by the manufacturer, the calibration value at wavelength λ is given by

$$J_o(\lambda) = J_s(\lambda) \; [\, F_o(\lambda) / \; F_s(\lambda) \,] \; . \tag{4}$$

On this matter, Shaw (ref. 3) suggests the use of several lamps to determine an average calibration value $\bar{J}_o(\lambda)$ at each wavelength. The comparison of these values with those given by the Langley method is a useful procedure in discovering casual errors produced by gaseous absorption features (in these cases, eq.(2) may not be valid) and, more generally, in controlling the reliability of Langley results. Moreover, standard lamps are a useful tool in checking instrumental responsivity and, hence, in discovering time-changes of $J_o(\lambda)$, if any.

3.3 Comparison with calibrated sun-photometers

Calibrated sun-photometers can be used to calibrate a sun-photometer with a comparable field of view and similar spectral characteristics of the bandpass filters. In order to obtain satisfactory results, it is convenient to take the calibration measurements at mountain stations, with clear sky conditions, by using all the sun-photometers simultaneously for rather large elevation angles of the sun. Assuming that the ratio between $J(\lambda)$ and $R\,J_o(\lambda)$ as obtained for a calibrated sun-photometer at wavelength λ, is a precise measurement of the monochromatic atmospheric transmission along the sun-path, the calibration values of the new sun-photometer can be easily determined.

4 THE EVALUATION OF PARTICULATE MATTER OPTICAL THICKNESS

Following the above procedures a sun-photometer can be accurately calibrated. Thus, the total optical thickness of the atmosphere can be evaluated at each measurement time as the ratio

$$\tau(\lambda) = \ln\left[R\, J_o(\lambda)/J(\lambda)\right]/m(h). \tag{5}$$

The optical thickness $\tau(\lambda)$ is produced by various attenuation processes, each due to an atmospheric constituent. Therefore, $\tau(\lambda)$ can be expressed at window wavelengths as the sum of different contributions, that is

$$\tau(\lambda) = \tau_p(\lambda) + \tau_R(\lambda) + \tau_w(\lambda) + \tau_z(\lambda) + \tau_g(\lambda)\ , \tag{6}$$

where subscripts p, R, w, z, and g refer to particulate matter, Rayleigh scattering, water vapour, ozone, and other minor atmospheric gases, respectively. These terms can be calculated as follows:

1) Rayleigh optical thickness $\tau_R(\lambda)$ is given by

$$\tau_R(\lambda) = (p_o/1013)\ u_R(\lambda), \tag{7}$$

where p_o is the air pressure at ground level (measured in mb) and $u_R(\lambda)$ is the optical thickness of the Rayleigh atmosphere reduced to a ground-level pressure of 1013 mb. Calculations of $u_R(\lambda)$ were made by Fröhlich and Shaw (ref. 12) for different atmospheric models, showing that $u_R(\lambda)$ is subject to small variations depending on both season and latitude. Recently, Young (ref. 13) pointed out that correct evaluations of $u_R(\lambda)$ can be obtained by multiplying the values of $u_R(\lambda)$ calculated by Fröhlich and Shaw (ref. 12) by a factor equal to 1.031 due to depolarization effects. The values of $u_R(\lambda)$ at the selected window wavelengths are given in Table 1 for two seasonal models at 45°N latitude.

2) optical thickness $\tau_w(\lambda)$ at window wavelengths of the visible and near infrared spectral range is produced by both isolated absorption lines and wings of water vapour bands, so that it can be correctly evaluated in terms of weak absorption in the form

$$\tau_w(\lambda) = c(\lambda)\, w, \tag{8}$$

TABLE 1

Values of the absorption coefficients $C(\lambda)$, $\chi(\lambda)$, and $\xi(\lambda)$ by atmospheric water vapour, ozone, and nitrogen dioxide, and values of Rayleigh optical thickness $u_R(\lambda)$, as obtained multiplying the values calculated by Fröhlich and Shaw (ref. 12) by 1.031, as suggested by Young (ref. 13).

Wavelength	$C(\lambda)$	$\chi(\lambda)$	$\xi(\lambda)$	$u_R(\lambda)$	
λ, μm	(g^{-1} cm^2)	$(cm\ STP)^{-1}$	$(cm\ STP)^{-1}$	45°N,Winter	45°N,Summer
0.38	0.0049	-	16.7	0.44504	0.44541
0.40	0.0076	-	17.2	0.35950	0.35981
0.44	0.0121	0.0022	13.6	0.24235	0.24254
0.50	0.0187	0.0294	4.7	0.14333	0.14344
0.55	0.0220	0.0836	1.0	0.09710	0.09718
0.64	0.0233	0.0780	-	0.05243	0.05247
0.67	0.0246	0.0441	-	0.04355	0.04358
0.75	0.0222	0.0097	-	0.02759	0.02761
0.87	0.0156	-	-	0.01516	0.01517
1.04	0.0157	-	-	0.00739	0.00739

where W is the water vapour content in the vertical atmospheric column (called precipitable water). The values of the weak absorption coefficient $C(\lambda)$ can be evaluated from the results obtained by Eldridge (ref. 14), Fraser (ref. 15), Tomasi (refs. 16 and 17), and Shaw (ref. 18). Values of $C(\lambda)$ at the window wavelengths are given in Table 1.

3) optical thickness $\tau_z(\lambda)$ is due to the Chappuis band of ozone, which presents continuous features of weak absorption from 0.44 μm to about 0.75 μm wavelength. Thus, optical thickness $\tau_z(\lambda)$ can be calculated as the product

$$\tau_z(\lambda) = \chi(\lambda)\, q, \qquad (9)$$

where $\chi(\lambda)$ is the ozone absorption coefficient and q is the atmospheric content of ozone along the vertical path, measured in cm STP. Values of $\chi(\lambda)$ are given in Table 1 at the selected wavelengths, according to Vigroux (ref. 19) and Selby et al. (ref 20). The vertical content of ozone can be evaluated by means of radiosonde measurements or with optical techniques. When measurements of q are not available, average seasonal values of q can be used in eq.(9) taking into account that q closely depends on latitude. Monthly average values of q at 45°N latitude usually range between 0.28 cm STP (October-November) and 0.40 cm STP (March-April).

4) optical thickness $\tau_g(\lambda)$ by minor atmospheric gases is mainly due to a nitrogen dioxide absorption band, which consists of a semicontinuum with a superimposed structure of weak lines from 0.25 μm to about 0.60 μm wave-

length. Thus, optical thickness $\tau_g(\lambda)$ can be calculated as the product

$$\tau_g(\lambda) = \xi(\lambda)\, s \, , \tag{10}$$

where $\xi(\lambda)$ is the NO_2 absorption coefficient and s is the vertical atmospheric content of NO_2, measured in cm STP. Values of $\xi(\lambda)$ are given in Table 1 according to the results found by Hall and Blacet (ref. 21). In unpolluted areas, the vertical content of NO_2 was found to appreciably vary in the course of the day with average values of 4 10^{-4} cm STP at sunrise, 3 10^{-3} cm STP at noon, and 6 10^{-4} cm STP at sunset (refs. 22 and 23).

Following the above suggestions, the monochromatic terms $\tau_R(\lambda)$, $\tau_w(\lambda)$, $\tau_z(\lambda)$, and $\tau_g(\lambda)$ can be calculated at each window wavelengths and at each measurement time. Thus, particulate matter optical thickness $\tau_p(\lambda)$ can be evaluated by subtracting the above terms from $\tau(\lambda)$. With this procedure, a set of values of $\tau_p(\lambda)$ at visible and near infrared wavelengths can be obtained from a spectral series of sun-photometer measurements. If the logarithms of $\tau_p(\lambda)$ are plotted as a function of wavelength λ, as shown in Fig. 3

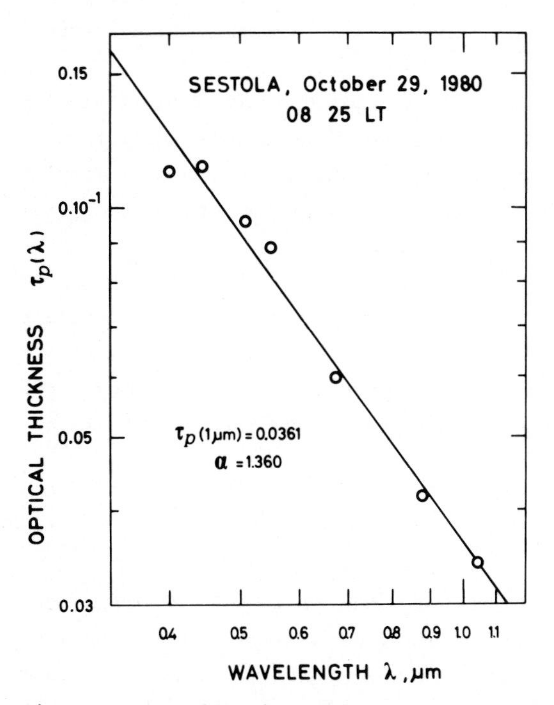

Fig. 3. Spectral series of optical thickness $\tau_p(\lambda)$, as obtained from measurements taken by use of the sun-photometer, FISBAT model (ref. 24), at Sestola on October 29, 1980 (08:25 local time). The best-fit line is obtained in terms of eq.(11).

for measurements taken at Sestola (940 m a.m.s.l.) in the Appennine chain (55 km south-west of Bologna), a best-fit line can be found in terms of Ångström's (ref. 25) formula

$$\tau_p(\lambda) = \tau_p(1\ \mu m)\ \lambda^{-\alpha} \qquad (11)$$

with λ measured in μm. Optical thickness $\tau_p(1\ \mu m)$ and exponent α are commonly called turbidity parameters. Since the particulate matter extinction generally follows the wavelenght dependence form in eq.(11), the turbidity parameters fully describe the particle attenuation conditions along the vertical atmospheric path.

5 EVALUATION OF THE VERTICAL MASS LOADING OF AEROSOL PARTICLES WITH KNOWN METHODS

Several methods were proposed to evaluate the vertical mass loading of atmospheric particles, which are all based on the determination of the turbidity parameters. Each method makes use of a particulate extinction model characterized by particular size distribution curves and by appropriate values of both complex refractive index and mean density of the aerosol particles:

1) examining measurements of the particle scattering coefficient at visible wavelengths, Charlson et al. (ref. 26) evaluated the average particle mass content along the sensing path by simply dividing the measured scattering coefficient by the mass extinction coefficient of particle populations having Junge-type size distribution curves in the form

$$n(r) = C\ r^{-(\nu + 1)} \qquad (12)$$

within the $0.04 \leqslant r \leqslant 2\ \mu m$ radius range, with ν ranging from 2 to 4. Therefore, vertical mass loading M of aerosol particles can be evaluated as the following ratio

$$M = \tau_p(0.50\ \mu m)/b(0.50\ \mu m), \qquad (13)$$

where $b(0.50\ \mu m)$ is the mass extinction coefficient at 0.50 μm wavelength. Thus, we considered a particle population with the size distribution form proposed by Charlson et al. (ref. 26), particle density $\varrho = 1.786$ g cm^{-3}, and a refractive index equal to that obtained by Hänel (ref. 27) for his model 6 of continental particles at relative humidity f = 36% (increasing). For this extinction model, which is hereinafter called model C, we calculated the mass

TABLE 2

Values of mass extinction coefficient $b(0.50\ \mu m)$ for particulate extinction models A and C.

Parameter ν	Mass extinction coefficient $b(0.50\ \mu m)$, $g^{-1}\ m^2$	
	Model A	Model C
1.9	0.404	1.536
2.1	0.533	1.715
2.3	0.714	1.904
2.5	0.947	2.081
2.7	1.200	2.243
2.9	1.396	2.357
3.1	1.447	2.407
3.3	1.334	2.387
3.5	1.119	2.302
3.7	0.885	2.168
3.9	0.681	2.006
4.1	0.523	1.833

extinction coefficient $b(0.50\ \mu m)$ for different values of ν ranging between 1.9 and 4.1, by using a computational program based on the exact Mie theory. The results are given in Table 2.

In order to examine the measurements shown in Fig. 3 and giving $\alpha = 1.3597$, we assumed that ν is equal to 3.3597 according to the relationship $\alpha = \nu - 2$. Making use of the coefficients given in Table 2 for model C, we calculated the most appropriate value of $b(0.50\ \mu m)$ by interpolating the logarithms of $b(0.50\ \mu m)$ in ν and found $M = 0.0392\ g\ m^{-2}$.

2) Griggs (ref. 28) proposed an extinction model having the radius distribution curve given by eq.(12) for $\nu = 3$, radius range from 0.1 to 20 μm, $\varrho = 1.448\ g\ cm^{-3}$, and total refractive index $n = 1.486$ at 0.50 μm wavelength. Mass extinction coefficient $b(0.50\ \mu m)$ is equal to 2.381 $g^{-1}\ m^2$, so that $M = 0.0389\ g\ m^{-2}$ was obtained from eq.(13) for the case shown in Fig. 3.

3) Shaw et al. (ref 29) defined an extinction model based on Junge-type radius distributions with radius range from 0.01 to 10 μm and ν varying between 2 and 4. However, the values of the mass extinction coefficient found by Shaw et al. (ref. 29) appear to be rather low, as unrealistically large values of M are obtained by using these results. Therefore, we decided to compute the values of $b(0.50\ \mu m)$ for an extinction model (which has been called A) having radius distribution curves as those proposed by Shaw et al. (ref. 29) and particle density and refractive index equal to those adopted for model C. The results are given in Table 2.

Examining the measurements shown in Fig. 3 in terms of model A, we calculated the value of $b(0.50\ \mu m)$ for $\nu = 3.3597$ by interpolation between the logarithms of the coefficients given in Table 2, from which $M = 0.0731$ g m^{-2} was obtained.

4) The method proposed by Box and Lo (ref. 30) is based on particle size distribution curves having the form of Deirmendjian's (ref. 31) modified gamma distribution function of type H. Thus, the vertical mass loading M can be evaluated through the following equation

$$M = 80\,\pi\, a\, \varrho\, B^{-3}, \qquad (14)$$

while a and B are Deirmendjian's parameters which can be determined from the measurements of optical thickness $\tau_p(\lambda)$. In particular, parameter B is twice the inverse of mode radius r_c. In order to represent the wavelength dependence curve of $\tau_p(\lambda)$, Box and Lo (ref. 30) considered both the exponential form and the power form given in eq.(11). Only this latter procedure is herein described, as it seems to be more realistic than the former.

The procedure begins with the determination of Ångström's parameters $\tau_p(1\ \mu m)$ and α. Parameter B is determined from the value of α by using appropriate relationship curves defined for different values of both real and imaginary parts of the refractive index, as shown in Fig. 4. When the value of B has been found, a second parameter β_o^{-1} can be determined by using other relationship curves. Fig. 4 shows some of these curves, as given by Box and Lo (ref. 30). When β_o^{-1} has been determined, parameter a can be calculated as the product of $\tau_p(1\ \mu m)$ by β_o^{-1}, so that M can be evaluated from eq.(14). Examining the case shown in Fig. 3 with $\alpha = 1.3597$ and using the curves given in Fig. 4 for $n = 1.5$ and $k = 0.05$, we obtained $B = 31.97\ \mu m^{-1}$ and, hence, $r_c = 0.063\ \mu m$. Since $\beta_o^{-1} = 45.94\ \mu m^{-2}$, we found $a = 1.656\ \mu m^{-2}$ and $M = 0.0228$ g m^{-2}.

The above extinction models are based on size distribution curves which may be unrealistic in many cases. The comparison between the values of $b(0.50\ \mu m)$ given by models A and C, which have different radius limits but equal values of particle density and of refractive index, shows that it is very important to appropriately choose the size range of the particle population. For this reason, we defined a set of extinction models with different features of the size distribution curve to characterize particle populations of different origin.

6 JUNGE-TYPE EXTINCTION MODELS

In order to define size distributions suited to represent different particle populations, we examined a large number of Junge-type size spectra,

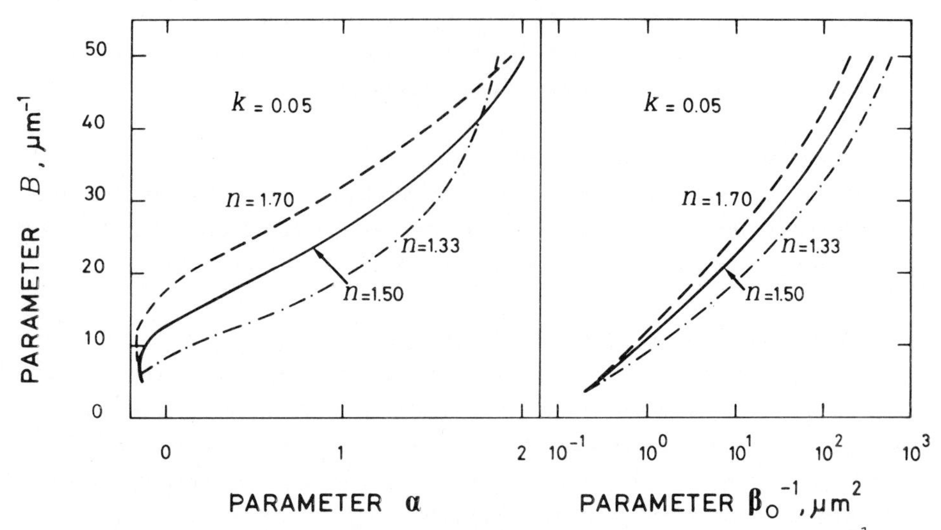

Fig. 4. Relationship curves between parameter B and parameters α and β_o^{-1}, as calculated by Box and Lo (ref. 30) for different values of real part n and for imaginary part $k = 0.05$ of the particulate matter refractive index.

which were measured by various authors in different parts of our planet. From these measurements, we determined the average profiles of the size distribution curves, paying particular attention to the radius intervals of the small and giant particles. For the large particles, we adopted variable Junge-type size distribution profiles having the form in eq.(12) with ν varying between 1.9 and 4.3. On this basis, we defined seven extinction models all having the radius range from 0.003 to 20 μm, as follows:

6.1 Model CR of continental and rural particles

The particle number $n(r)$ per cubic centimeter of air and unit radius interval is proportional to r^5 for $0.003 \leqslant r \leqslant 0.01$ μm, constant for $0.01 \leqslant r \leqslant 0.03$ μm, proportional to r^{-1} for $0.03 \leqslant r \leqslant 0.1$ μm, to $r^{-(\nu + 1)}$ for $0.1 \leqslant r \leqslant 3$ μm, to $r^{-5.5}$ for $3 \leqslant r \leqslant 10$ μm, and to r^{-7} for $10 \leqslant r \leqslant 20$ μm. The particle density and the complex refractive index were taken equal to those of model 6 of continental particles, as proposed by Hänel (ref. 27) at different values of the decreasing relative humidity.

6.2 Model CU of continental and urban particles

The particle number $n(r)$ is proportional to r^{-1} for $0.003 \leqslant r \leqslant 0.03$ μm, to r^{-2} for $0.03 \leqslant r \leqslant 0.1$ μm, to $r^{-(\nu + 1)}$ for $0.1 \leqslant r \leqslant 6$ μm, to $r^{-5.5}$ for $6 \leqslant r \leqslant 10$ μm, and to r^{-7} for $10 \leqslant r \leqslant 20$ μm. For particles with radius smaller than 0.1 μm, the values of the density and of the complex refractive index are assumed to be equal to those of model CR at $f = 36\%$ (decreasing). For $r \geqslant 0.1$ μm, both particle density and refractive index are equal to those adopted by Hänel (ref. 27) in his model 5 of urban aerosol particles.

6.3 Model AU of all urban particles

The size distribution curves are those adopted for model CU. The particle density and the refractive index are assumed to be equal to those of Hänel's model 5 within the whole radius range from 0.003 to 20 μm.

6.4 Model UCA of urban particles with carbonaceous substances

The size distribution curves are equal to those adopted in models CU and AU. The density and the refractive index vary as functions of the relative humidity according to Shettle and Fenn (ref. 32) who defined the urban particles as a mixture of 56% of water soluble substances (mainly ammonium and calcium sulfate and organic compounds), 24% of dust-like aerosol particles, and 20% of soot particles with carbonaceous substances.

6.5 Model S of continental and sea-salt particles

The particle number $n(r)$ is proportional to r^5 for $0.003 \leqslant r \leqslant 0.01$ μm, constant for $0.01 \leqslant r \leqslant 0.03$ μm, proportional to r^{-1} for $0.03 \leqslant r \leqslant 0.1$ μm, to r^{-2} for $0.1 \leqslant r \leqslant 0.3$ μm, to $r^{-(\nu + 1)}$ for $0.3 \leqslant r \leqslant 5$ μm, to r^{-6} for $5 \leqslant r \leqslant 10$ μm, and to r^{-7} for $10 \leqslant r \leqslant 20$ μm. For particles with radius $r \leqslant 0.1$ μm, the density and the refractive index are equal to those adopted in model CR at $f = 36\%$. For $r \geqslant 0.1$ μm, the density and the refractive index are equal to those proposed by Hänel (ref. 27) for his model 2 of sea-spray aerosol particles.

6.6 Model M of maritime particles

The size distribution curves are equal to those defined for model S. The particle density and the refractive index are equal to those proposed by Hänel (ref. 27) for his model 3 of maritime aerosol particles.

6.7 Model MD of maritime and desertic particles

The particle number $n(r)$ has the form used in model CR for $r \leqslant 0.1$ μm and is proportional to r^{-1} for $0.1 \leqslant r \leqslant 0.2$ μm, to $r^{-(\nu + 1)}$ for $0.2 \leqslant r \leqslant 4$ μm, and to r^{-7} for $4 \leqslant r \leqslant 20$ μm. For particle radius $r \leqslant 0.1$ μm, the density and the refractive index are equal to those adopted for model CR at $f = 36\%$ (decreasing). For $r \geqslant 0.1$ μm, the density and the refractive index were taken from model 4 proposed by Hänel (ref. 27).

6.8 Extinction coefficients

The size distribution curves of the above models, as obtained for $\nu = 3$, are shown in Fig. 5. For each of the Junge-type models, we calculated the values of the particle density and of the refractive index at dry air conditions and at values of relative humidity f varying between 30% and about 90%. For these values, the volume extinction coefficients $\beta(\lambda)$ were computed at the wavelengths of 0.40, 0.50, 0.55, 0.60, 0.70, 0.80 and 1.00 μm, by varying parameter ν from 1.9 to 4.3, by steps of 0.2. Each spectral series of seven monochromatic values of $\beta(\lambda)$, as found for each value of parameter ν, was examined in terms of eq.(11) to find the best-fit value of exponent α. The relationship curves between exponent α and parameter ν were found to present considerably different features from those of the commonly used relationship curve having the form $\alpha = \nu - 2$. This result was found for all the extinction models and for all the values of the relative humidity. The relationship curves found for different values of f between 52% and 79% are shown in Fig. 6. These results clearly indicate that more precise estimates of vertical mass loading M can be obtained by determining the mass extinction coefficient $b(0.50\ \mu\text{m})$ by interpolation in α rather than in ν, the value of α

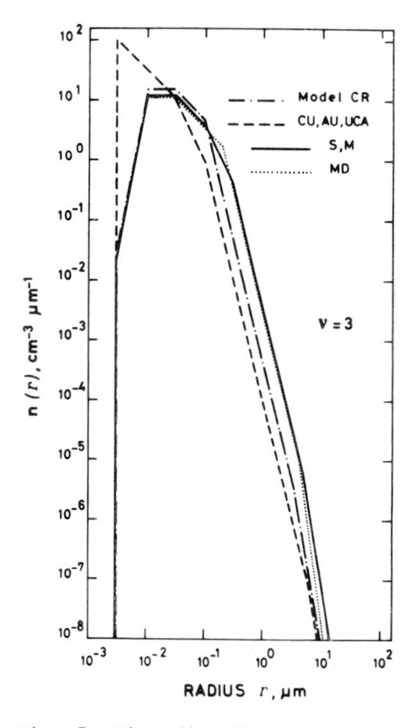

Fig. 5. Size distribution curves used for the various Junge-type extinction models, as obtained for $\nu = 3$.

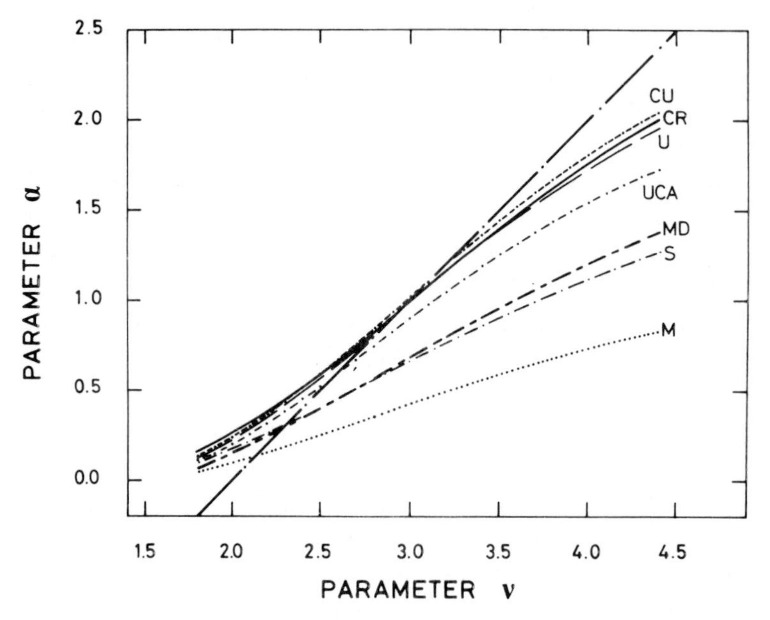

Fig. 6. Relationship curves between parameters α and ν, as found for the present Junge-type extinction models at relative humidity f = 58% (model CR), f = 79% (models CU and AU), f = 70% (model UCA), f = 65% (model S), f = 52% (model M), and f = 62% (model MD).

Table 3
Value of mass extinction coefficient $b(0.50\ \mu m)$, measured in $g^{-1}\ m^2$, and of corresponding parameter α (in parenthesis), as obtained for extinction models CR, CU, AU and UCA, at different values of relative humidity f.

	Model CR f = 58%	Model CU f = 79%	Model AU f = 79%	Model UCA f = 70%
1.9	0.7640 (0.2082)	0.4937 (0.1875)	0.4926 (0.1811)	0.5042 (0.1560)
2.1	0.9251 (0.3216)	0.6270 (0.3013)	0.6251 (0.2915)	0.6517 (0.2584)
2.3	1.1314 (0.4498)	0.8101 (0.4390)	0.8068 (0.4248)	0.8611 (0.3830)
2.5	1.3849 (0.5913)	1.0491 (0.5943)	1.0441 (0.5748)	1.1392 (0.5226)
2.7	1.6784 (0.7468)	1.3372 (0.7611)	1.3308 (0.7354)	1.5045 (0.6730)
2.9	1.9928 (0.9086)	1.6482 (0.9339)	1.6418 (0.9017)	1.9160 (0.8247)
3.1	2.2991 (1.0724)	1.9399 (1.1077)	1.9359 (1.0676)	2.3344 (0.9745)
3.3	2.5663 (1.2347)	2.1708 (1.2786)	2.1714 (1.2307)	2.7056 (1.1180)
3.5	2.7711 (1.3926)	2.3183 (1.4416)	2.3243 (1.3855)	2.9904 (1.2527)
3.7	2.9046 (1.5439)	2.3838 (1.5952)	2.3940 (1.5306)	3.1772 (1.3770)
3.9	2.9711 (1.6872)	2.3845 (1.7348)	2.3967 (1.6615)	3.2766 (1.4901)
4.1	2.9828 (1.8215)	2.3419 (1.8693)	2.3532 (1.7874)	3.3092 (1.5915)
4.3	2.9543 (1.9459)	2.2742 (1.9881)	2.2823 (1.8975)	3.2960 (1.6816)

being obtained from the sun-photometer measurements of $\tau_p(\lambda)$. We calculated the values of $b(0.50\ \mu m)$ for each model and for each value of ν and determined the relationship curves between $b(0.50\ \mu m)$ and parameter α. The results obtained for the seven models at different values of f are given in Tables 3

Table 4
Values of mass extinction coefficient $b(0.50\ \mu m)$, measured in $g^{-1}\ m^2$, and of corresponding parameter α (in parenthesis), as obtained for extinction models S, M and MD, at different values of relative humidity f.

	Model S $f = 65\%$	Model M $f = 52\%$	Model MD $f = 62\%$
1.9	0.6962 (0.1360)	0.4038 (0.0714)	0.5350 (0.1159)
2.1	0.8498 (0.2160)	0.4987 (0.1244)	0.6543 (0.1954)
2.3	1.0491 (0.3058)	0.6241 (0.1852)	0.8098 (0.2971)
2.5	1.2993 (0.4014)	0.7847 (0.2516)	1.0058 (0.4010)
2.7	1.5999 (0.5070)	0.9819 (0.3206)	1.2428 (0.5106)
2.9	1.9426 (0.6110)	1.2124 (0.3905)	1.5151 (0.6242)
3.1	2.3126 (0.7146)	1.4676 (0.4592)	1.8108 (0.7379)
3.3	2.6947 (0.8114)	1.7349 (0.5257)	2.1137 (0.8495)
3.5	3.0577 (0.9066)	2.0005 (0.5880)	2.4072 (0.9574)
3.7	3.3905 (0.9977)	2.2526 (0.6468)	2.6776 (1.0607)
3.9	3.6841 (1.0828)	2.4834 (0.7066)	2.8505 (1.1498)
4.1	3.9201 (1.1609)	2.6888 (0.7507)	3.1202 (1.2513)
4.3	4.1260 (1.2338)	2.8684 (0.8092)	3.2901 (1.3382)

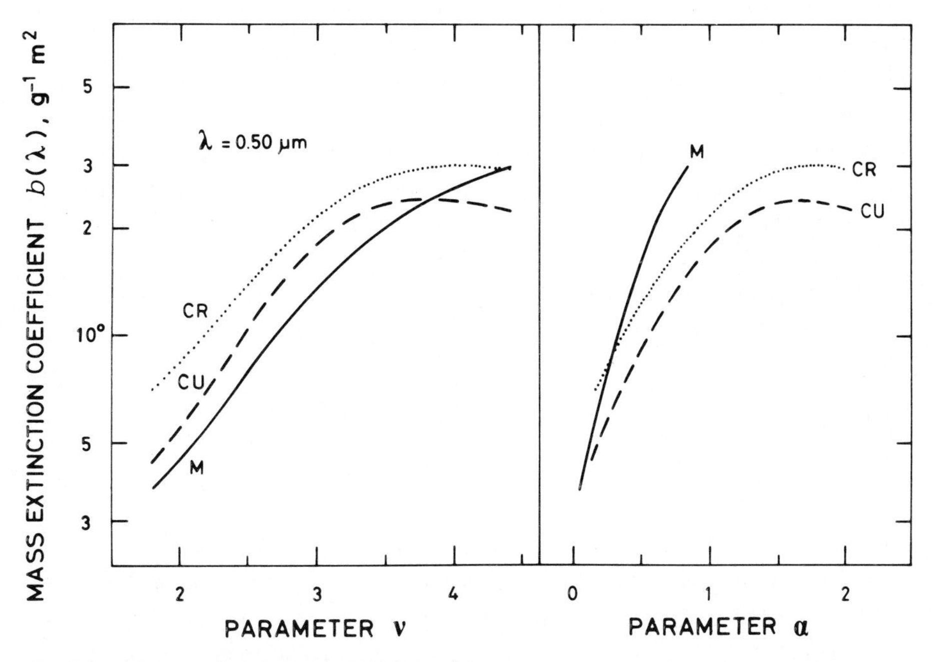

Fig. 7. Mass extinction coefficient $b(0.50\ \mu m)$ as a function of parameter ν and of turbidity parameter α for extinction models CR (continental and rural particles at $f = 58\%$), CU (continental and urban particles at $f = 79\%$), and M (maritime particles at $f = 52\%$).

and 4. Fig. 7 shows $b(0.50\ \mu m)$ as a function of both parameters ν and α for three extinction models. As can be seen, the mass extinction coefficient can vary by almost an order of magnitude as α increases from 0 to 2.

In order to find the value of the vertical mass loading, one determines the most appropriate value of $b(0.50\ \mu m)$ by interpolating the logarithms of the mass extinction coefficients in α. For this purpose, the values of $b(0.50\ \mu m)$ calculated at relative humidity conditions similar to those occurring in the ground layer during the measurements can be appropriately used. Thus, M can be calculated from eq.(13). Considering that the measurements shown in Fig. 3 were taken at a mountain station and at relative humidity close to 30%, the examination of these data was made in terms of model CR at f = 36% (decreasing). For $\alpha = 1.3597$, we found $b(0.50\ \mu m) = 2.717\ g^{-1}\ m^{2}$ and $M = 0.0340\ g\ m^{-2}$, which is a considerably higher value than that found with the procedure suggested by Box and Lo (ref. 30).

7 BIMODAL EXTINCTION MODELS

Particle size distributions of continental, maritime, and urban origin frequently present multimodal features related to natural formation processes or to human activities. At parity of particulate extinction, the presence of rather large and giant particle modes generally implies higher values of the vertical mass loading of aerosol particles, due to the fact that the rather large and giant particles produce appreciably lower values of the mass extinction coefficient than the relatively small ones. Taking into account these remarks, we propose the following five particulate extinction models obtained on the basis of models defined by Shettle and Fenn (ref. 32):

7.1 Tropospheric monomodal model

The size distribution is given by a log-normal curve in the form

$$n(r) = \frac{N}{\ln(10)\ \sigma \sqrt{2\pi}\ r} \exp\left[-\frac{(\log r - \log \bar{r})^2}{2\sigma^2}\right] \qquad (15)$$

where the cumulative particle number density N is taken equal to 1 cm^{-3}, standard deviation σ is equal to 0.35, median radius $\bar{r}$ ranges from 0.027 to 0.039 μm as relative humidity f increases from 0% to 90%, and ln and log are the natural and decimal logarithm, respectively. The tropospheric particles are assumed to be composed of a mixture of 70% of water soluble substances and 30% of dust-like aerosols. The real and imaginay parts of the refractive index and their variations as functions of the relative humidity are calculated according to Shettle and Fenn (ref. 32). The particle density varies from 1.791 to 1.266 g cm^{-3} as f increases from 0% to 90%.

7.2 Rural monomodal model

The size distribution function $n(r)$ is given by a log-normal curve in the form of eq.(15) with $N = 1\ cm^{-3}$, $\sigma = 0.40$, and $\bar{r}$ ranging from 0.43 to 0.65 μm as f increases from 0% to 90%. The composition and the complex refractive index of the particles at dry air conditions are equal to those adopted for the tropospheric model, whereas their variations related to the relative humidity conditions have been calculated taking into account the different growth rates of the large particles from the small ones. So the particle density decreases from 1.791 to 1.233 g cm^{-3} as f increases from 0% to 90%.

7.3 Oceanic monomodal model

The size distribution curve is given by a log-normal curve having the form given in eq.(15) with $N = 1\ cm^{-3}$, $\sigma = 0.40$, and $\bar{r}$ varying from 0.16 to 0.38 μm as f increases from 0% to 90%. These aerosols are largely sea-salt particles. The refractive index and its variations with the relative humidity are taken according to Shettle and Fenn (ref. 32). The particle density varies from 2.250 g cm^{-3} at f = 0% to 1.093 g cm^{-3} at f = 90%.

7.4 Coastal bimodal model

According to the results found by Tomasi and Prodi (ref. 33) from coastal aerosol samples taken in Red Sea area and along the Somalian Coast, this model was assumed to be composed of 50% of rural particles and of 50% of oceanic particles characterized by features equal to those of the above proposed models. Therefore, the size distribution curves at different relative humidity values consist of two modes, each having the form in eq.(15) with $N = 0.5\ cm^{-3}$, while the average particle density decreases from 1.814 to 1.209 g cm^{-3} as f increases from 0% to 90%.

7.5 Urban bimodal model

The size distribution curve consists of two modes. The first modal curve is given by eq.(15) with $N = 0.999875\ cm^{-3}$, $\sigma = 0.35$, and $\bar{r}$ ranging from 0.025 to 0.042 μm as f varies from 0% to 90%. The second modal curve is given by eq.(15) with $N = 1.25\ 10^{-3}\ cm^{-3}$, $\sigma = 0.40$ and $\bar{r}$ ranging from 0.40 to 0.71 μm as f increases from 0% to 90%. These urban particles are composed of a mixture equal to that assumed in Junge-type model UCA, so that the complex refractive index and its variations with f are taken according to Shettle and

Fenn (ref. 32). The particle density of the small and large particles varies from the common value of 1.889 g cm^{-3} at f = 0% to 1.189 and 1.161 g cm^{-3}, respectively, at f = 90%.

7.6 Extinction coefficients

Making use of a computational program based on the exact Mie theory, we calculated the volume extinction coefficient $\beta(\lambda)$ at the nine wavelengths listed in Table 5 for the five particles extinction models, as defined at f = 0, 50, 70, 80, and 90%. The results obtained at f = 70% are given in Table 5 together with the values of mean particle mass $\bar{m}$. Fig. 8 shows the volume extinction coefficient as a function of the wavelength for the five models corresponding to f = 0% and f = 90%. The tropospheric and urban models present spectral dependence curves which appreciably decrease as a function of the wavelength. The tropospheric extinction coefficients closely follow the spectral dependence form proposed by Ångström (ref. 25) with values of exponent α ranging between 1.428 (f = 0%) and 1.312 (f = 90%). The values of exponent α found for the urban extinction models are all very similar and vary between 1.055 and 1.083. On the other hand, the rural, oceanic and coastal models present spectral dependence curves which are almost constant or slowly increasing with the wavelength for any value of the relative humidity.

TABLE 5
Values of volume extinction coefficient $\beta(\lambda)$, measured in cm^{-1}, as obtained at f = 70% for the five extinction models based on log-normal radius frequency distributions normalized to unit number density per cubic centimeter of air. The values of mean particle mass $\bar{m}$, measured in g, are given in parenthesis.

Wavelength λ, μm	Tropospheric model	Rural model	Urban model	Oceanic model	Coastal model
	($3.01\ 10^{-15}$)	($3.02\ 10^{-11}$)	($6.95\ 10^{-15}$)	($2.60\ 10^{-12}$)	($1.64\ 10^{-11}$)
0.380	$1.450\ 10^{-10}$	$7.916\ 10^{-8}$	$1.639\ 10^{-10}$	$1.712\ 10^{-8}$	$4.814\ 10^{-8}$
0.400	$1.374\ 10^{-10}$	$7.947\ 10^{-8}$	$1.567\ 10^{-10}$	$1.720\ 10^{-8}$	$4.834\ 10^{-8}$
0.500	$1.089\ 10^{-10}$	$8.080\ 10^{-8}$	$1.291\ 10^{-10}$	$1.756\ 10^{-8}$	$4.918\ 10^{-8}$
0.550	$9.864\ 10^{-11}$	$8.138\ 10^{-8}$	$1.189\ 10^{-10}$	$1.771\ 10^{-8}$	$4.954\ 10^{-8}$
0.671	$7.495\ 10^{-11}$	$8.295\ 10^{-8}$	$9.570\ 10^{-11}$	$1.797\ 10^{-8}$	$5.046\ 10^{-8}$
0.777	$5.912\ 10^{-11}$	$8.443\ 10^{-8}$	$8.052\ 10^{-11}$	$1.794\ 10^{-8}$	$5.108\ 10^{-8}$
0.865	$4.933\ 10^{-11}$	$8.520\ 10^{-8}$	$7.076\ 10^{-11}$	$1.786\ 10^{-8}$	$5.153\ 10^{-8}$
1.000	$3.862\ 10^{-11}$	$8.653\ 10^{-8}$	$5.943\ 10^{-11}$	$1.776\ 10^{-8}$	$5.215\ 10^{-8}$
1.048	$3.568\ 10^{-11}$	$8.697\ 10^{-8}$	$5.616\ 10^{-11}$	$1.773\ 10^{-8}$	$5.235\ 10^{-8}$

Since the measurements of $\tau_p(\lambda)$ are usually found to follow the wavelength dependence form given in eq.(11) with exponent α ranging for the most part between 1.5 and 0.3, we propose the use of bimodal extinction models to examine the measurements of the particulate matter optical thickness. When the measurements are taken in rural, or maritime or coastal areas, these bimodal models can be a linear combination of the tropospheric model with a secondary model chosen among the three models for large particle extinction. For measurements taken in urban areas and giving a value of exponent α smaller than 1.05, the bimodal urban model can be combined with the rural model to give form to a trimodal extinction model. The procedure is that followed by Tomasi and Prodi (ref. 33): from the determination of turbidity parameters $\tau_p(1\ \mu m)$ and α, the particulate matter optical thickness $\tau_p(0.50\ \mu m)$ can be evaluated from eq.(11). Each pair of optical thicknesses ($\tau_p(0.50\ \mu m)$, $\tau_p(1.00\ \mu m)$) can be examined in terms of the bimodal extinction model which best describes the particle population features in the measurement area. With regard to this, Fig. 9 shows that the mass extinction coefficient does not present considerable variations as the relative humidity increases from 50% to 90%. This suggests that the values of $\beta(\lambda)$ given in Table 5 and calculated at f equal to 70% can be reliably used for all the normally occurring conditions of relative humidity in the low atmosphere, when sun-photometric measurements can be carried out.

Indicating with subscripts 1 and 2 the basic model of small particles and the secondary extinction model of large particles, respectively, the following system is obtained:

$$\begin{cases} N_1\ \beta_1(0.50\ \mu m) + N_2\ \beta_2(0.50\ \mu m) = \tau_p(0.50\ \mu m) \\ N_1\ \beta_1(1.00\ \mu m) + N_2\ \beta_2(1.00\ \mu m) = \tau_p(1.00\ \mu m) \end{cases} \qquad (16)$$

from which positive values of N_1 and N_2 can be found in most cases. Since $\beta_1(\lambda)$ and $\beta_2(\lambda)$ are evaluated for unit particle number density along the path of 1 cm length, N_1 and N_2 are the total numbers of particles in the vertical atmospheric column of 1 cm^2 section, pertinent to the two selected models. Then, the vertical mass loading of aerosol particles is given by the sum

$$M = 10^4\ (N_1 \bar{m}_1 + N_2 \bar{m}_2), \qquad (17)$$

where $\bar{m}_1$ and $\bar{m}_2$ are the values of the mean particle mass for the two selected models.

Examining the measurements shown in Fig. 3 in terms of the linear combination of the tropospheric and rural models, we found $N_1 = 8.68\ 10^8\ cm^{-2}$

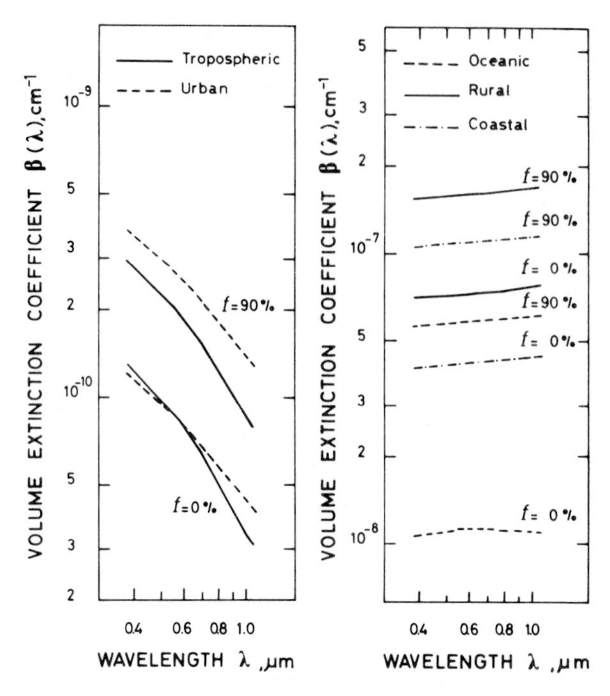

Fig. 8. Volume extinction coefficient $\beta(\lambda)$ as a function of the wavelength, as obtained for the five extinction models based on log-normal size distribution curves corresponding to relative humidity values of 0% and 90%.

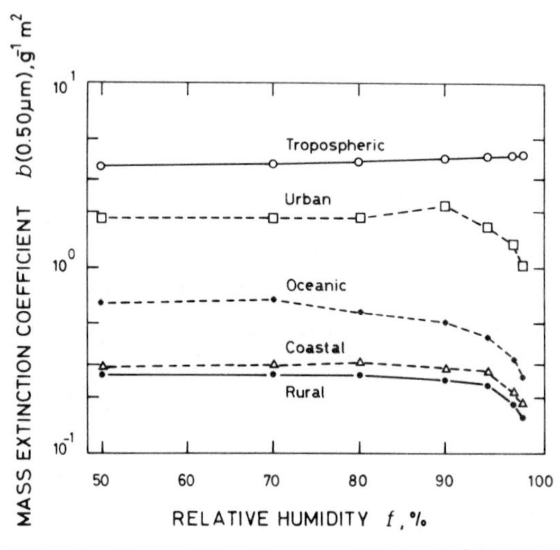

Fig. 9. Mass extinction coefficient $b(0.50\ \mu m)$ as a function of relative humidity f, as found for the five extinction models based on log-normal size distribution curves.

for the tropospheric mode, $N_2 = 6.32\ 10^4\ cm^{-2}$ for the rural mode, and $M = 0.0422\ g\ m^{-2}$.

8 DISCUSSION AND CONCLUSION

Based as they are on different size distribution curves, the various extinction models proposed herein give different estimates of the vertical mass loading of aerosol particles. The estimates of M obtained by examining the measurements shown in Fig. 3 were found to range from $0.0228\ g\ m^{-2}$ (haze H model proposed by Box and Lo, ref. 30) to $0.0731\ g\ m^{-2}$ (model A). In order to examine the merits and the demerits of these models, we calculated the radius distribution curves of volume V occupied by the particles in the vertical atmospheric column, as obtained for the solutions found by using the various extinction models. These curves are shown in Fig. 10.

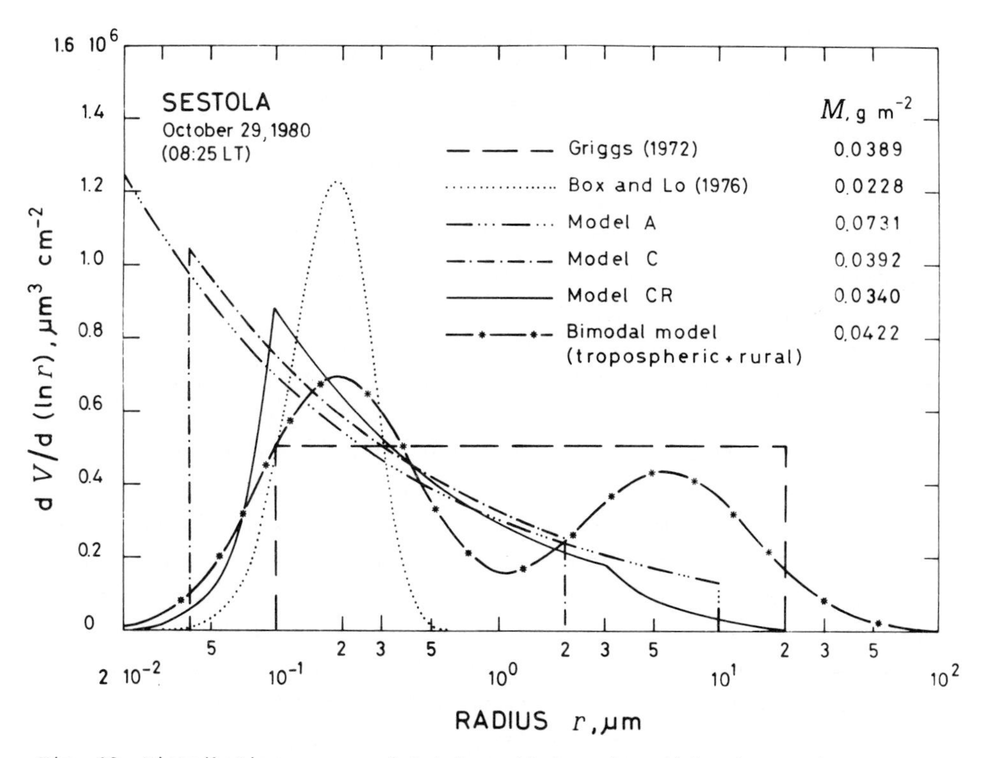

Fig. 10. Distribution curves of total particle volume V in the vertical atmospheric column, as given by the solutions found by examining the particulate matter optical thickness measurements shown in Fig. 3 in terms of the various extinction models proposed herein. The corresponding estimates of mass loading M are also given.

The extinction model proposed by Griggs (ref. 28) to represent continental and urban particle populations is unrealistic in all the cases in which parameter ν appreciably differs from 3. Considering that the radius range has an upper limit of 20 μm, the estimates of M are expected to be overestimated when the measurements of exponent α are larger than about 1. In fact, for these turbidity conditions, giant particles contribute poorly to the vertical extinction and to the cumulative particle volume.

The haze model H proposed by Box and Lo (ref. 30) generally attributes the extinction to particle populations largely composed of small particles and, hence, inadequately accounts for the contributions made by large and giant particles. Since the mass extinction coefficients produced by rather large and giant particles are lower than those given by relatively small particles in all the cases, the estimates of M are expected to be underestimated when exponent α presents relatively low values.

Model A based on the size distribution model proposed by Shaw et al. (ref. 29) generally gives overestimated values of M because it gives an unrealistically large volume of particles for any value of parameter ν. Model C having radius range from 0.04 to 2 μm, according to Charlson et al. (ref. 26), attributes the most part of volume V to the large particles. Moreover, it gives rather a large volume of small particles and, at the same time, rather a small volume of giant particles. However, these overestimation and underestimation effects should almost completely cancel each other out when the particle size distribution curve presents average features.

The volume distribution curve obtained in Fig. 10 by using one of the Junge-type extinction models follows the increasing profile of the haze H model (Box and Lo, ref. 30) in the radius range below 0.1 μm and decreases as a function of the radius throughout the large particle range since the value of turbidity parameter α is related to a value of ν smaller than 3. When ν is equal to 3, the volume distribution curve is constant with the radius, as found for the model proposed by Griggs (ref. 28). When ν is smaller than 3, as frequently occurs for maritime particle populations, the volume distribution curve increases throughout the whole large particle radius range, in which the size distribution curve has the form given in eq.(12), and reaches its maximum at the upper limit. The volume distribution curve shown in Fig. 10 rapidly decreases as a function of the radius throughout the whole radius range of the giant particles. This behaviour is common to all the present Junge-type extinction models.

Considering that each Junge-type model is based on a set of size distribution curves characterized by realistic features, accurate estimates of the vertical mass loading can be only obtained if use is made of appropriate

Junge-type models. These can be chosen by taking into account the origin of the particulate matter, the relative humidity conditions of the low atmospheric layer as well as the spectral features of the atmospheric attenuation, as measured by the sun-photometer.

The bimodal extinction models presented herein are obtained from the widespread aerosol particle models proposed by Shettle and Fenn (ref. 32). However, Fig. 10 clearly shows that the radius distribution curve of the particle volume produced by these bimodal extinction models has radius range without a realistic upper limit and presents very high volume contributions by large and giant particles. This is a serious defect in all the bimodal models made by using log-normal size distribution curves to represent large particle populations. As pointed out by Twomey (ref. 34), the size distributions of natural particles do not actually exhibit the symmetry found in log-normal distribution curves but generally have short-tailed wings in the upper radius range. Therefore, the multimodal extinction models can give accurate results only when size distribution curves skewed to the right are adopted to represent the average features of the large and giant particle size spectra. Thus, particle extinction models based on log-normal size distribution curves generally give overestimated values of the vertical mass loading of particulate matter.

Figure 10 clearly shows that the mass extinction coefficient depends closely on the upper limit of the particle radius range as well as on the right wing shape of the particle radius distribution curve. Therefore, Junge-type extinction models can be reliably used when the giant particle mass fraction in the vertical atmospheric column is relatively limited, the predominant contribution being made by large particles. In cases in which the extinction by giant particles along the sunpath is rather important, bimodal extinction models appear to be very realistic. However, Junge-type extinction models can also give realistic estimates of the particulate mass loading. In fact, when the particle population is characterized by a size distribution curve with a rather wide right wing, which gives greater values of particle number density than those given by our Junge-type size distribution models, considerably smaller values of spectral exponent α than 1 are usually obtained from the sun-photometer measurements. Correspondingly, Junge parameter ν is smaller than 3 so that our extinction models produce a peaked shape of the volume distribution function with the maximum located in the upper limit of the radius range throughout which the Junge parameter is made to vary. With these features, our extinction models give relatively low values of the mass extinction coefficients so as to compensate for a good part the defect of not taking sufficient account of the extinction produced by giant particles of relatively large sizes.

However, in order to select the most appropriate size distribution models, spectral measurements of the skylight aureole around the sun (ref. 7) taken with the sun-photometer as well as measurements of particle size spectra from both aircraft and ground-level samplings can be of great usefulness. Moreover, realistic values of both complex refractive index and density of the particulate matter can be chosen by thinking of the origin of the particle population and the meteorological conditions of the low atmosphere.

REFERENCES

1 F.E. Volz, Arch. Meteorol. Geophys. Bioklimatol., B 10 (1959) 100-131.
2 F.E. Volz, Appl. Opt., 13 (1974) 1732-1733.
3 G.E. Shaw, Pure Appl. Geophys., 114 (1976) 1-14.
4 C. Tomasi and R. Guzzi, J. Phys. E: Sci. Instrum., 7 (1974) 647-649.
5 D.E. Pitts, W.E. McAllum, M. Heidt, K. Jeske, J.T. Lee, D. DeMonbrun, A. Morgan and J. Potter, J. Appl. Meteor., 16 (1977) 1312-1321.
6 R.E. Bird and R.L. Hulstrom, J. Appl. Meteor., 21 (1982) 1196-1201.
7 D. Deirmendjian, Use of scattering techniques in cloud microphysics research. I. The aureole method, Report R-590-PR, USAF Project Rand, Santa Monica, California, 1970, 39 pp.
8 K.L. Coulson, Solar and Terrestrial Radiation, Academic Press, New York-London, 1975, pp. 235-249.
9 F. Kasten, Arch. Meteorol. Geophys. Bioklimatol. B 14 (1966) 206-223.
10 K.Ya. Kondratyev, Radiation in the Atmosphere, Academic Press, New York-London, 1969, pp. 870-872.
11 M.P. Thekaekara, Solar electromagnetic radiation, NASA Space Vehicle Design Criteria (Environment), NASA SP-8005, Washington, D.C., 1971.
12 C. Fröhlich and G.E. Shaw, Appl. Opt., 19 (1980) 1773-1775.
13 A.T. Young, J. Appl. Meteor., 20 (1981) 328-330.
14 R.G. Eldridge, Appl. Opt., 6 (1967) 709-713.
15 R.S. Fraser, J. Appl. Meteor., 14 (1975) 1187-1196.
16 C. Tomasi, Quart. J. Roy. Meteor. Soc., 105 (1979) 1027-1040.
17 C. Tomasi, Il Nuovo Cimento, 2C (1979) 511-526.
18 G.E. Shaw, Appl. Opt., 19 (1980) 480-482.
19 E. Vigroux, Ann. de Phys., 8 (1953) 709-763.
20 J.E.A. Selby, E.P. Shettle and R.A. McClatchey, Atmospheric transmittance from 0.25 to 28.5 μm: supplement LOWTRAN 3B (1976), Environm. Res. Papers, No. 587, AFGL-TR-76-0258, Hanscom AFB, Massachusetts, 1976, 79 pp.
21 T.C., Jr., Hall and F.E. Blacet, J. Chem. Phys., 20 (1952) 1745-1749.
22 A.W. Brewer, C.T. McElroy and J.B. Kerr, Nature, 246 (1973) 129-133.
23 G.E. Shaw, J. Geophys. Res., 81 (1976) 5791-5792.
24 C. Tomasi, F. Prodi, M. Sentimenti and G. Cesari, Appl. Opt., 22 (1983) 622-630.
25 A. Angström, Geogr. Ann., 11 (1929) 156-166.
26 R.J. Charlson, N.C. Ahlquist and H. Horvath, Atmos. Environm., 2 (1968) 455-464.
27 G. Hänel, Advances in Geophysics, 19 (1976) 73-188.
28 M. Griggs, J. Air Pollution Control Ass., 22 (1972) 356-358.
29 G.E. Shaw, J.A. Reagan and B.M. Herman, J. Appl. Meteor., 12 (1973) 374-380.
30 M.A. Box and S.Y. Lo, J. Appl. Meteor., 15 (1976) 1068-1076.

31 D. Deirmendjian, Electromagnetic Scattering on Spherical Polydispersions, Elsevier, New York, 1969, pp. 75-119.
32 E.P. Shettle and R.W. Fenn, Models for the aerosols of the lower atmosphere and the effects of humidity variations on their optical properties, Environm. Res. Papers, No. 676, AFGL-TR-79-0214, Hanscom AFB, Massachusetts, 1979, 94 pp.
33 C. Tomasi and F. Prodi, J. Geophys. Res., 87 (1982) 1279-1286.
34 S. Twomey, Atmospheric Aerosols, Elsevier, Amsterdam, 1977, pp. 4-14.

Optical Remote Sensing of Air Pollution,
Lectures of a course held at the Joint Research Centre, Ispra, Italy, 12–15 April 1983,
P. Camagni and S. Sandroni (Eds). 329–350

LASER-FLUORESCENCE TECHNIQUES

N. OMENETTO

1. GENERAL CONSIDERATIONS

Nowadays, it is amply recognized that there is an increasing demand and an effective need for sensitive and instantaneous detection methods able to quantify several atmospheric pollutants over a large concentration range. As an example, one can simply recall the increasing attention given to radicals such as NO and SO_2 because of their important role in the problem known as "acid rain", NO and SO_2 being converted to nitric and sulphuric acids, respectively. Therefore, it is not surprising that the technique of laser induced fluorescence has been considered and tested in the field of remote sensing, in view of its specificity and, above all, its sensitivity which, on theoretical grounds, appears to be much higher than that achievable with most other monitoring methods (refs. 1-8). This can be deduced from the values collected in Table I, which shows order of magnitude estimates of the differential cross sections pertaining to several methods which could be and/or have already been put to test in remote sensing experiments.

The interaction processes are usually characterized by the value of their wavelength dependent cross section (cm^2) rather than in terms of the familiar decadic extinction coefficient for absorption, i.e. $\epsilon(\lambda)$ (liters $mole^{-1} cm^{-1}$). The relation between these quantities as well as several other expressions and/or definitions which are thought to be useful for the considerations presented in this lecture are collected in Table II.

It is worth noting that the term "Resonant Raman Scattering" has been omitted from Table I, the reason being that the terms Resonant Raman and Resonance Fluorescence refer to the same physical process where the excitation wavelength must match a molecular electronic transition, i.e. a process where the energy of the photon corresponds to the energy difference between the initial and final states of the molecule (1, 6, 9, 10). It is therefore clear that the fluorescence technique, in contrast with the other scattering techniques, needs <u>spectral tunability</u> over the absorption profile of the species to be detected.

TABLE I

Typical range of differential cross sections for several interaction processes pertinent to remote sensing of atmospheric pollutants

Interaction process	Differential cross section ($d\sigma/d\Omega$), $cm^2\ sr^{-1}$
Mie scattering	$10^{-8} - 10^{-10}$
Rayleigh scattering	$10^{-26} - 10^{-27}$
Raman scattering	$10^{-29} - 10^{-30}$
Differential absorption and scattering	$10^{-18} - 10^{-23}$
Atomic fluorescence	$10^{-14} - 10^{-15}$
Quenched atomic fluorescence	$10^{-16} - 10^{-17}$
Molecular fluorescence	$10^{-19} - 10^{-21}$
Quenched molecular fluorescence	$10^{-24} - 10^{-26}$

TABLE II

Useful quantities and relationships

$\sigma(\lambda)$	=	wavelength dependent absorption cross section; cm^2
$k(\lambda)$	=	wavelength dependent absorption coefficient; cm^{-1}
n_i	=	number density of species i; cm^{-3}
$\epsilon(\lambda)$	=	decadic molar extinction coefficient; liters $mole^{-1}$ cm^{-1}
– $k_i(\lambda)$	=	$\sigma_i(\lambda)\, n_i$
– $\epsilon(\lambda)$	=	$2.62 \times 10^{20}\ \sigma(\lambda)$
– $\sigma(\lambda)$	=	$3.81 \times 10^{-21}\ \epsilon(\lambda)$
– 1 ppmv	=	2.45×10^{13} molecules cm^{-3} (298 K, 1 atm)
– 1 ppmv	=	$\frac{\text{Molecular weight}}{24.5}$ (g) x 10^3 $\mu g\ m^{-3}$
– Photon energy	=	$\frac{1.98 \times 10^{-16}}{\lambda(nm)}$; J
– E_ℓ	=	Energy per laser pulse = Peak power (W) x pulse duration (s); J
– Photons per laser pulse	=	$\frac{E_\ell\ \lambda\ (nm)}{1.98 \times 10^{-16}}$
– Average power	=	Peak power x duty cycle, W (square pulse)
– Duty cycle	=	Pulse duration (s) x Repetition frequency (Hz)

The interaction processes reported in Table I are schematically depicted in Fig. 1. For the sake of completeness, the laser induces optogalvanic and optoacoustic effects are also shown. Indeed, the use of the PADAR technique (Photoacoustic Detection and Ranging) has been recently discussed in the literature with regard to the remote detection of gases (ref. 11).

From the Table and the Figure, one can conclude that the fluorescence technique is advantageous over the other processes, mainly because: (i) it is selective since both excitation and emission wavelengths are characteristic of the constituent under investigation; (ii) it is sensitive, because of the large cross section; (iii) it is spectrally free from interferences due to the unspecific, on-line scattering signals.

However, as seen from the values given in Table I, quenching is rather severe both in atomic and molecular fluorescence, resulting in a significant reduction of the signals strength as well as of the effective lifetime of the excited level. The pertinent cross section for the fluorescence process can be expressed as follows (ref. 1)

$$\sigma_F(\lambda,\lambda_\ell) = \sigma_F(\lambda_\ell) \cdot S^F(\lambda) \qquad (1)$$

where

$$\sigma_F(\lambda_\ell) = \lambda_\ell \sigma^{abs}(\lambda_\ell)\, \phi_F/\lambda \qquad (2)$$

Here, λ_ℓ stands for the laser excitation wavelength, $\sigma^{abs}(\lambda_\ell)$ is the absorption cross section at λ_ℓ, $S^F(\lambda)$ represents the fraction of the total fluorescence emitted into the wavelength interval $(\lambda, \Delta\lambda)$, and ϕ^F is the quantum efficiency or quantum yield of the fluorescence process, i.e. the ratio between the number of photons emitted and absorbed or the fractional probability of radiative de-excitation with respect to the total probability of de-excitation of the level. $\sigma^F(\lambda_\ell)$ is therefore the spectrally integrated cross section for the fluorescence excited at λ_ℓ. Assuming that the fluorescence is isotropic and that the measurements are taken on resonance $(\lambda = \lambda_\ell)$, we have

$$\frac{d\sigma(\lambda_\ell)}{d\Omega} = \frac{\sigma^{abs}(\lambda_\ell)}{4\pi}\left(\frac{\tau}{\tau_{rad}}\right) \qquad (3)$$

where ϕ^F has now been expressed as the ratio between the lifetime of the level in the presence of collisions and the radiative lifetime. Ba taking $\sim 10^{10}\ s^{-1}$ as being the collision rate at atmospheric pressure and 10 μs for the molecular

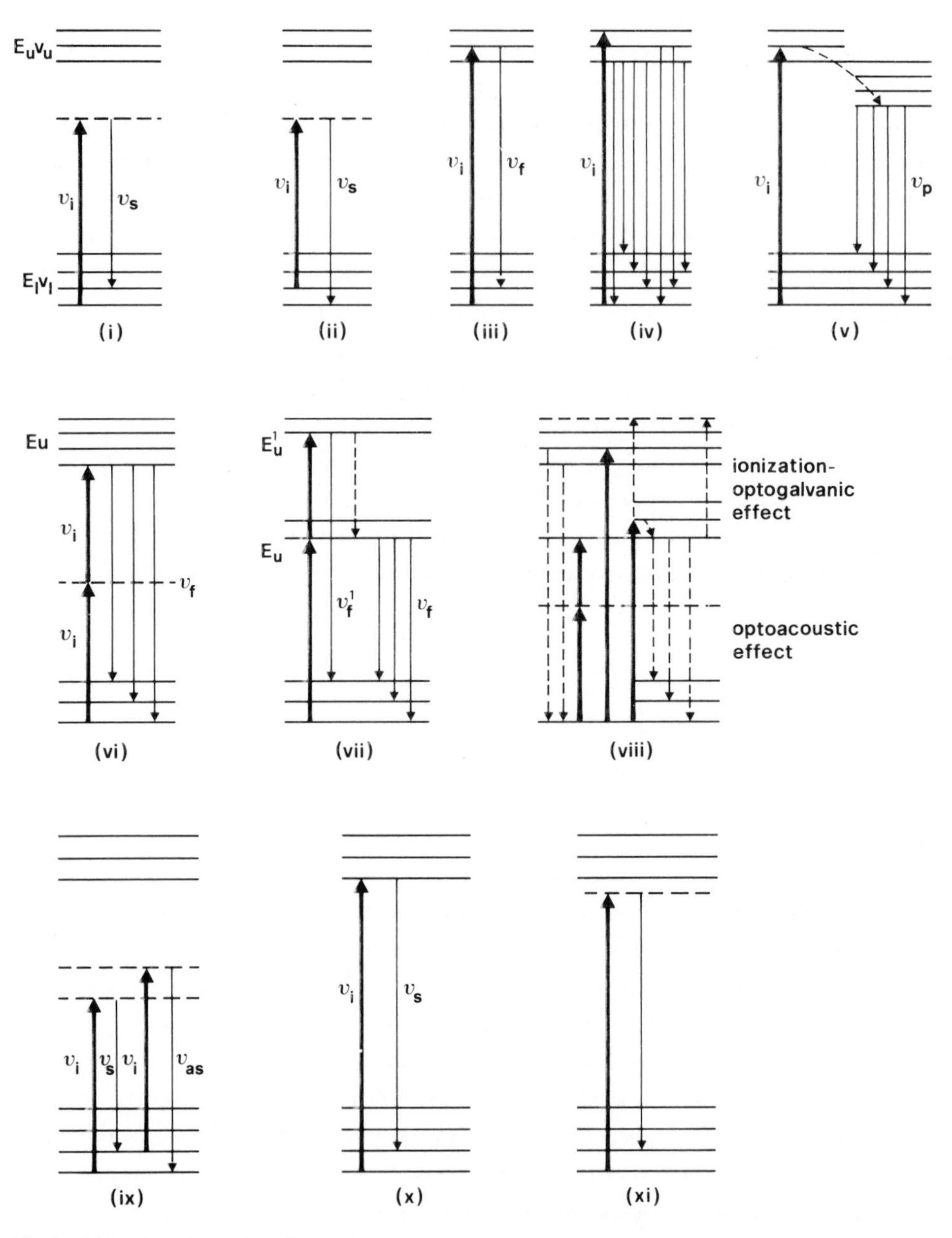

Fig. 1 - Schematic representation of molecular transitions in optical spectrometry. From N. Omenetto and J.D. Winefordner: CRC Crit. Rev. Anal. Chem., 13, 59 (1981). Reproduced by permission. (i) Stokes spontaneous Raman; (ii) Anti-Stokes spontaneous Raman; (iii) Resonance and Stokes narrow band fluorescence; (iv) Broad band fluorescence; (v) Broad band phosphorescence; (vi) 2-photon absorption (virtual level) fluorescence; (vii) Multiple photon absorption with two lasers (real levels) fluorescence; (viii) Optogalvanic and optoacoustic effects; (ix) Anti-Stokes coherent Raman scatter; (x) Resonance Raman; (xi) Near-resonance Raman.

radiative lifetime, one can see that the quenching factor is 10^{-5}.

It would therefore be advantageous to excite fluorescence transitions from a level whose radiative lifetime is much shorter than a microsecond so that the fluorescence process can effectively compete with the collisional quenching process. For example, SO_2 fluorescence would give much better sensitivity if excited in the spectral region from 2000 to 2300 Å because of the radiative decay time of approximately 9 ns of the excited state produced (ref. 12).

The number of species that can in principle be detected by means of the fluorescence technique is reported in Table III. This long list includes atomic as well as polyatomic species, with the exception of atoms such as sodium whose fluorescence has been observed at night at 90 km of altitude with a dye laser tuned at one of the D lines (ref. 13). Also, a Shuttle-borne Lidar system based on the detection of backscattered resonance fluorescence radiation has been recently modelled (ref. 14) and applied to the measurements of sodium and potassium number densities in the upper atmosphere (80-110 km). The same authors (ref. 15) have simulated numerically similar measurements of magnesium ion number density in the ionosphere (80-500 km).

The presence of many radicals in the list (CH, CN, OH, etc.) stems from the great attractiveness and potential of the laser induced fluorescence technique in the field of combustion diagnostics (ref. 16). In fact, flames and similar hostile environments do benefit from the use of non-intrusive optical techniques. As a result, there is an enhanced understanding of the complex dynamics of combustion processes.

In addition to the species of Table III, Table IV reports several compounds which have received detailed laboratory investigation because of their potential presence as atmospheric particulates and therefore also subjected to fluorescence under laser irradiation. These materials were grouped in several categories such as: (i) fly ash from coal-burning plants; (ii) calcium and ammonium sulphates from heating sources; (iii) lead, copper, cadmium and mercury compounds from smelters, incinerators, alloy plants and steel mills; (iv) fluorides from aluminium reduction plants or phosphorous plants; (v) other particles of general interest including sulphates, chlorides, oxides and sulphides of lead, zinc, cadmium, mercury, copper, nickel, chromium, vanadium, arsenic and beryllium; (vi) coal sample as a typical organic material (ref. 17). As shown in the Table, nine out of thirtyone tested samples gave sizeable fluorescence signals.

It can therefore be anticipated that the aerosol fluorescence, when present, can mask or be superimposed on the fluorescence return of the particular

TABLE III

Atomic, diatomic and polyatomic species amenable to fluoresce under laser escitation[a)]

Species	Spectral absorption region (A°)	Observations
C	1560	not practical
C_2	4000 – 6000	
CH	3600 – 5000	
CH_2 (Methylene)	1415	not practical
CH_3 (Methyl)	2160	not practical
CH_4 (Methane)	500 – 1455	not practical
C_2H_2 (Acetylene)	2100 – 2370	
C_2H_4 (Ethylene)	2600 – 3400	
C_6H_6 (Benzene)	2270 – 2670; 3000 – 3400	high sensitivity
$C_{10}H_8$ (Naphtalene)	2400 – 3000; 3000 – 3800	high sensitivity
$C_{14}H_{10}$ (Anthracene)	3000 – 3800; 3800 – 4800	high sensitivity
CH_2O (Formaldehyde)	2300 – 3500; 3600 – 4000	
Cl	120; 133 – 140	not practical
Cl_2	visible – 5560	
ClO	2600 – 3000	
ClO_2	2700 – 4800	
CN	3500 – 4600	
CO	1500 – 2400	
CO_2	1400 – 1700	not practical
CS	3500 – 2700	
CS_2		multiphoton excitation scheme
H	1215 (Lyman)	not practical
H_2	1200	not practical
HCl	< 2300	not practical
HCN	1600 – 1915	not practical
HNO_2	3000 – 4000	very dissociative absorption bands. Multiphoton excitation scheme.
HNO_3	1200 – 3000	very dissociative absorption bands.
H_2O_2		multiphoton excitation scheme
H_2S	1900 – 2700	very dissociative absorption bands.
N	1135	not practical
N_2	vacuum uv	not practical
NH	3360	
NH_2	4300 – 9000	
NH_3	1700 – 2170	not practical
NO	1950 – 3400	high sensitivity
NO_2	3200 – 10,000	high sensitivity
NO_3		multiphoton excitation scheme
O	1300	not practical
O_2	1700 – 2200	not practical
OH	2400 – 4000	high sensitivity
S	1807	not practical
SO_2	2600 – 3400; 3400 – 3900	high sensitivity

a) from refs. 2, 16, 20

TABLE IV

Spectral characteristics of several inorganic and organic compounds potentially present as by-products of chemical and industrial plants[a)]

Species	Observations
AlF_3	
$Al_2(SO_4)_3$	
CaF_2	Fluorescence observed for all these compounds,
$CaSO_4$	at excitation wavelengths ranging from
$CuSO_4$	3000 to 3800 A°.
Cryolite	
Coal sample	
$HgSO_4$	
Phosphate rock sample	
CdO	
$CdSO_4$	
$CuCl_2$	
CuO	
Fly ash	No significant fluorescence could be observed
H_2SO_4	for these compounds[b)].
HgO	
HgS	
$PbCl_2$	
PbO	
PbS	
$PbSO_4$	

a) from ref. 17.

b) The reader should be aware that the laser induced fluorescence of some of these compounds has been investigated and reported in the literature.

species investigated, resulting in a serious degradation, if not in the impossibility of gathering meaningful information from the backscattered signal. Indeed, a "fluorescent particle lidar" system has been proposed (ref. 18) to study the pollution plume transport and diffusion and demonstrated (ref. 19) by observing the return signal from fluorescent particles sprayed into the base of a 30-m tall ventilating stack emitting an airflow of 112 $m^{-3}min^{-1}$. The particles contained a fluorescent dye encapsulated in a thermoplastic material, were characterized by an ellipsoidal shape and had a 0.3 μm mean diameter (ref. 19).

As shown in Table III, many atomic as well as molecular species have their strongest absorption lines or bands below 2000 Å, i.e. in the vacuum uv region. As well known, in Lidar applications, the strong absorption of radiation due to atmospheric oxygen negates the practical use of wavelengths shorter than 2300 Å for pathlengths greater than 100 m and below 1850 Å the atmosphere is totally opaque (ref. 5). One should also recall that at an ozone concentration of 0.1 ppm, the absorption at 2500 Å ($\sim$200 cm^{-1}) compares with that due to oxygen (ref. 5). This of course does not preclude the use of balloon-borne lidar systems equipped with a suitable sampling apparatus.

As a general comment, as Davis et al (ref. 20) state, one could say that for any gaseous molecular species being present in the atmosphere in a concentration range of 10^6 molecules per cm^3 (sub-pptv), having a bonding excited state with a radiative lifetime of 1 μs and an absorption cross section of 10^{-18} cm^2 in the near uv-visible region, one should challenge the laser excited fluorescence as a selective and sensitive monitoring technique. One can also rely upon the use of multiphoton laser induced fluorescence techniques (refs. 20, 21) especially in those cases where bonding excited states do not exist but the molecule can be photolysed at a given wavelength to give a molecular fragment in a vibrationally excited state which could then be pumped into the first excited electronic state by another wavelength. The advantage of this scheme is that the fluorescence can subsequently be observed at a third wavelength which is (much) lower than the other two, therefore allowing for spectral rejection of the noise generated at the pumping frequencies (ref. 20).

On the other hand, despite the great attractiveness of the fluorescence technique, it should be clearly stressed that the significant amount of data published in the literature refer, to the best of our knowledge, either to laboratory investigations or to data obtained with airborne instrumentation and sampling platforms with independent calibration apparatus. We are therefore still lacking an experimental demonstration of the real feasibility of laser induced fluorescence for the remote sensing of atmospheric pollutants at the typical operating ranges of a Lidar instrument and in a typical atmospheric environment where quenching effects, signal losses due to scattering and other interfering fluorescence sources must be carefully considered.

2. FLUORESCENCE STUDIES OF OH, NO, NO_2, SO_2 and AEROSOLS

2.1 The hydroxyl radical fluorescence

This radical, being not in itself a pollutant, could however play a crucial role in controlling daytime photochemical processes. It can be safely stated that the OH fluorescence has been investigated as deeply as the atomic fluorescence of sodium, whose characteristics have been described in a large number of publications.

A balloon-borne lidar system has been built at the Goddard Space Flight Center (NASA) and flown in the stratosphere where an OH density of 5×10^6 cm^{-3} could be detected (ref. 22). In addition, by observing the fluorescence from two excited vibrational states, pressure and temperature, respectively, can be remotely monitored (ref. 23).

A schematic diagram showing the excitation and fluorescence wavelengths of OH is shown in Fig. 2. In this scheme (ref. 20), the optical pumping takes place at 2819 Å while the fluorescence detection is set at 3095 Å. A complete airborne lidar system capable of detecting OH concentration ranging from 30 parts-perquadrillion (3.7×10^5 molecules cm^{-3}) at altitudes of 6 km to 0.8 parts-pertrillion (2.0×10^7 molecules cm^{-3}) at 0.5 km has been described (ref. 20). The laser consists of a Nd-YAG pumped dye laser equipped with a Fabry-Perot etalon to provide a spectral excitation bandwidth of 0.013 Å. In order to minimize interferences from the presence of broad band fluorescence compounds and in order to clearly establish the fluorescence noise level in the vicinity of the absorption line, measurements are taken alternately on and off the absorption peak.This procedure has recently received detailed scrutiny in view of an optimization of the attainable signal-to-noise ratio (ref. 24).

2.2 NO fluorescence

A detailed study of the fluorescence characteristics of this molecule has been recently published by Bradshaw et al. (ref. 25). As shown in the schematic diagram of Fig. 3, NO is excited into the v'=0 manifold of the $A^2\Sigma^+$ electronic state by a laser photon centered at 2260 Å, while shifted fluorescence is observed at 2590 Å. As excitation source, a Nd-YAG pumped dye laser, equipped with a frequency doubling crystal and a frequency mixing crystal to

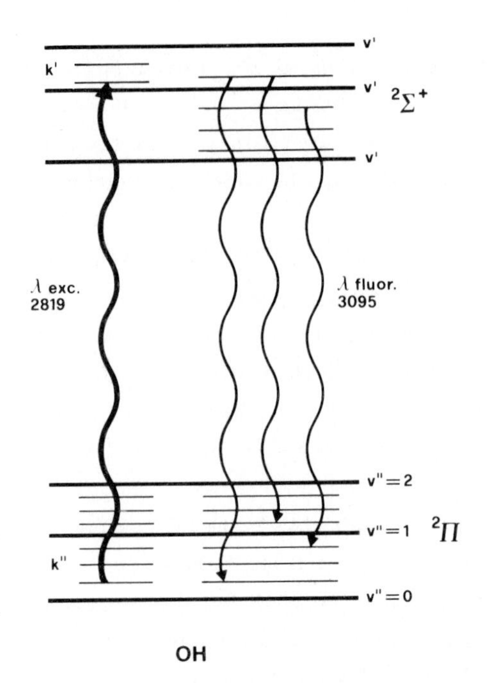

Fig. 2 - Schematic diagram of the energy levels pertinent to OH fluorescence. Redrawn from ref. 20.

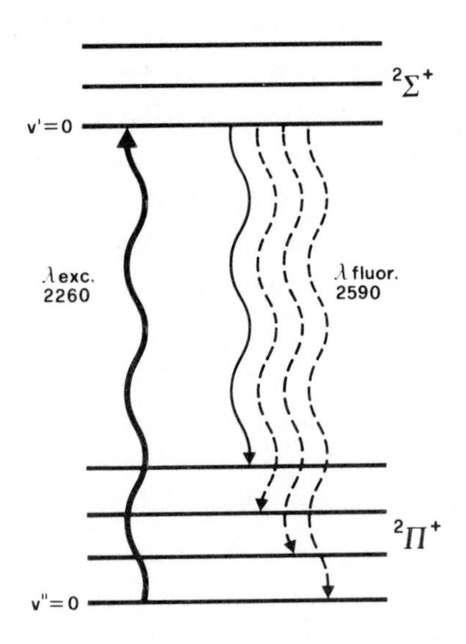

Fig. 3 - Schematic diagram of the energy levels pertinent to NO fluorescence. Redrawn from ref. 25.

achieve the 2260 Å transition, is used. The limit of detection for airborne field experiments is estimated at 8 pptv for an integration time of 20 minutes. This level appears to be well within the normal concentration range expected for natural tropospheric levels (3-300 pptv). It was also shown (ref. 25) that the interference expected from NO_2 and SO_2 in tropospheric conditions should be negligible.

2.3 NO_2 fluorescence

Ambient atmospheric concentrations of NO_2 vary widely depending upon location but should be less than 10 ppbv. Since photodissociation of the molecule occurs at wavelengths shorter than 4100 Å, the fluorescence is usually excited in the blue region, where cw argon ion lasers (4880 Å and 4579 Å) and He-Cd laser (4420 Å) are easily available (refs. 26-29). Special care was taken by Tucker et al. (ref. 28) in designing a laser NO_2 monitor based on the 4880 Å excitation and photon counting detection. Several filters and baffles were used to minimize scattered light. In an improved version (ref. 29) using an excitation wavelength of 4420 Å, a hundredfold greater sensitivity was achieved. Interferences due to the presence of other gases such as NO, SO_2, ozone and water vapour were not detectable. However, quenching is severe at atmospheric pressure due to the long radiative lifetime of the excited level(s). The limit of detection for an 80 sec integration time was found to be approximately 0.2 ppbv. Calibration was performed by flowing in the fluorescence cells known concentrations of NO_2 in nitrogen or air using a calibrated NO_2 permeation tube.

2.4 SO_2 fluorescence

Electronic absorption and fluorescence spectra of SO_2 have been widely studied in the past and recently (refs. 30-32, 12, 25). When excited within the absorption band located at 2600-3200 Å, the radiative lifetime is 50 μs and therefore, as for NO_2, strong quenching is effective. As previously reported (ref. 12), the absorption band at 1800-2300 Å should offer a much improved detection. In fact, Bradshaw et al. (ref. 25) using excitation at 2220 Å and detection at 2600 Å, predicted a detection limit in a field sampling system of 1.6 pptv for a 20 minutes integration time, this limit being more than adequate for typical free tropospheric levels with range between 50 and 90 pptv.

It should be borne in mind that the SO_2 absorption spectrum shows fine structure. For example, Woods et al. (ref. 33) in their high resolution study of the absorption of SO_2 at atmospheric pressure and at temperatures of 100°C have shown that a considerable error (~20%) will result in concentration measurements if the spectral fine structure is not accounted for.

When excited in the region 2650-3000 Å, the fluorescence has a broad spectral distribution, peaked around 350 nm and extending with a long wavelength tail up to 4500 Å (see Fig. 4). Phosphorescence appears also clearly at higher pressures and is spectrally structured.

2.5 Aerosol fluorescence

As stated previously, the fluorescence of aerosols is a potential interference in all cases considered above. This type of fluorescence was first noted by Gelbwachs et al. (refs. 34, 35) during the investigation of the NO_2 fluorescence since the presence of aerosols in the air stream analysed resulted in a broad band fluorescence that was more intense than that of NO_2. This interference disappeared, in the "in situ" apparatus, by an appropriate aerosol filter. Allegrini and Omenetto (ref. 36) have investigated the potential interference form aerosol fluorescence on the detection of NO_2 and SO_2 by both Raman and fluorescence techniques. A simple generator was used to produce desolvated aerosols of fluoranthene and tar, which are likely to be present in the atmosphere. The aerosols were directed into a fluorescence cell traversed by a pulsed dye laser beam.

Figures 5 and 6 show the fluorescence spectra obtained with gated photon counting electronics. Typical conditions for the measurements were: 10^5 cm^{-3} particle density in the cell; 0.4 μm mean particle diameter; solvent cyclohexane. The emission spectrum of fluoranthene (Fig. 5) is clearly visible in the blue region. At an excitation wavelength of 2600 Å, the fluorescence starts approximately at 4000 Å and peaks in the 4500-5000 Å region. Tar fluoresces very weakly at this excitation, while its emission is easily observable when the laser is tuned, as for NO_2, at 4200 Å. In this case, emission extends approximately up to 6500 Å (Fig. 6). An important application of this technique would be the identification and the quantification of the aerosol material, but this seems to be an exceedingly difficult task.

As indicated by Gelbwachs et al. (refs. 34, 35), the measurements of the decay time of the fluorescence is important for assessing the feasibility of a temporal discrimination of the interference signal. Preliminary measurements (ref. 36) have shown that aerosol fluorescence decay time is of the order of a microsecond. Some theoretical calculations based upon the signal-to-noise expression pertinent to lidar operation will be given in the subsequent section.

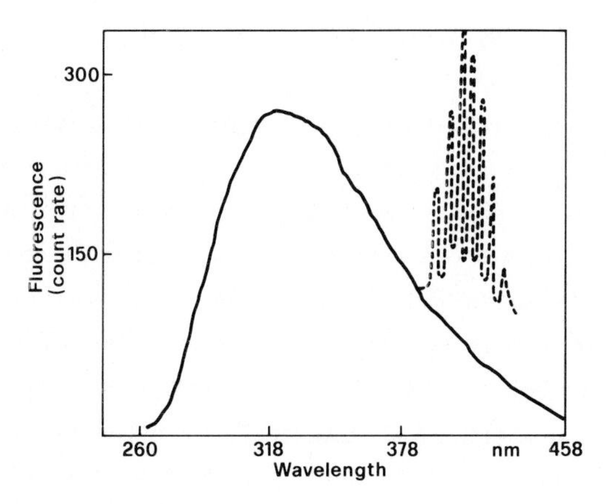

Fig. 4 - Low resolution fluorescence spectrum of SO_2 (9 mTorr). The excitation wavelength is set at 2650 Å. Indicated in the figure is also the structured phosphorescence in the 4000 Å region, which appears at higher pressures.

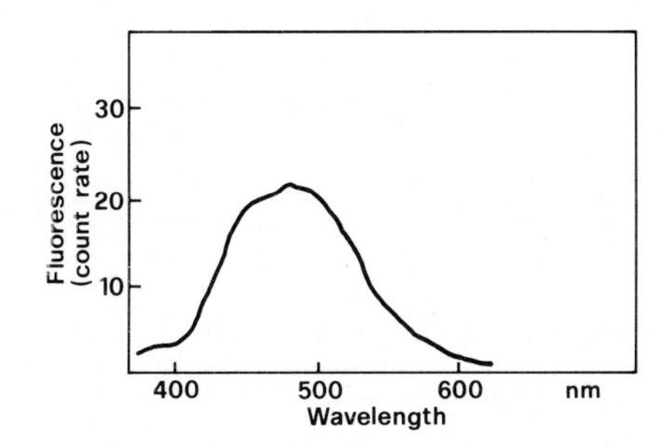

Fig. 5 - Fluorescence spectrum of fluoranthene aerosols excited at 2650 Å by a frequency doubled, nitrogen pumped dye laser. Redrawn from ref. 36.

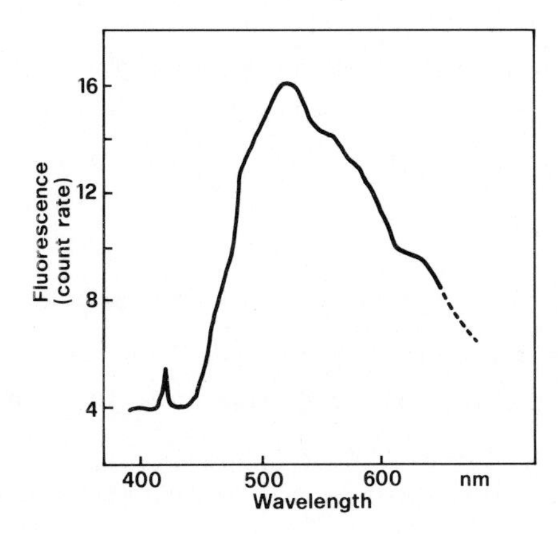

Fig. 6 - Fluorescence spectrum of tar aerosols excited at 4210 Å by a nitrogen pumped dye laser. Redrawn from ref.36.

3. FLUORESCENCE LIDAR EQUATION: EFFECT OF FLUORESCENCE DECAY TIME

The following treatment applies to the usual monostatic configuration of an environment lidar (laser and receiver close to each other). The backscattered radiation from the target is spectrally analysed to discriminate it from other radiations and background. As recognized by Kildal and Byer (ref. 5) and thoroughly discussed by Measures (refs. 1, 37), in the case of a fluorescent target, finite relaxation effects have to be taken into consideration. In fact, the interaction between the laser beam and the medium can be expressed as follows (ref. 37). In the case of a purely scattering medium, the laser induced spectral radiance, $W(\lambda, R)$, ($W\ cm^{-4}\ sr^{-1}$) per unit range increment at wavelength λ and range R, is given by the relation

$$W(\lambda,R) = \sum_i \{n_i(R) \frac{d\sigma_i^{sc}(\lambda_\ell)}{d\Omega} S^{sc}(\lambda) \frac{E_\ell T(\lambda_\ell,R)}{\tau_\ell A_E(R)}\} \tag{4}$$

where

$n_i(R)$ = number density of the i-scatterer; cm^{-3}

$\frac{d\sigma^{sc}(\lambda_\ell)}{d\Omega}$ = differential cross section at the laser wavelengths; $cm^2\ sr^{-1}$

$S^{sc}(\lambda)$ = fraction of scattered radiation collected; cm^{-1}

$T(\lambda_\ell, R)$ = atmospheric transmission at λ_ℓ and range R

E_ℓ = energy of laser pulse; J

τ_ℓ = duration of the rectangularly shaped laser pulse; s

$A_E(R)$ = area of irradiation by laser beam at range R; cm^2

On the other hand, if the medium possesses fluorescence properties, we have:

$$W(\lambda,R) = \sum_i \{n_i^*(R) \frac{hc}{\lambda} \frac{1}{\tau_{rad}} \frac{1}{4\pi} S_i^F(\lambda)\} \tag{5}$$

where

$n_i^*(R)$ = number density of laser excited molecules; cm^{-3}

(hc/λ) = energy of the photon at wavelength λ; J

$S_i^F(\lambda)$ = fraction of fluorescence radiation collected; cm^{-1}

This different interaction therefore leads to a different Lidar equation. The well documented Lidar equation for a purely scattering medium can be written in terms of the increment of radiative energy received by the detector during the interval (t, τ_d), where τ_d is the integration period for the detector and $t = (2R/c)$. We then have

$$E(\lambda,R) = E_\ell T_r(\lambda)T(R)(\frac{A_r}{R^2})n(R)\frac{\sigma^{sc}(\lambda_\ell)}{4\pi}\frac{c\tau_d}{2} \qquad (6)$$

where

$T_r(\lambda)$ = optical efficiency of the detection system

$T(R) \equiv T(\lambda_\ell,R)T(\lambda,R) = \exp\{-\int_0^R [\epsilon(\lambda_\ell,R) + \epsilon(\lambda,R)]\,dR\}$

= total atmospheric transmission factor, where $\epsilon(\lambda_\ell, R)$ and $\epsilon(\lambda, R)$ are the atmospheric attenuation coefficients at the laser and detection wavelengths, respectively

A_r = effective aperture of receiving optics; cm^2

R = range; cm

The effective range resolution for this system is given by $c[\tau_d + \tau_\ell)/2]$. It should be emphasized that Eq. (6) holds for:

i) a rectangularly shaped laser pulse of duration τ_ℓ;
ii) one scattering species in a homogeneous scattering medium;
iii) isotropic scattering cross section;
iv) spectrally narrow-band returned radiation;
v) range of interest R much greater $(c\tau_\ell/2)$; and finally
vi) integration period of the detector much smaller than $(2R/c)$.

In the case of a fluorescent medium, the laser radiation will be able to excite a given fraction of molecules from the initial ground state to a specific excited level. The <u>response time</u> of the system plays an important role here. If the radiation density is such that optical saturation effects can safely be neglected, the time variation of the excited state population (dn^*/dt) can be given by the following expression (ref. 37):

$$\frac{dn^*(R,t)}{dt} = \frac{\sigma^{abs}(\lambda_\ell)\lambda_\ell}{hc}\, n(R,t)\, I(R,t) - \frac{n^*(R,t)}{\tau} \qquad (7)$$

where

$n(R, t)$ = number density of molecules in the ground state; cm^{-3}

$I(R, t)$ = intensity of laser beam at range R and time t as given by the last factor in Eq. (4); W cm^{-2}

τ = effective lifetime of the excited level, which includes both radiative and collisional deactivation; s

With the assumption that the ground state number density prior to irradiation is $n_o(R)$ and that $n^*(R,0) = 0$, Eq. (7) can be solved to give

$$n^*(R,t) = \frac{n_0(R)\sigma^{abs}(\lambda_\ell)\lambda_\ell}{hc} \exp(-t/\tau) \int_0^t I(R,x)\, e^{x/\tau}\, dx \tag{8}$$

where $I(R,x)$ represents the laser intensity at range R and at a time x after the leading edge of the laser pulse reaches this location. If one considers that the target medium at a range $R' < R$ has been exposed to laser pumping radiation for a period $(R-R')/c$, then by substitution of Eq. (8) into Eq. (5) and by making use of Eqs. (2) and (3), we obtain

$$W(\lambda,R') = \frac{n_0(R')\, S^F(\lambda)}{4\pi} \frac{\sigma^F(\lambda)}{\tau} e^{-t'/\tau} \int_0^{t'} I(R',x)\, e^{x/\tau}\, dx \tag{9}$$

where $t' \equiv 2(R'-R)/c$.

If we now assume a rectangularly shaped laser pulse as before and that the target boundary lies at R_o, we see that in the range interval $(ct/2)$ to $c(t-\tau_\ell)/2$ the number of excited molecules increases, while in the range interval $c(t-\tau_\ell)/2$ to R_o the excited state density decays with its characteristic decay time τ. The growth in the excited number density will be represented by the integral

$$\int_{\frac{c(t-\tau_\ell)}{2}}^{ct/2} [1 - e^{-(t-\frac{2R'}{c})/\tau}]\, e^{-\epsilon(R'-R_0)} \frac{dR'}{R'^2} \tag{10}$$

while the decay will be represented by the integral

$$\int_{R_0}^{c(t-\tau_\ell)/2} [1 - e^{-(t-\tau_\ell)/\tau}]\, e^{-\epsilon(R'-R_0)}\, e^{-(t-\tau_\ell-\frac{2R'}{c})/\tau} \frac{dR'}{R'^2} \tag{11}$$

The range resolution in this case will be given by $c(\tau_d+\tau_\ell+\tau)/2$, as seen from Fig. 7. When the integrals are solved and with the assumptions that the spectral window of the detection system processes only a small fraction of the fluorescence spectrum and that the ground state population density is constant, the lidar equation for fluorescence can be written as

$$E(\lambda,R) = E_\ell T_r(\lambda) T(R_0) (\frac{A_r}{R^2}) (\frac{n_0 \sigma^F(\lambda_\ell) S^F(\lambda)}{4\pi}) \cdot (\frac{c\tau_d}{2}) \{\gamma(R)\, e^{-\epsilon(R-R_0)}\} \tag{12}$$

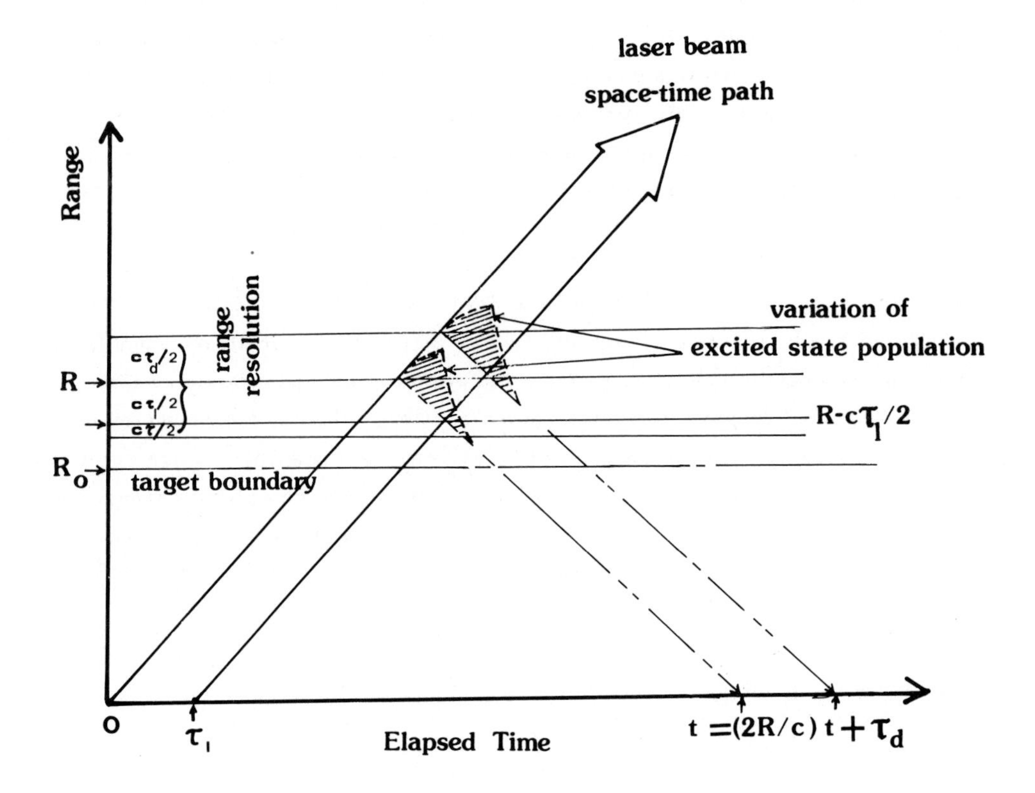

Fig. 7 - Space-time view of laser pulse propagation and excitation of fluorescent target medium for a rectangular laser pulse. From Measures (37), reproduced by permission.

It can be seen that this equation closely resembles Eq. (6), apart from the last factor in braces. Measures (refs. 1, 37) has called $\gamma(R)$ a "correction factor" that takes into account the effects of finite lifetime of the fluorescence, the laser pulse duration and shape and the detector integration period. As a representative example, Fig. 8 shows that such correction factor approaches unity for large penetration depths in an optically thin target medium, as would be typical for atmospheric applications. It is clear, however, that the use of Eq. (6) instead of Eq. (12) in the case of a fluorescent target can seriously overestimate the expected signal in situations where the return arises close to a sharp boundary, leading to large concentration errors.

4. INTERFERENCE EFFECTS DUE TO OTHER INTERACTION PROCESSES

We shall now evaluate the effect of other potential backscattered signals on the fluorescence signal expected from the pollutant sought. The detection system is assumed to work in a gated photon counting scheme, i. e. a photomultiplier is used as optoelectrical transducer, and the signal pulses are collected and averaged over a gate width of variable duration and delay with respect to a suitable trigger pulse that enables the periodic start of data processing. The average rate, R_m, of signal pulses for a given process characterized by a decay time τ will be given by

$$R_m = \frac{Np}{\tau} e^{-t/\tau} \tag{13}$$

where N_p is the number of photons per laser pulse and is given by a simple modification of one of the appropriate equations derived before, expressed in number of photons rather than in energy or power units. For example, in the case of fluorescence with $\gamma(R) = 1$, the number of fluorescence photons per laser pulse, $N_{p,F}$ will be given by

$$N_{p,F} = n_F V P_\ell \sigma_F(\lambda) \eta_T S^F(\lambda) \tag{14}$$

where V is now the volume of sampling, p_ℓ is the number of photons per cm^2 per pulse of the exciting beam; $\sigma_F(\lambda) \equiv \sigma^{abs}(\lambda)\phi$ is the cross section of the fluorescence; n_T is an overall efficiency factor and $S^F(\lambda)$ is the fraction of fluorescence radiation collected.

During the gate opening time, t_g, the detection system is supposed to collect all the signal pulses, i. e.

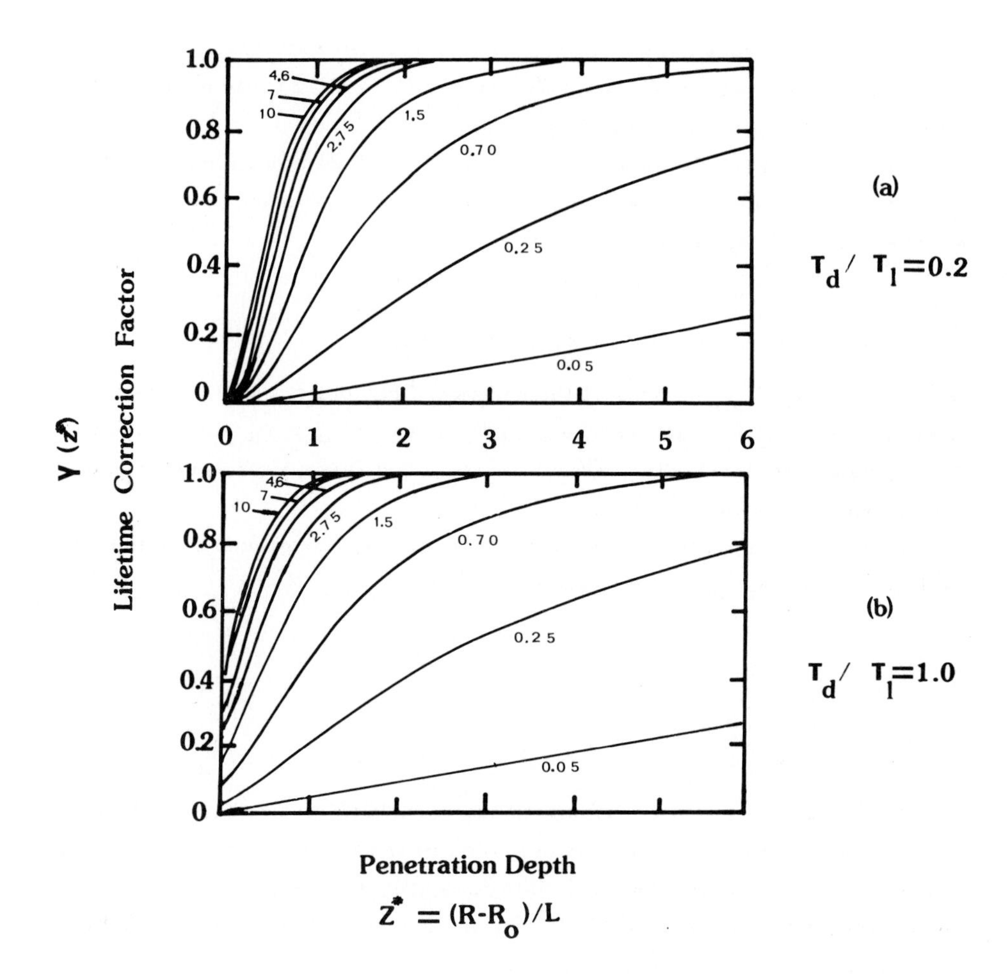

Fig. 8 - Variation of correction factor, $\gamma(Z^*)$ with normalized penetration Z^* into a fluorescent target for several values of T^* (ratio of pulse duration to fluorescence lifetime). L is the laser pulse length. From Measures (1), reproduced by permission.

$$N_F = \int_0^{t_g} \frac{N_{p,F}}{\tau_F} e^{-t/\tau_F} = N_{p,F}(1 - e^{-t_g/\tau_F}) \tag{15}$$

The number of photons collected over a measuring time Δt, and at a laser repetition frequency f, will be $N_F(\Delta tf)$. We can derive similar expression for the Raman signals and the aerosol fluorescence.

By considering the fluorescence of a selected pollutant being measured in the presence of both Raman and aerosol fluorescence returns, the overall signal-to-noise ratio for the measurement will be

$$\frac{S}{N} = \frac{N_{p,F}(1-e^{-t_g/\tau_F})(f\Delta t)^{1/2}}{\{N_{p,F}(1-e^{-t_g/\tau_F}) + 2N_{p,R}(1-e^{-t_g/\tau_R}) + 2N_{p,a}(1-e^{-t_g/\tau_a}) + 2Dt_g\}^{1/2}} \tag{16}$$

where $N_{p,R}$ and $N_{p,a}$ represent the number of photons per laser pulse due to Raman and aerosol fluorescence signals, respectively; τ_R and τ_a are the decay times associated with the two processes and D is comprehensive of dark counts and background radiation.

If the Raman signal is spectrally rejected and the background contribution dark counts are negligible, we have

$$\frac{S}{N} = \frac{\{N_{p,F}(1-e^{-t_g/\tau_F})\}^{1/2}(f\Delta t)^{1/2}}{\{1 + 2\dfrac{N_{p,a}(1-e^{-t_g/\tau_a})}{N_{p,F}(1-e^{t_g/\tau_F})}\}^{1/2}} \tag{17}$$

We can therefore see that the signal-to-noise ratio obtainable for the fluorescence signal due to the pollutant is degraded by the presence of aerosol fluorescence. This can be quantified by defining a degradation factor, DF, which is given by the ratio between the (S/N) obtained without aerosols to the (S/N) obtained in the presence of aerosols, i.e.

$$\text{D.F.} = \{1 + 2\frac{n_a \sigma_a^{abs} \phi_a (1-e^{-t_g/\tau_a})}{n_F \sigma_F^{abs} \phi_F (1-e^{-t_g/\tau_F})}\}^{1/2} \tag{18}$$

where subscripts a and F stand for aerosol and fluorescence, respectively.

If the gate width t_g is made (much) larger than τ_F, so as to collect all fluorescence photons, Eq. (18) reduces to

$$\text{D.F.} = \{1 + 2\frac{n_a \sigma_a^{abs} \phi_a (1-e^{-t_g/\tau_a})}{n_F \sigma_F^{abs} \phi_F}\} \tag{19}$$

Equation (19) clearly shows the important role played by the decay time of the aerosol fluorescence. If, as indicated by the preliminary laboratory data reported in ref. 36, this lifetime is indeed in the microsecond range while the decay time of the quenched fluorescence is in the nanosecond range, then t_g will be much less than τ_a, so that Eq. (19) can be simplified to give

$$\text{D.F.} = \{1 + 2\left(\frac{n_a \sigma_a^{abs} \phi_a}{n_F \sigma_F^{abs} \phi_F}\right)\left(\frac{t_g}{\tau_a}\right)\}^{1/2} \tag{20}$$

This equation can now be evaluated for two specific aerosol loads and in the case of SO_2 fluorescence.

Case 1 - Typical heavy load conditions

$n_a = 10^5\ cm^{-3}$; $\sigma_a^{abs}\ (\equiv \frac{\pi d^2}{4}) = 7.8x10^{-9}\ cm^2$ for a particle diameter

$d = 1\ \mu m$; $\phi_a \cong 0.5$; $n_{SO_2} = 2.45x10^{13}\ (ppm^{-1})\ C_{SO_2}\ (ppm)$; $\sigma_{SO_2}^{abs} = 10^{-19}\ cm^2$;

$\phi_{SO_2} = 10^{-5}$; $t_g = 5x10^{-9}\ s$; $\tau_a = 2x10^{-6}\ s$.

Inserting these values in Eq. (20) we get

$$\text{D.F.} = \{1 + \frac{7.8 \times 10^4}{C_{SO_2}}\}^{1/2} \tag{21}$$

which is clearly unacceptable for normal concentrations of SO_2 in the atmosphere.

Case 2 - In this case, the particle diameter is taken as being 0.05 μm, so that

$\sigma_a^{abs} = 2x10^{-11}\ cm^2$ and $n_a = 10^2\ cm^{-3}$.

We then obtain

$$\text{D.F.} = \{1 + \frac{0.1}{C_{SO_2}}\}^{1/2} \tag{22}$$

and the fluorescence technique results in a viable approach for SO_2 concentrations equal or greater than 0.01 ppm.

LITERATURE

1 R. M. Measures, Chapter 6 in "Analytical Laser Spectroscopy", N. Omenetto, Editor, J. Wiley and Sons, N. Y. (1979).
2 K. Schofield, J. Quant. Spect. Radiat. Transfer, 17, 13 (1977).
3 B. L. Sharp, Chemistry in Britain, 342 (1982).
4 R. L. Byer, Opt. Quant. Electro., 7, 147 (1975).
5 H. Kildal and R. L. Byer, Proc. of the IEEE, 59, 1644 (1971).
6 H. Rosen, P. Robrish and O. Chamberlain, Appl. Optics, 14, 2703 (1975).
7 R. M. Measures and G. Pilon, Opto-electronics, 4, 141 (1972).
8 M. Birnbaum, "Laser-excited fluorescence techniques in air pollution monitoring", Chapter 5.
9 P. F. Williams, D. L. Rousseau and S. H. Dworetsky, Phys. Rev. Lett., 32, 196 (1974).
10 J. G. Hochenbleicher, W. Kiefer and J. Brandmüller, Appl. Spectroscopy, 30, 528 (1976).
11 D. J. Brassington, J. Phys. D: Appl. Phys., 15, 219 (1982).
12 H. Okabe, J. Amer. Chem. Soc., 93, 7095 (1971).
13 M. R. Bowman, A. J. Gibson and M. C. W. Sanford, Nature, 221, 456 (1969).
14 S. D. Yeh and E. V. Browell, Appl. Optics, 21, 2365 (1982).
15 S. D. Yeh and E. V. Browell, Appl. Optics, 21, 2373 (1982).
16 A. C. Eckbreth, P. A. Bonczyk and J. F. Verdieck, Appl. Spectr. Reviews, 13, 15 (1978).
17 M. L. Wright, EPA-report R2-73-219 (1973).
18 B. G. Schuster and T. G. Kyle, Appl. Optics, 19, 2524 (1980).
19 T. G. Kyle, S. Barr and W. E. Clements, Appl. Optics, 21, 14 (1982).
20 D. D. Davis, W. S. Heaps, D. Philen, M. Rodgers, T. McGee, A. Nelson and A. J. Moriarty, Rev. Scient. Inst., 50, 1505 (1979).
21 J. Bradshaw and D. D. Davis, Opt. Letters, 7, 224 (1982).
22 W. S. Heaps, T. J. McGee, R. D. Hudson and L. O. Candill, NASA report No. X-963-81-27, August 1981.
23 T. J. McGee and T. J. McIlrath, Appl. Optics, 18, 1710 (1979).
24 W. S. Heaps, Appl. Optics, 20, 583 (1981).
25 J. D. Bradshaw, M. O. Rogers and D. D. Davis, Appl. Optics, 21, 2493 (1982).
26 K. Sakurai and H. P. Broida, J. Chem. Phys., 50, 2404 (1969).
27 P. B. Sackett and J. T. Yardley, Chem. Phys. lett., 9, 612 (1971).
28 A. W. Tucker, A. W. Petersen and M. Birnbaum, Appl. Optics, 12, 2036 (1973).
29 A. W. Tucker, M. Birnbaum and C. L. Fincher, Appl. Optics, 14, 1418 (1975).
30 H. D. Mettee, J. Phys. Chem., 73, 1071 (1969).
31 F. P. Schwarz, H. Okabe and J. H. Whittaker, Anal. Chem., 46, 1024 (1974).
32 F. P. Schwarz and H. Okabe, Anal. Chem., 47, 703 (1975).
33 P. T. Woods, B. W. Jolliffe and B. R. Marx, Opt. Comm., 33, 281 (1980).
34 J. A. Gelbwachs, M. Birnbaum, A. W. Tucker and C. L. Fincher, Opto-Electron., 4, 155 (1972).
35 J. Gelbwachs and M. Birnbaum, Appl. Optics, 12, 2442 (1973).
36 I. Allegrini and N. Omenetto, Environ. Science and Techn., 13, 345 (1979).
37 R. M. Measures, Appl. Optics, 16, 1092 (1977).

Optical Remote Sensing of Air Pollution,
Lectures of a course held at the Joint Research Centre, Ispra, Italy, 12—15 April 1983,
P. Camagni and S. Sandroni (Eds). 351—361

LONG-PATH MONITORING OF TROPOSPHERIC OH BY UV-LASER ABSORPTION *

R. ZELLNER and J. HÄGELE

1 INTRODUCTION

The essential role of the hydroxyl radical in the chemistry of the troposphere was proposed more than a decade ago (ref. 1,2). Today, OH is generally accepted as the perhaps most important driving agent of chemical conversion and transformation in the troposphere. This is based on its predicted concentration level (ref. 3-5) and its reactivity. Intensive laboratory kinetic investigations have revealed, that OH reacts very rapidly with most known tropospheric trace constituents; e. g. SO_2, NO_x, CO, alkanes, alkenes, aromatics and oxygenated and S- and N-containing hydrocarbons (for a recent review of OH rate constants see e. g. ref. 6). This has two significant consequences:

a) The chemical lifetime (identical with the atmospheric residence time) of most trace constituents is determined by their reactivity towards OH. Writing the chemical loss process of a constituent A as

(i) $$OH + A \longrightarrow products$$

its rate law will be given by $-d[A]/dt = k_i[OH][A]$ which - assuming a steady state concentration for OH - then defines the lifetime of A as $\tau_A = (k_i[OH])^{-1}$.

b) OH is coupled to other reactive trace constituents, e. g. CO, CO_2, NO_x. The nature of this coupling and the corresponding interconversion of individual species of the HO_x family (= H, OH, HO_2) is shown in Fig. 1. Since larger fractions of CO, SO_2, and NO_x are anthropogenic, OH again plays a major role in establishing their concentrations and, inversely, the level of OH is affected by changing concentrations of these pollutants.

---

* work supported by "Bundesministerium für Forschung und Technologie (BMFT)"

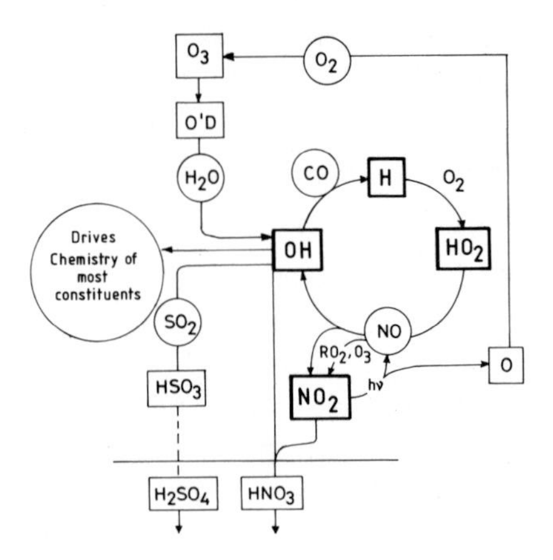

Fig. 1 Main tropospheric cycle of HO_x and its coupling with CO, NO_x and SO_2.

2 TROPOSPHERIC OH LEVEL AND GENERAL DESCRIPTION OF ITS MEASUREMENT

Tropospheric OH is generated and consumed in the sequence of the following reactions:

(1) $O_3 + h\nu \quad (\lambda \leq 310 \text{ nm}) \longrightarrow O(^1D) + O_2$

(2) $O(^1D) + H_2O \longrightarrow 2OH$

(3) $O(^1D) + M \longrightarrow O(^3P) + M; \quad M = N_2, O_2$

(4) $OH + CO \longrightarrow CO_2 + H$

(5) $OH + CH_4 \longrightarrow H_2O + CH_3$

(6) $OH + O_3 \longrightarrow HO_2 + O_2$

from which the steady state concentration of OH can be calculated to be

$$[OH] = 2J_1k_2[O_3][H_2O] \Big/ \left\{ (k_2[H_2O] + k_3[M]) \cdot (k_4[CO] + k_5[CH_4] + k_6[O_3]) \right\}$$

where J_1 is the first order rate coefficient for O_3 photodissociation integrated over all wavelength of tropospheric ozone absorption, $J_1 = \int I(\lambda) \cdot \sigma(\lambda) \phi(\lambda) d\lambda$. Using the accepted rate coefficients for (1) to (6) and observations of the average distributions of O_3, H_2O, and CO, the above equation predicts a global tropospheric 24-hour average OH concentration of 1.3×10^5 cm^{-3} (ref.6), corresponding to a mixing ratio of 5×10^{-3} ppt. Due to the photolysis term (J_1), OH is of course expected to show strong diurnal and annual variations.

Its chemical lifetime, approximately equal to $(k_4[CO])^{-1}$, is estimated to be ~ 2.5 s for 100 ppb CO.

Not all the OH consumed in reactions (4) to (6) is, however, lost from the atmosphere. HO_2, formed in (6) and - more importantly - in a consecutive reaction of H atoms ($H + O_2(+M) \longrightarrow HO_2(+M)$) recycles OH via the reactions

(7) $HO_2 + O_3 \longrightarrow OH + 2O_2$

(8) $HO_2 + NO \longrightarrow OH + NO_2$

with the important consequence, that the annual average OH concentration during a 24-hour period is ~ 5 x 10^5 cm^{-3}. Most likely ranges are between (3 - 10) 10^5 cm^{-3} (ref. 6). Regional day-time concentrations may be expected in the order of several 10^6 cm^{-3}. The ability to measure ambient OH concentrations would allow the accuracy and applicability of photochemical models of the troposphere to be tested. However, its direct observation has become a very controversial topic.

The $X^2\Pi_{3/2,1/2} \rightarrow A^2\Sigma^+$ optical transition of the OH radical (Fig. 2) is spectroscopically well characterized (see e. g. ref. 7). From the values of the Franck-Condon factors the two vibrational band transitions ($v'' = 0 \rightarrow v' = 0,1$) at 306 and 282 nm with easily resolvable rotational structure are the most important. To monitor atmospheric OH, either absorption or fluorescence measurements can be made.

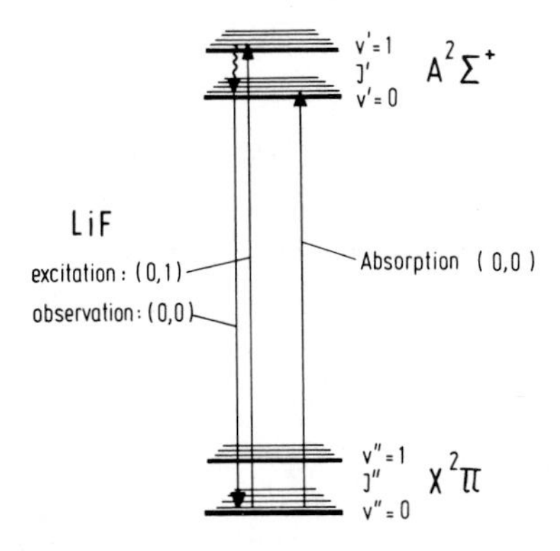

Fig. 2 A-X optical spectrum of the OH radical showing usual transitions used in absorption or fluorescence measurements.

The technique of laser induced fluorescence (LIF), a potentially very sensitive method, was the first to be applied to this problem (ref. 8,9), yet results obtained have been less than satisfactory. Some of the problems have been insufficient detection sensitivity and interference from other atmospheric constituents, notably ozone (ref. 10, 11). The laser radiation at 282 nm used in the LIF method to excite OH dissociates ambient O_3 to produce $O(^1D)$ which in turn generates OH via reaction (2). At ground level the characteristic time of artificial OH generation is about 1 ns. Hence, OH is generated, excited, and detected with the same laser pulse (pulse length typically 7 ns). Quantitative considerations of this effect (ref. 12) lead to the conclusion that for laser intensities of $> 50\ \mu J/cm^2$ generated OH levels exceed typical ambient levels, namely $10^6\ cm^{-3}$. Hence, meaningful LIF techniques have to apply drastically reduced pulse intensities. Two such improved approaches, one using a high repetition (17 kHz) atomic copper vapour laser (ref. 13) and the other applying an unfocussed laser beam with drastically increased field of view (ref. 14) are now being pursued.

An alternative method to the LIF technique for measurements of tropospheric OH is the use of long path laser absorption. This technique, first applied by Perner et al. (ref. 15,16) is devoid of any OH generation problems since i) the laser beam can be expanded to reduce the beam intensity (typically $\sim 1\ \mu J/cm^2$) and beam divergence and ii) measurements are made in the (0,0) band transition (usually: $Q_1(2)$ line at 307.995 nm) where ozone absorption is considerably weaker. However, due to the low optical densities of OH, pathlengths of several km have to be used.

The expected absorptions can be calculated using the fundamental laws of line absorption (see e. g. ref. 17). For a transition ($A^2\Sigma^+$, v'=0,J" $\leftarrow$ $X^2\Pi$, v"=0,J") we have for the number density in the absorbing state

$$\text{(ii)} \qquad [OH]_{J''} = (mc^2/\pi e^2) \cdot \int \alpha_\nu d\nu \,/ f_{J'J''}$$

where $f_{J'J''}$ is the line oscillator strength and $\int \alpha_\nu d\nu$ is the integrated absorption. $[OH]_{J''}$ is then related to the total OH density by

(iii) $[OH] = [OH]_{J''} \cdot Q_{rot}/(2J''+1)\ \exp(-E_{J''}/RT)$

where Q_{rot} is the rotational partition function and $E_{J''}$ is the energy of the J" absorbing level. Depending on the linewidth of the laser, eqs. (ii) and (iii) can be utilized in two different ways:

1. by directly measuring $\int \alpha_\nu d\nu$ using a broad band laser light source and tuning across the line by means of a high resolution monochromator (ref. 15,16)

2. by using a laser system with bandwidth small compared to the bandwidth of the OH line (ref. 18 and below).

In the latter case no monochromator is essential. However, the absorption coefficient in the line center (α_o) has to be known. For tropospheric conditions the frequency distribution of the absorption ($\alpha_\nu = [OH]_{J''} \cdot \varepsilon_\nu$) will not simply be of Doppler shape but will be pressure broadened. Calculations for the $Q_1(2)$ line yield $\varepsilon_o = 2.3 \times 10^{-15}$ cm^2. Hence, for OH concentrations of 1×10^6 cm^{-3} (corresponding to $[OH]_{J''} = 1.1 \times 10^5$ cm^{-3} in the J" = 2.5 level) the expected absorption, $A = \varepsilon_o \cdot [OH]_{J''} \cdot \ell$, is in the order of 2.5×10^{-4} for a pathlength of ℓ = 10 km. This can only be determined by suitable averaging techniques. In the following we will describe details of a double beam absorption technique used in our own experiments.

3. DOUBLE BEAM ABSORPTION TECHNIQUE

3.1 General Considerations

Laser beam transmission through the atmosphere is dependent on particle scattering, molecular absorption, and turbulence fluctuation effects. Particle scattering, mainly caused by air humidity and air pollution, can lead to an attenuation of light intensity of more than 20 dB/km. For 3 dB/km the transmission range is reduced to about 10 km.

Atmospheric turbulence causes local changes in density and refraction index, resulting in laser beam displacement or widening. Consequently strong intensity fluctuations with a frequency distribution of $\lesssim$ 1 kHz arise, which have to be compensated by either a reference beam or by taking average values of increasing numbers of measurements. Using a reference beam which probes the same atmosphere as the analysis beam reduces the standard deviation of a single measurement to the accuracy of the detection system. Neglecting OH-absorption by the reference beam and all other molecular absorptions, the intensity ratio of the outgoing analysis and reference beam, I_{out}/I^R_{out}, is only attenuated by OH-absorption. According to Lambert-Beer's law for weak absorption the transmitted intensity ratio I_{in}/I^R_{in} may be expressed as:

$$\text{(iv)} \quad I_{in}/I^R_{in} = (I_{out}/I^R_{out})(1 - \varepsilon_{OH} \cdot [OH] \cdot \ell)$$

Our method makes use of two short laser pulses of 0.0006 nm bandwidth (analysis) and 0.1 nm bandwidth (reference), respectively. The spectral conditions are shown schematically in Fig. 3.

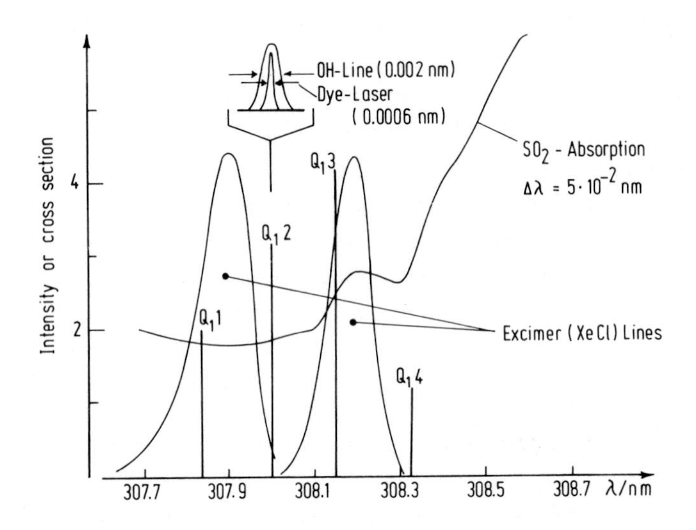

Fig. 3 Atmospheric and laser spectral conditions in double beam UV-absorption experiment.

At the $Q_1(2)$-absorption wavelength of OH (308 nm), other tropospheric substances (SO_2, CH_2O) also show strong absorptions. Fig. 3 contains a medium resolution spectrum of SO_2 (ref. 20). Measurements at higher resolution (see e. g. ref. 16,18) indicate that even at atmospheric pressure there will be considerably more structure in the SO_2 bands, whereupon serious interferences with the analysis of OH absorption data according to eq. (iv) may occur. This interference depends on the absolute SO_2 level. Its absorption will become comparable to that of OH for $[SO_2] \sim 5$ ppb. In order to reduce field measurement data for OH for this effect, measurements have to be taken with the analysis laser on and off the OH line.

3.2 Experimental Set-up

The essential components of the experimental set-up are shown schematically in Fig. 4. A photograph is provided in Fig. 5. An excimer laser (LAMBDA, EMG 102, E $\sim$ 150 mJ) serves as pump source for a frequency-doubled dye laser (LAMBDA, FL 2002E). A splitted portion of the direct excimer laser beam is collimated to nearly 1 mrad, delayed over 100 ns ($\hat{=}$ 30 m) and joined to the output of the frequency-doubled dye laser. This "double-beam", consisting of two light pulses of the same duration ($\sim$10 ns), the same divergence (1 mrad) and nearly the same energy (1 mJ), is directed to a 10x beam expander and further to a 15x Cassegrainian telescope expanding the beam to nearly 30 cm in diameter. Table 1 summarizes the properties of the outgoing "double-beam".

TABLE 1

	Analysis Beam	Reference Beam
wavelength	307.995 nm	308 nm
bandwidth	0.0006 nm	0.1 nm
pulse length	10 ns	10 ns
energy	1 mJ	1 mJ
divergence (of the extended beam)	7 μ rad	7 μ rad

A 50 cm plane mirror with a surface flatness of $\lambda/10$ serves as a reflector in a distance of 6.4 km and 15.7 km, respectively. The reflected beam (d $\approx$ 40 - 50 cm) is collected by a 65 cm Gregorian telescope and directed through an interference filter (SCHOTT, UV-DIL, $\Delta\lambda$ = 6.9 nm) to a photomultiplier (EMI, QB 9815), which also detects a splitted portion of the outgoing beam. In order to control the wavelength stability of the dye laser, a fraction of the frequency-doubled light is used for OH-fluorescence excitation in a reference cell. The fluorescence arising from this is detected by a second photomultiplier.

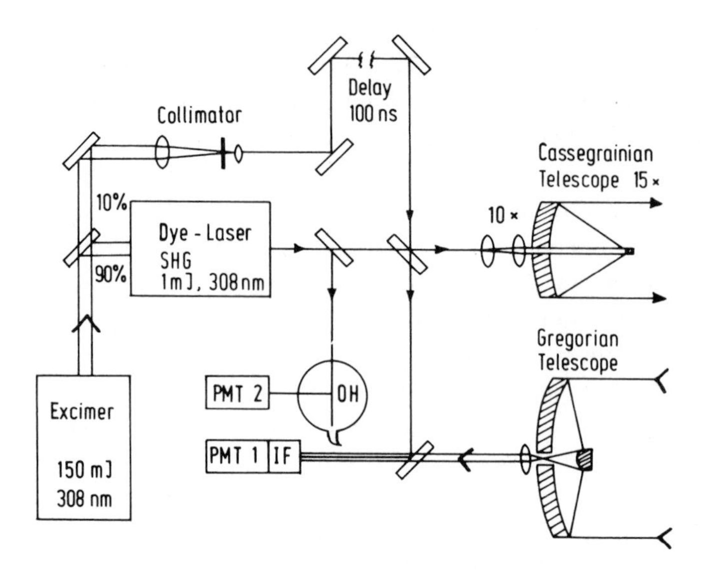

Fig. 4 Schematic representation of experimental set-up

Fig. 5 Photograph showing experimental arrangement

3.3 Pulse Timing and Intensity Measurement

Fig. 6 shows the chronological order of the detected pulses. The pulse length is about 10 ns, the interval between analysis pulse and reference pulse is 100 ns, the interval between outgoing and incoming double-pulse is 43 µs and 105 µs, respectively. As the excimer laser normally runs at 20 Hz, the next pulse sequence follows 50 ms later.

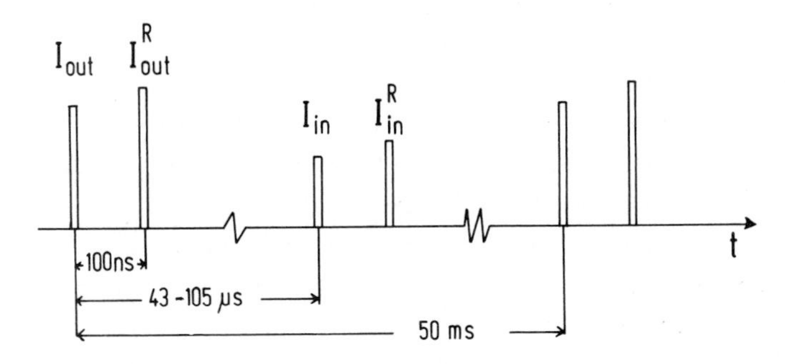

Fig. 6 Pulse timing diagram

The large number of pulses arriving at the detector during a period of up to several hours requires careful analog data processing prior to digital computing (Fig. 7).

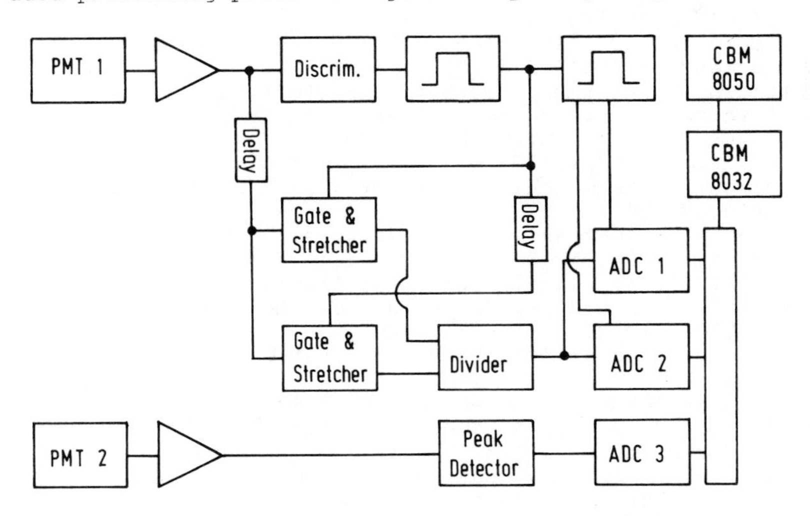

Fig. 7 Signal processing equipment

The incoming first two pulses (Δt = 100 ns) are separately streched to 3 μs (ORTEC LG 105) and their intensity ratio is transferred to an analog-to-digital converter ADC 1 (ORTEC 800). The second two pulses of a sequence pass through the same modules but their ratio is given to ADC 2. Gate-setting generators started by the discriminated first pulse of a sequence provide for correct timing. The accuracy of this electronic circuit is better than 0.5 %, the digital resolution can be chosen up to 12 Bit. The reference signal measured by PMT 2 is given to ADC 3. The digitized data are processed by a 64-kByte microcomputer (CBM 8032) and transferred to a floppy disc (CBM 8050). The computer also resets the ADCs after a data transfer and is used for simultaneous and later data analysis. For the condition that the signal intensities show random fluctuations, averaging over 2500 signals leads to an accuracy of 10^{-4} corresponding to an OH-concentration of $2 \cdot 10^{-5}$ cm^{-3}. Since beam displacements cause data losses up to estimated 80 % and absorption measurements will be made on and off the OH-line, a single measurement period takes about 20 minutes.

REFERENCES

1 H. Levy II, Science 173 (1971), 141
2 B. Weinstock, Science 166 (1969), 224; B. Weinstock, H. Niki, Science 176 (1972), 290
3 P. J. Crutzen, J. Fishman, Geophys. Res. Lett. 4 (1977), 321; J. Fishman, P. J. Crutzen, Atmospheric Science Paper No. 284, Colorado State Univ. (1978)
4 P. Warneck, Planet. Space Sci. 23 (1975), 1507
5 P. J. Crutzen, in "Atmospheric Chemistry", E. D. Goldberg (Ed.), Springer 1982
6 R. Atkinson, K. R. Darnall, A. C. Lloyd, A. M. Winer, J. N. Pitts jr., Adv. Photochemistry 11 (1979), 375
7 D. R. Crosley, R. K. Lengel, Quant. Spectrosc. Radiat. Transfer 15 (1975), 579; K. R. German, J. Chem. Phys. 63 (1975), 5252; ibid. 64 (1976), 4065
8 C. C. Wang, L. I. Davis jr., Phys. Rev. Lett. 32 (1974), 349 - 352
9 D. D. Davis, W. Heaps, T. McGee, Geophys. Res. Lett. 3 (1976), 331; D. D. Davis, W. Heaps, D. Philen, T. McGee, Atmospher. Environment 13 (1979), 1197; D. D. Davis, W. S. Heaps, D. Philen, M. Rodgers, T. McGee, A. Nelson, A. J. Moriarty, Rev. Sci. Instruments 50 (1979), 1505
10 D. D. Davis, M. O. Rodgers, S. D. Fischer, K. Asai, Geophys. Res. Lett. 8 (1981), 69; ibid. 8 (1981), 73
11 M. Hanabusa, C. C. Wang, S. Japar, D. K. Killinger, W. Fischer, J. Chem. Phys. 66 (1977), 2118

12 G. Ortgies, K. H. Gericke, F. J. Comes, Z. Naturforsch. 36a (1981), 177
13 R. M. Stimpfle, J. G. Anderson, 2nd Symp. on Composition of the Non-Urban Troposphere, Williamsburgh, Va. 1982
14 L. I. Davis, C. C. Wang, X. Tang, H. Niki, B. Weinstock, 2nd Symp. on Composition of the Non-Urban Troposphere, Williamsburgh, Va. 1982
15 D. Perner, D. H. Ehhalt, H. W. Pätz, U. Platt, E. P. Roth, A. Volz, Geophys. Res. Lett. 3 (1976), 466
16 G. Hübler, D. Perner, U. Platt, A. Toennissen, D. H. Ehhalt, 2nd Symp. on Composition of the Non-Urban Troposphere, Williamsburgh, Va. 1982
17 A. C. G. Mitchell, M. W. Zemansky, "Resonance Radiation and Excited Atoms", Cambridge University Press (1961)
18 G. Ortgies, F. J. Comes, to be published
19 R. Gruß, Nachrichtentech. Z. 22 (1969), 184
20 D. J. Brassington, Appl. Opt. 20 (1981, 374

Optical Remote Sensing of Air Pollution,
Lectures of a course held at the Joint Research Centre, Ispra, Italy, 12—15 April 1983,
P. Camagni and S. Sandroni (Eds). 363–380

OPTIMISATION OF MONITORING NETWORKS BY REMOTE SENSING TECHNIQUES

D.ONDERDELINDEN

INTRODUCTION

The Dutch Air Quality Monitoring System was installed about 10 years ago to gather information on air pollutant concentrations in space and time. The behaviour of hourly as well as daily averaged values over periods of a year generally is expressed as 50, 95 and 98 percentiles, which then are compared with levels set by public health standards. The geographical pattern of the pollutant levels can be derived from the measurement results by means of interpolation procedures. The accuracy of the obtained pattern depends on network density and statistical spatial characteristics of the component under study. A second objective is the determination of the contribution of sources or source categories to the pollutant levels in order to enable an emission oriented strategy for decreasing concentrations. Due to the generally moderate spatial resolution of networks additional information to be obtained from mobile measurements of groundconcentration as well as vertical profile is necessary. The relation between emissions, transport as determined by meteorological conditions, vertical concentration profile or gasburden, as measured by remote sensing techniques and groundconcentration is to be given by short term meso-scale transport models. In the ideal case the model uses routinely available meteorological parameters making extrapolation to long term description possible. In this paper the significance of the three items, fixed monitoring networks, dynamic monitoring with remote sensors and thirdly modelling will be discussed.

FIXED STATION NETWORKS

In the Netherlands, situated in between the largest industrial areas of Western Europe (the associated SO_2-pollutant emissions are presented in Fig. 1) the ambient levels of SO_2 are monitored continuously in a network of 200 fixed stations, with interstation distance of 20 km in rural and 3-10 km in urban-industrial areas.

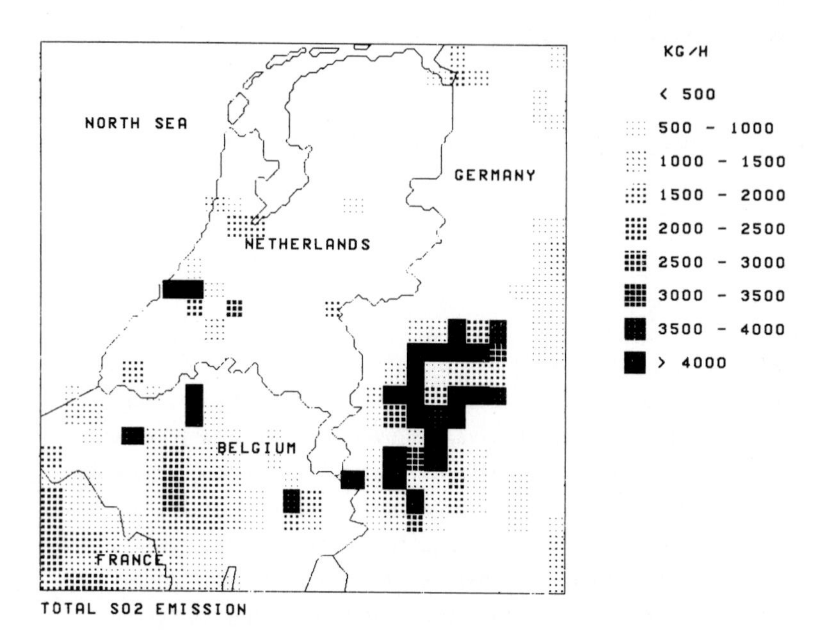

Fig. 1. SO_2-emissions in the 400 x 400 km² surroundings of the Netherlands (in kg/h).

NO_x and O_3 are measured in subsets of 92 and 35 stations respectively (Fig. 2). Information about the spatial correlation and representativety is derived from the measurement results of a regular baseline-grid comprising 108 stations for SO_2, 44 stations for NO_x and 28 stations for O_3. The measurement area (200 x 250 km²) covers the total area of the Netherlands. The concentrations are measured contineously (~1 minute) and after data validation hourly values are formed. The following instruments are used:

SO_2	Philips PW 9700	coulometry
NO_x	Philips PW 9762/00	chemoluminescense
O_3	Philips PW 9771/00	chemoluminescense

The instruments for SO_2 and NO_x are calibrated every 24 h and the O_3-monitor every measuring cycle of 80 s.

The results of this set of fixed monitoring stations can be considered as:

- concentration levels at individual isolated locations, without inter-relations to other positions in the area under study; for example at locations where maximum levels are expected to be used for a worst case study or for short distance model validation.
- concentrations related to the observations at other stations, enabling extrapolation beyond the spatial argument of the station itself to other locations in the area with a known confidence level.

In the second case the spatial pattern can be deduced from empirical, statistical relationships between the observations. To evaluate the significance of such configurations the spatial representativity of the observations needs to be quantified.

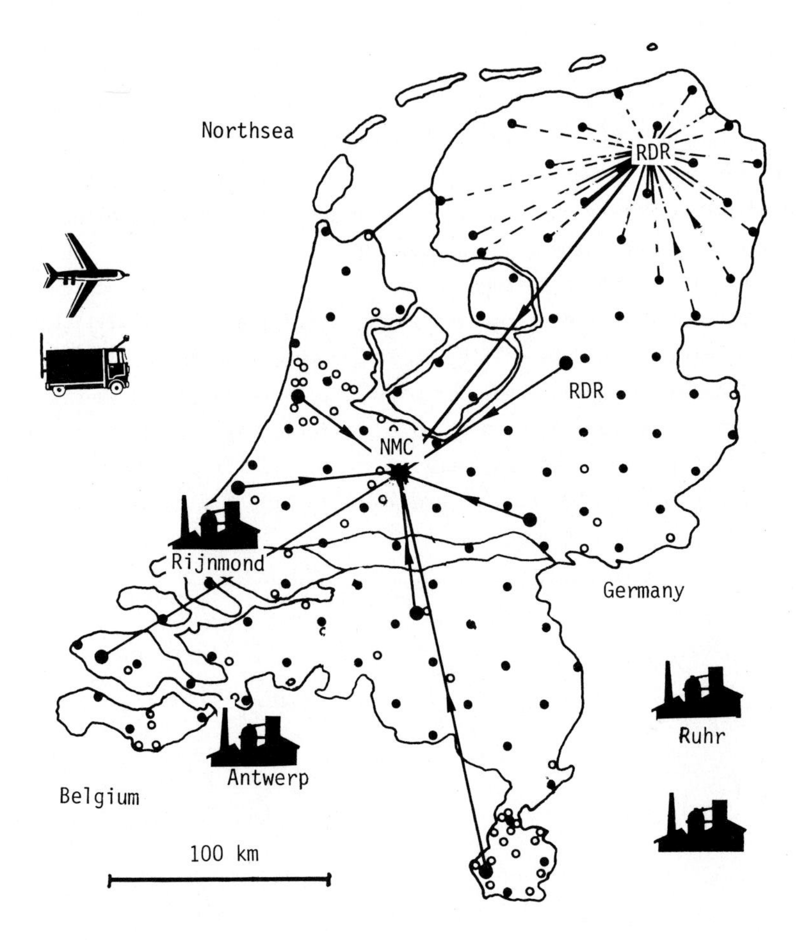

Fig. 2. The Dutch national air quality monitoring system.

Spatial correlation

If the concentration at time t of station i at position x_i is denoted by C_{it} the measured correlation $r^m(x)$, averaged over the area considered, is obtained as:

$$r^m(x) = \overline{<(C_{it} - M)\ (C_{jt} - M)>} \ / \ \overline{<(C_{it} - M)^2>} \quad (1)$$

where the averaging is over time (bar) and space (< >) such that $x = |x_i - x_j|$. The average (space and time) momentary correlation function $r_{.t}(x)$ is obtained by putting $M = M_{.t}$, the spatial average of the observed concentrations at time t. The long term correlation function ($r_{..}(x)$ is obtained by putting $M = M_{..}$, the spatial and time average of the measured concentrations. It can be shown (ref. 2) that

$$r_{.t}\ Var_{.t} = r_{..}\ Var_{..} - Var(M_{.t}) \tag{2}$$

elucidating the addition of systematic time behaviour of the whole field considered (the variance of the $M_{.t}$-plane) and the spatial momentary gradients (the variance of the momentary concentrations with respect to the $M_{.t}$-plane: $Var_{.t}$) into the variance with respect to the time and space average $M_{..}$. The distribution of individual concentrations in space and time cannot be considered as a normal distribution. At high concentrations spatial gradients will be high as well, at low levels gradients will be small. Therefore an appropriate procedure to arrive at normal distributions, for which the above mentioned averaging procedure is more meaningful, is considered to be formed by procedures based on the logarithm of the observed concentrations. Empirical logarithmic correlation functions r^m for NO_2 and NO (ref. 4) are shown in Fig. 3 and Fig. 4.

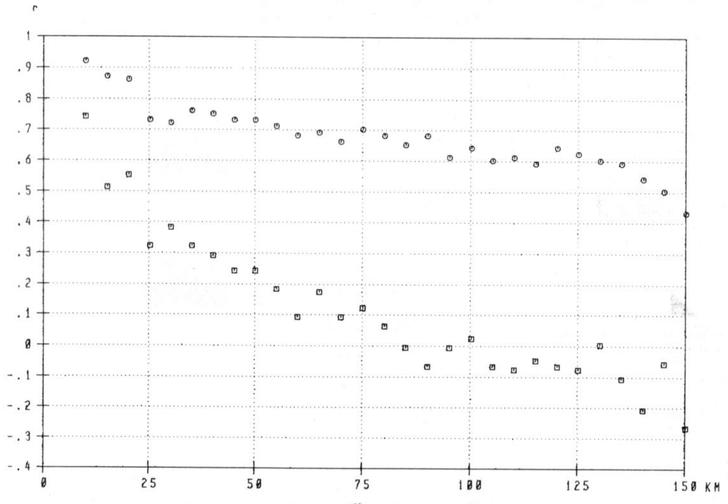

Fig. 3 Correlation functions $r^m_{..}$ (o) and $r^m_{.t}$ (□) from hourly NO_2-concentrations (logarithmic); 10-24 Jan. 1982; 35 stations.

- All correlations exhibit values, at extrapolation to zero interstation distance, r^m_o smaller than 1. This is attributed to a difference between "true" concentration, representative for a larger area and measured concentration which might be influenced by local sources. If it is assumed that this difference h_f possesses no spatial correlation and has the same variance for all stations it follows that:

$$\overline{h_f^2} = (1 - r_o^m)\ Var^m \qquad (3)$$

where Var^m the variance of the observed (logarithmic) concentrations with respect to the chosen first moment ($M_{.t}$ or $M_{..}$) for the calculation of r^m. From the four curves of Fig. 3 and Fig. 4 h_f is estimated as 10% for NO_2 and 30% for NO and as such show the influence of local sources.

- The difference in the observed correlation distance of the two types of correlation is quite significant. For NO_2 that distance is about 150 km in the $r_{..}$-case and some 35 km in the $r_{.t}$-case. The difference can be explained from relation (2) from which in fact an asymptatic behaviour of $r_{..}$ towards the ratio $Var(M_{.t})/Var_{..}$ could be expected.

In considering representativity and spatial interpolation of momentary hourly averaged network data it is clear from the discussion above that the function $r_{.t}$ must be taken into account.

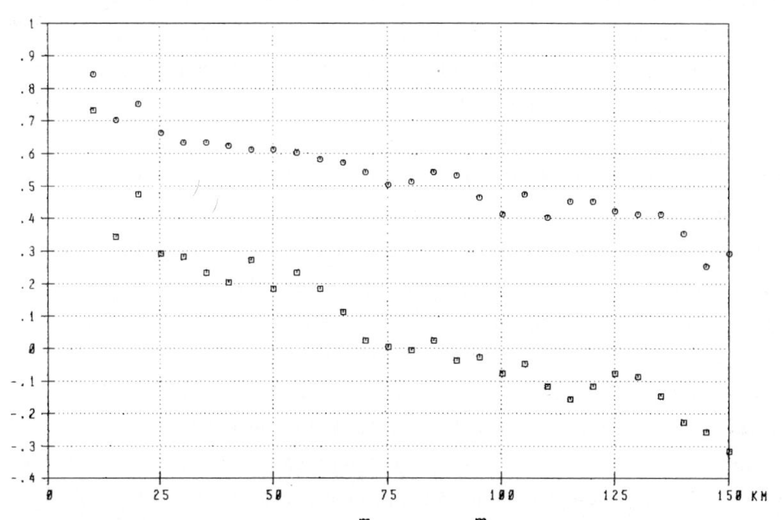

Fig. 4. Correlation functions $r_{..}^m$ (○) and $r_{.t}^m$ (□) from hourly NO-concentrations (logarithmic); 10-24 Jan. 1982; 35 stations.

Interpolation

The concentration at a given position and time C_{ot} can be estimated from the observed concentration C_{it} at N surrounding stations (ref. 9) from the relation:

$$C_{ot} = \sum_{i=1}^{N} p_{io}\,(C_{it} + h_{it}) + E_{ot} \qquad (4)$$

where p_{io} weighting coefficient for station i

h_{it} the earlier defined local influence and measuring error

E_{ot} the interpolation error, i.e. the discrepancy between the true concentration C_{ot} and its estimate. The coefficients p_{io} can be optimized by minimization of the mean square errors E^2_{ot} over a large number of realisations, for example a sequence of T hourly values. The coefficients then follow from the equation:

$$p_o = (R + \eta^2 I)^{-1} r_o \qquad (5)$$

where p_o the vector of elements p_{io}

R the matrix (NxN) of correlations r_{ij}

r_o the vector of expected correlations r_{io}

η^2 the normalized local effects: $\eta^2 = h^2 / <(C_{it} - M)^2>$

From empirical correlation functions as presented in Fig. 3 and Fig. 4 for NO_x relative interpolation errors F (logarithmic procedure) can be computed numerically for various network densities. For an interstation distance of 40 km the interpolation errors for NO_2 and NO were found to be 15% and 40% respectively. In Fig. 5 the relation between relative interpolation error F and interstation distance x_o is plotted for SO_2 as deduced from earlier measurements (jan./febr. 1979). It can be concluded that at smaller interstation distance the marginal reduction of F is strongly reduced; F tends to an asymptotic value which depends on the relative local disturbance h_f. From stations which are situated close together this "error" was estimated from relation (3) as h_f = 17% with an upper limit of 25%. The curve for h_f = .25 gives smaller interpolation error because part of the difference between interpolated values and measured values is ascribed to local disturbance, thus decreasing the interpolation errors. It should be remarked that the SO_2-correlation functions and consequently the results presented in Fig. 5 apply to the SO_2-pattern in the Netherlands where the meso-scale contributions are relatively large due to nation wide use of low sulphur containing gas for space heating. At the present density of x_o = 20 km the interpolation error at which the hourly SO_2-pattern is reconstructed amounts about 20%, implying a 95% confidence interval of about 40%.
As a further increase in network density is ineffective, mobile dynamic monitoring systems are used to realize a very high spatial resolution over a short time interval. The interpretative link between the two approaches is given by operational models; one of these models the lagrangian PUFF-model will be described below.

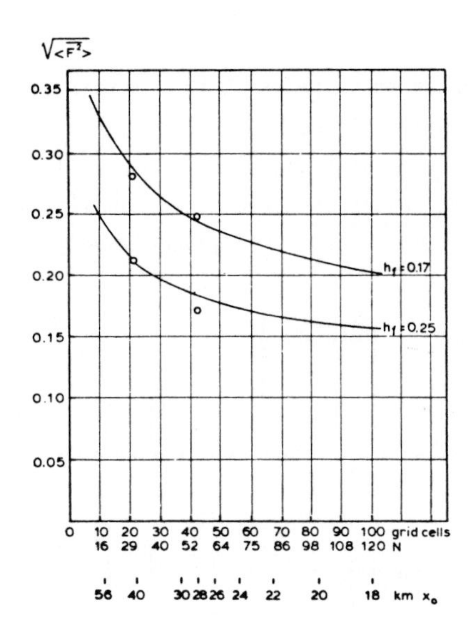

Fig. 5. Relative interpolation error as a function of SO_2-network density (interstation distance x_0).

MESO-SCALE PUFF-MODEL

This lagrangian model (ref. 7) describes the transport and dispersion over the 400 x 400 km² surrounding of the Netherlands. The used emissions are of the type as given by Fig. 1, together with information on source height (ref. 10, ref. 11). The emissions are presented (and used) for grid squares of 15 km side, effective plume height for these areas are computed as a mean of the individual source heights, weighted according to the emission strength. Within the Netherlands a number of point sources are taken into account. The transport is described with a large number of Gaussian puffs with horizontal concentration distribution:

$$c(r) = f \frac{M}{2\pi\sigma_r^2\, h} \exp\left(-r^2/2\sigma_r^2\right) \qquad (6)$$

where r the distance from the puff centre

σ_r the standard deviation

M the mass represented by the puff

f a shape factor for the vertical profile near the surface, resulting from dry deposition at the surface

h height of the layer modelled by the puff

The degree of sophistication of the model is adapted to the routinely available

meteorological input data:
- wind speed and direction at 40 stations at 10 m height and 5 stations (TV-towers) at levels between 150 and 300 m.
- solar radiation measured at 3 stations
- diurnal profile of mixing height measured by an acoustic sounder or lidar.

In the model three atmospheric layers are distinguished as illustrated in Fig. 6.

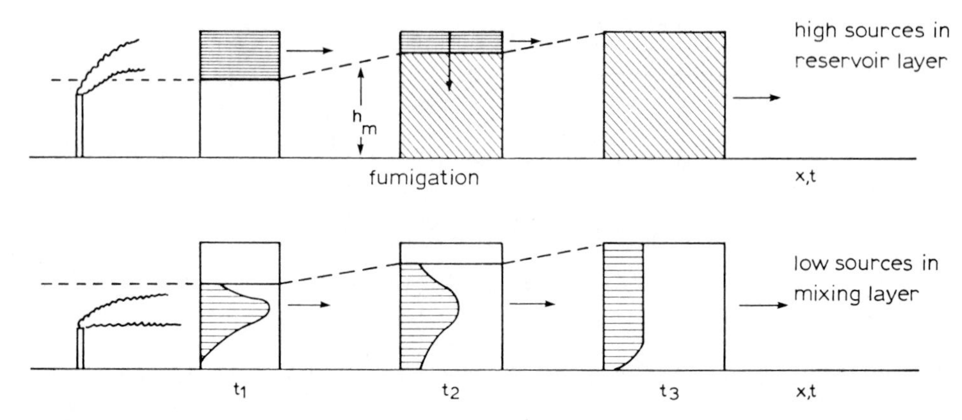

Fig. 6. Vertical stratification of pollution in reservoir and mixing layer and downward transport during fumigation.

- mixing layer, assumed to be constant during the night and increasing during daytime as a result of solar heating. The initial height and rising speed are derived from the acoustic sounder measurements. Close to the source the vertical concentration profile is assumed to be gaussian. The associated dispersion parameter σ_z is chosen in the usual way, where stability class is derived from surface layer parameters.
- surface layer of 50 m, being the lowest part of the mixing layer. From windspeed, radiation and surface roughness the surface layer parameter Obukhof length L and friction velocity u_* are derived. These parameters in turn are used to estimate the aerodynamic resistance of the entire layer, which together with the surface resistance determine a quasi-steady state concentration gradient.
- reservoir layer, in which the pollution emitted by high sources is transported. The layer is situated above the mixing layer and simulated by the upper compartment of the puffs with f=1. During the morning hours the mixing height increases and puffs in the reservoir layer will be mixed downward to the mixing layer (fumigation), depending on their initial height. Generally this process will result in higher ground concentrations.

Horizontal dispersion is modelled by increasing σ_r according to:

$$\sigma_r^2 (t + \Delta t) = \sigma_r^2(t) + 2K_h \Delta t + 2K_s \Delta t \qquad (7)$$

Herein K_h is the apparent horizontal diffusivity and K_s the additional diffusivity resulting from wind shear over the mixing or reservoir layer (ref. 5). K_h is estimated from the standard deviation of the measured wind direction and an assumed lagrangian correlation function. Shear dispersion is derived from the wind shear in the following way. Assuming a constant wind shear S in rad/m over the extend z over the vertical pollutant profile the crosswind velocity of the pollutant is given by $v_s = Sz$. The shear displacement Δy_s after travel time Δt then is given by $\Delta y_s = Sz\ \Delta t$. On the other hand $y_s = \int_0^t v_s\ dt = S\int_0^t zdt$. The equivalent K_s then follows from:

$$K_s = \tfrac{1}{2} \frac{\overline{dy_s^2}}{dt} = \overline{y_s \frac{dy_s}{dt}} = S^2 \overline{\int_o^t z(t)z(t')dt'} = \tfrac{1}{2}\ \sigma_z^2\ S^2 t \qquad (8)$$

where the averaging is performed over all possible correlated z values at times t en t'. In this approximation σ_z^2 gives the vertical extend of the plume with an upperbound at complete mixing $\sigma_z^2 = h^2/_{12}$, i.e. the second moment at a constant concentration over layer depth h.

The puffs are released at "distances" of 10 km or less. The advection time step is generally one hour, but substantially smaller during the first hours of transport. The puffs are advected according to wind fields in the mixing or the reservoir layer. The wind fields are obtained by in - and extrapolation of the measured wind direction and speed. The vertical behaviour of the direction is extrapolated according to the Ekman spiral profile, the windspeed according to a power law.

The model computes hourly averaged concentration fields over the 400 x 400 km² model area and as such can be compared to the interpolated field as derived from the network measurement stations. Furthermore a spatial line profile of concentration and vertically integrated concentration (gasburden) along the route traversed by mobile measurement systems can be generated. This opens the possibility of validation of the model in between emission and ground level concentration.

DYNAMIC MONITORING; REMOTE SENSING

A third essential contribution to the measuring strategy is formed by the use of two dynamic monitoring systems. They are used to obtain high resolution spatial profiles of pollutants for validation of the model input parameters such as emission distribution and meteorological parameters. The measurements are made from moving laboratories; the data are combined with spatial coordinates, obtained from a navigational system and stored as profiles of concentration (SO_2, NO_x, O_3, b_{scat}) and gasburden (SO_2, NO_2-correlation spectrometry). The use of remote sensors in particular has the following advantages:

- the emission position, relative source contribution and absolute source strength can be determined without the consideration of the dispersion process.
- the overall decay in term of fluxes at several downwind traverses can be measured and decay parameters (deposition, chemical conversion) can be estimated.

Given these advantages of gasburden measurements the initial first estimate pollutant concentration field, especially near the foreign source areas can in principle be updated by iterative matching of emission, measured gasburden profile, modelled gasburden (and concentration) profile, modelled and measured network concentration field.

Meso-scale experiments

In this case mobile measurements are made over traverses in the order of 150 km and take about three hours. Time variability is not taken into account. The correlation spectrometer (Barringer Cospec IV) is used in the passive, upward looking mode. The total SO_2- or NO_2-flux is obtained by integration of the measured gasburden profile and additional information on windspeed.

Example 1

One of the earlier encounters (at our institute) of measuring and modelling results is given in Fig. 7 (ref. 3). The concentration field as reconstructed from the network by interpolation is given in the upper right map. Wind direction is indicated by small arrows. The model results are presented in the upper left, while the measured concentration and gasburden profiles are presented in the lower part of the figure, in combination with the modelled field. At an anti-cyclonic southerly circulation SO_2 is advected from source areas in Belgian and France (Fig. 1) into the Netherlands. The measured SO_2-levels of the network are of the order 100-150 $\mu g/m^3$, while in the dynamically measured concentration profile peaks of more than 300 $\mu g/m^3$ can be observed and as such give quantitative insight in the significance of the respective source areas. The meso-scale plume pattern as calculated by the model and affirmed by these profiles cannot be reconstructed from the network measurement results. In the Western part of the traverse a discrepancy between model and gasburden (as well as ground concentration) is observed. The fluxes therefore are due to unaccounted emissions probably in the Belgium Gent area and an emission area in the Northern part of France.

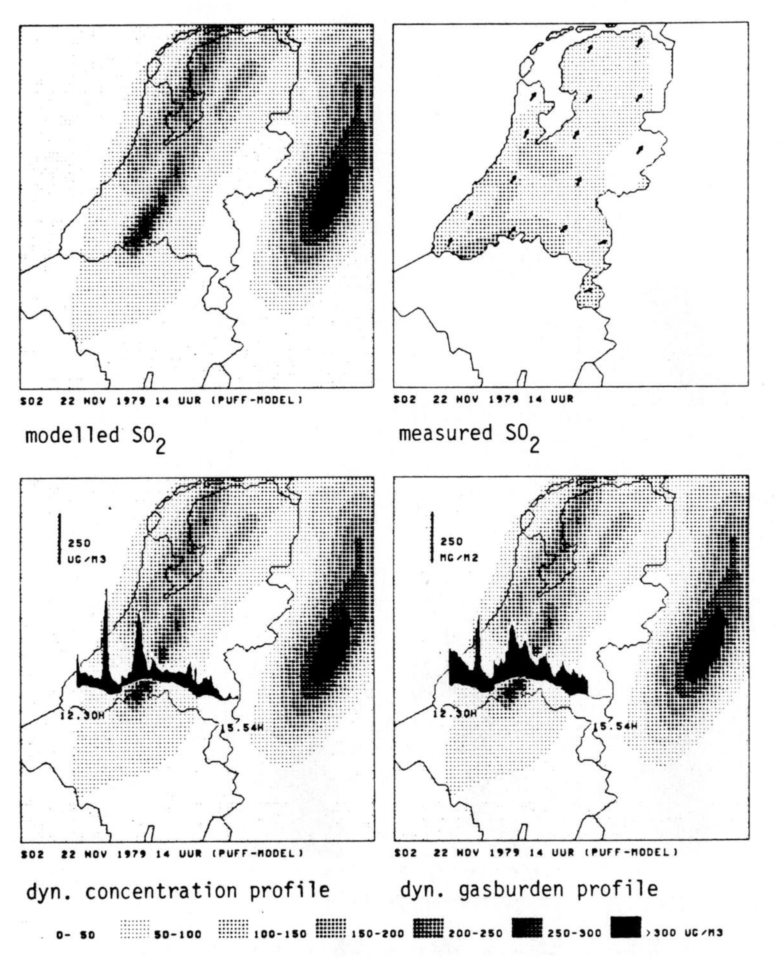

Fig. 7. Modelled and measured SO_2-concentrations for November 22nd 1979.

Example 2

Another example of source tracking is given in Fig. 8 (ref. 3). At northerly winds the emission from the Dutch Rijnmond industrial area is transported into Belgium. The gasburden profile indicates the plume of a source not incorporated in the emissions used for the model calculation. The source also affects ground concentration as measured dynamically. In the interpolated concentration field from the network this contribution is not observed. This unaccounted (power plant) emission thus has to be introduced into the model in order to obtain a more realistic field. To the east the gasburden increases due to apparently unaccounted emissions in the central part of the Netherlands.

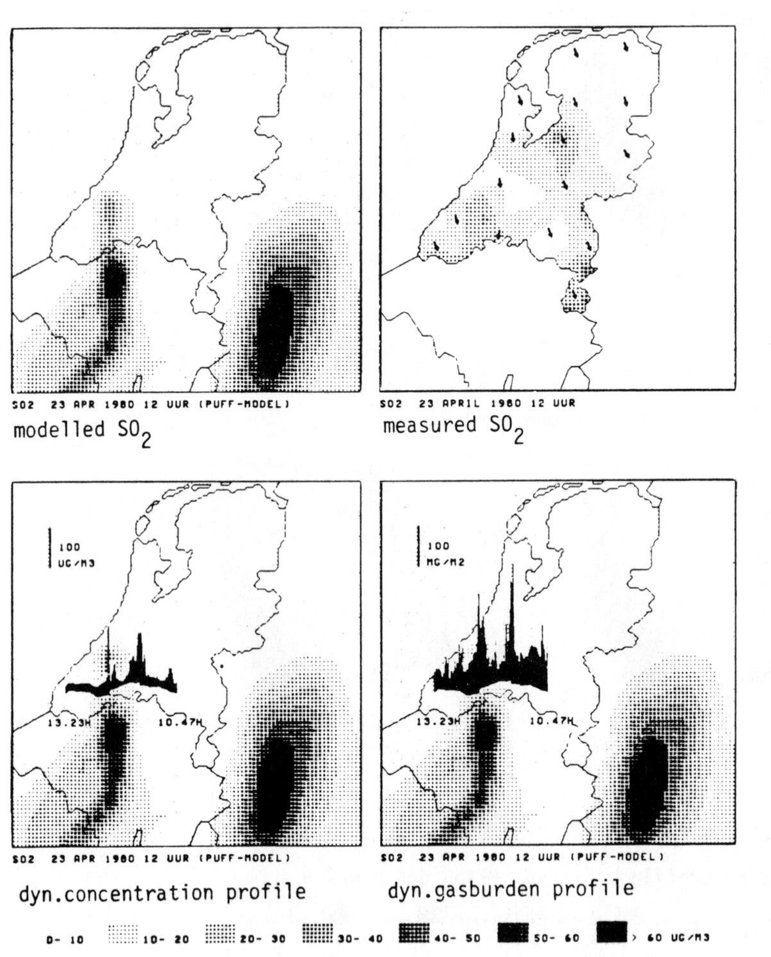

Fig. 8. Modelled and measured SO_2-concentrations for April 23rd 1980.

Example 3

An example in which dispersion characteristics can be validated by the availability of dynamic measurement is given in Fig. 9 and Fig. 10 (ref. 8) where the concentration field of february 3rd at 13.00 hours as deduced from the 100 baseline grid stations of the network is given in the upper right map. At a stable

south easterly anti-cyclonic circulation with relatively high windspeed (~6m/s) and low mixing depth (350m) the meso-scale plumes from sources areas in the Netherlands, Belgium and Germany dominate the model area. The dynamically measured concentration and gasburden profiles together with modelled results are given at the lower right part of the figures; the traverses all correspond to the measured concentration profile which is plotted over the modelled concentration profile in the lower part of the figures.
The modelling results presented in Fig. 9 apply to a $K_s = 0$ in equation 7. In Fig. 10 the corresponding results including the derived value for an extra dispersion due to shear are presented.

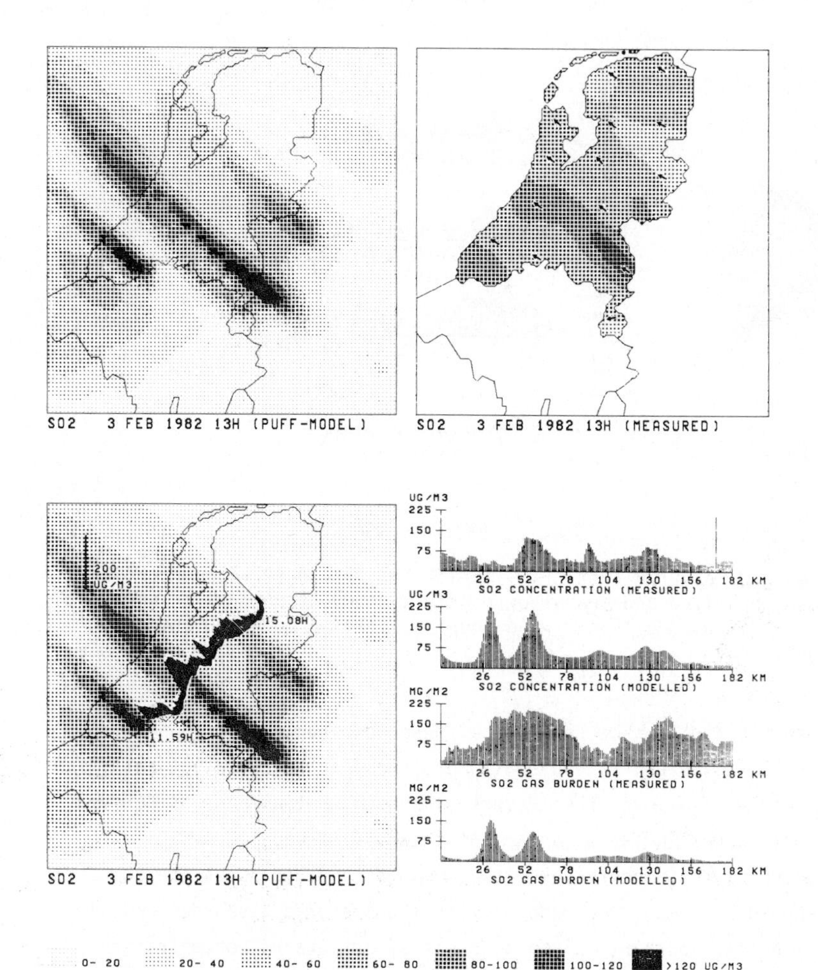

Fig. 9. Modelled and measured (network and mobile traverse) SO_2-concentration patterns; modelled and measured (mobile system) concentration and gas-burden profiles. Additional wind shear dispersion not modelled ($K_s = 0$).

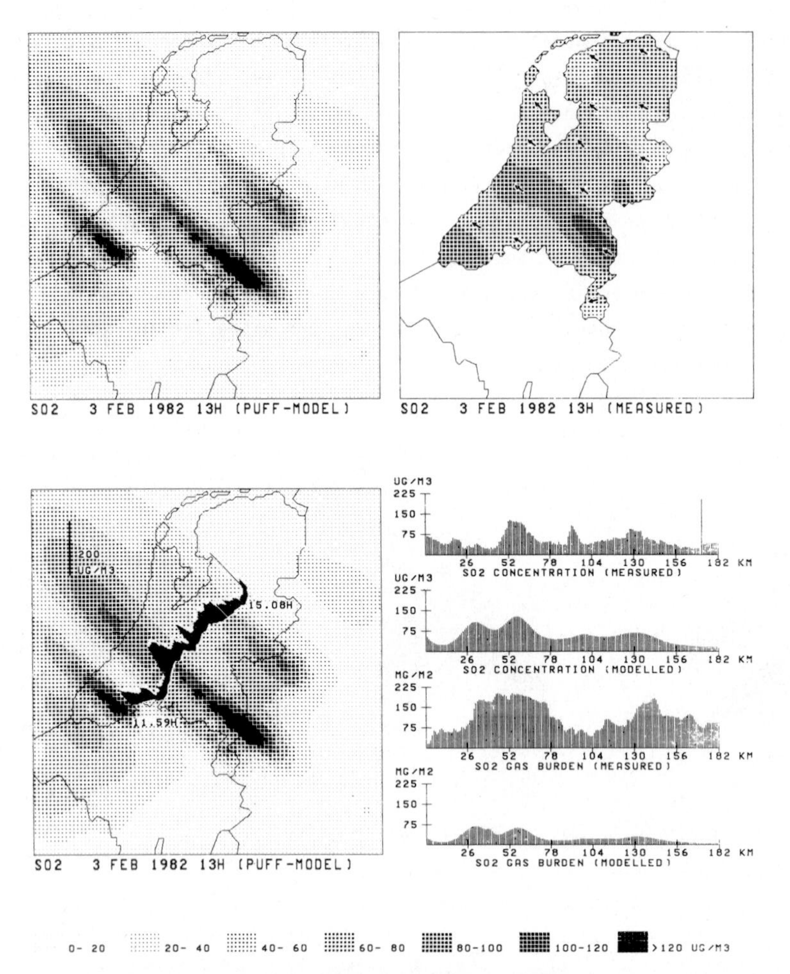

Fig. 10. Modelled and measured (network + mobile traverses) SO_2-concentration patterns; modelled and measured (mobile system) concentration and gas-burden profiles. Additional wind shear dispersion modelled according to equation 7.

The following conclusions can be drawn from these combined results:

1. The measured SO_2-gasburden ($\cong$ 400 tons/hour) seems to be unrealistically high. The modelled flux over the traverse amounts 150 tons/hour as derived from emission figures in the wind part of the model area.
2. On the southern part of the traverse two dominating plume zones are computed. The combined width seems to correspond with the measured gasburden profile (Fig. 10). The most southerly plume does not affect ground concentration. In the model the respective emission is erraneously allocated to the mixing layer; actually the plume appears to have penetrated through the inversion layer.

3. The spatial correlation between network and model results is rather insensitive for the variation of inputparameters, the correlation coefficient is found to vary from 0.48 to 0.52. Comparison of measured and modelled gasburden profiles provide a more accurate determination of dispersion parameters. However it must be borne in mind that the interpretation of gasburden profiles should be performed carefully as the broadening of the plumes as well as the measured flux over the traverse might be influenced by disturbing optical effects. Resulting from light scattering at lower atmospheric levels in particular on aerosols, under- and overestimation of the true gasburden might occur. As higher SO_2-levels in meso-scale plumes are mostly associated with high aerosol concentrations and the plume width is larger than its height overestimation of gasburden is expected. Computations (ref. 1 and ref. 6) suggest that overestimations of the order of 10% can occur under a wide plume and as such does not explain the difference between expected and measured flux. Another effect which might influence the calculated flux from the measurements is the choice of the zero-level of the instrument. The highest possible constant zero level is 40 mg/m^2 reducing the measured flux to 240 tons/h. The remaining difference then might be attributed to SO_2-emissions outside the model area.

ASSIMILATION OF MODEL DATA INTO NETWORK INTERPOLATION PROCEDURES

Having optimized the model performance by means of the results of mobile (remote sensing) measurements to a certain level, the concentration fields from the model can be intoduced into the network results (ref. 7). Instead of reconstructing the concentration field from a limited set of network stations with a statistically based interpolation procedure, information on the expected detailed structure of the concentration field as given by the model is used for this reconstruction. Because the concept of surface-, mixing- and reservoir layer allows a reasonable simulation of the diurnal variability of atmospheric stability and associated vertical concentration profiles, the spatial information available in the sequence of calculated hourly patterns over one day is thought to be suited for inclusion in the assimilation scheme.

From 24 modelled hourly concentration fields, after substraction of the hourly average spatial concentration $(c)_t$, characteristic fields f_h were computed. These characteristic fields in fact are time independent eigenvectors as determined from the covariance matric CC' , where elements of matric C are defined as: $c_{it} - (c)_t$, t = 1,2,- 24 , i = 1,2,- N, the number of spatial positions calculated. The vectors f_h are ordered according to decreasing explained variance and in practical applications the space time variability over a day to large extend is characterized by about 7 of such fields. The true (interpolated) field at a given hour is approximated by:

$$c^*_{it} = b_{ot}\ (c)_t + \sum_{h=1}^{7} b_{ht}\ f_{hi}\ , \quad i = 1 - N \tag{9}$$

The coefficients b_{ht} are estimated for that hour by multiple regression, minimizing the residuals e_j over a subset of M positions where the concentrations c^m are measured:

$$c^m_{jt} = b_{ot}\ (c)_t + \sum_{h=1}^{7} b_{ht}\ f_{hj} + e_j \quad j = 1 - M \tag{10}$$

Having estimated the coefficients b from restricted number of position where measurements are available, the estimated (interpolated) concentrations at the N positions are given by relation 9.
In the following example the procedure is illustrated. The results are preliminary and for an eventual final judgement of the suggested approach a larger number of case studies has to be made.

Example
The PUFF-model was run for february 3th 1982; computations were performed for 86 positions. In Fig. 11b the modelled field for 18.00 hours for the model area (400 x 400 km^2) is given. The field is dominated by the meso-scale plumes from the major source areas. The measured field, as interpolated with the statistical interpolation scheme from measurements at the same 86 positions, is given in Fig. 11a. The rms-discrepancy between model and measurements is 36 $\mu g/m^3$ at an averaged value of 56 $\mu g/m^3$. The procedure as given above (equation 10) was performed for a subset of 29 stations more or less equally distributed over the country. The reconstructed field from model and 29 stations is given in Fig. 11d. The rms-discrepancy with respect to the complete set of 86 stations amounts 22 $\mu g/m^3$. Statistical interpolation from the 29 stations gives a corresponding value of 27 $\mu g/m^3$. This field is presented in Fig. 11c.
The reconstruction by means of modelled information thus appears superior to the statistical interpolation. This implies that a trade off between number of measurement stations and model results (sustained by dynamic monitoring) has a potential for optimizing air pollution monitoring programs.

Fig. 11. SO_2-concentration fields in µg/m³ for February 3rd 1982, 18.00 hours.

Acknowledgements

Discussions with N.D.van Egmond and H.Kesseboom are greatfully acknowledged.

REFERENCES

1 M.M. Millan, Atm.Env. 14, 1980, pp. 1241-1253.
2 N.D. van Egmond and D. Onderdelinden, Atm. Env. 15, 1981, pp. 1035-1046.
3 N.D. van Egmond, Proc. 75th Annual Meeting of APCA, 1982, paper 82-631.
4 D. Onderdelinden and N.D. van Egmond, Air Pollution by Nitrogen Oxides, Elsevier, Amsterdam, editors T. Schneider and L. Grant, 1982, pp. 209-223.
5 N.D. van Egmond, D. Onderdelinden and H. Kesseboom, Proc. 13th Internat. Meeting Air Poll. Modelling and its Application, NATO-CCMS, Marseille, 1982, to be published.
6 A. van der Meulen and D. Onderdelinden, Atm. Env. 17, 1983, pp. 417-428.
7 N.D. van Egmond and H. Kesseboom, Atm. Env. 17, 1983, pp. 257-274.
8 N.D. van Egmond and D. Onderdelinden, VIth World Congress on Air Quality, 1983, paper C VI 14.
9 N.D. van Egmond and O. Tissing, Proc. IV Int. Clean Air Congress, Tokyo, Japan, 1977, pp. 658-662.

10 TNO, Rekensysteem Luchtverontreiniging II, P.O. Box 214, Delft, The Netherlands, 1979.
11 OECD-LRTAP, 2 rue André Pascal, 75775 Paris Cedex 16, France, 1979.

Optical Remote Sensing of Air Pollution,
Lectures of a course held at the Joint Research Centre, Ispra, Italy, 12—15 April 1983,
P. Camagni and S. Sandroni (Eds). 381—415

METEOROLOGICAL EFFECTS IN REMOTE-SENSING OPERATIONS

Millán M. Millán

1. INTRODUCTION: A PERSONAL VIEW

It is interesting to note that the application of spectroscopic techniques to the remote sensing of air pollutants or, in a more general context, to the study of air pollution problems in the lower atmosphere, is still in its infancy in spite of a great flurry of activity over the last decade or so. In the author's opinion, this seems to be the result of several factors, one of which could be that this field has been approached almost exclusively from the instrumentalist side.

Given this situation, solving the instrumental development problems first and being busily in the process of understanding its interactions with a real atmosphere second, leave little time to "sell" the results to the air pollution meteorological community at large.

The situation is further complicated by the fact that anyone approaching, or requiring support from, another field of science has a tendency to grasp quickly a set of concepts which can become his gospel truth, even though those concepts may be totally questionable by those working in that field. In this manner, instrumentalists have put their sensors to use in a much too simplified manner for the meteorologist, and, conversely, the latter have developed air pollution and plume models which seldom pass any instrumental test.

Before raising too much dust in this matter, it should be said that most applications of remote sensing devices to the study of air pollution take place at distances from a few metres to tens of kilometres from the source. Within these distances dispersion effects (diffusion and transport) fall within the realms of the micro and, particularly, mesometeorological scales and phenomena, and tend to be further complicated by either, or both, sources and topographical effects. All of these fall into a no man's land which is not (yet) adequately covered by conventional fluid mechanics or meteorology and their associated instrumental methods.

It should become quite obvious throughout this chapter that meteorological considerations play a very important role in the interpretation and processing of remotely sensed data and, conversely, that some meteorological aspects of air pollution can be better understood through the use of the remote sensing instruments. In the context of acquiring background information, assuming that the reader is familiar with some principles of fluid dynamics and thermodynamics, the author would suggest reading through the books or chapters indicated in refs. 1 to 13, in the order given. This includes introductory material on basic atmospheric behaviour, micrometeorology and turbulence (ref. 1-7), specialized concepts relating to air pollution meteorology (ref. 8 to 11) and more specialized monographs (ref. 12, 13).

The contents of this chapter are based almost exclusively on the experiences of the author and some of his colleagues using the COSPEC instruments and, correspondingly, it is the application and interpretation of COSPEC results that are emphasized. The text will be somewhat colloquial and the presentation is intended to illustrate, with some applications, the problems which can be encountered when comparing the behaviour of pollutants in the real atmosphere with some of the "simpler" conceptions about what is supposed to happen.

Finally, it should be emphasized that the development of a cohesive theory on atmospheric turbulence and meteorology as applied to air pollution is in the process of being developed at the present, and that it will necessarily have to include means of incorporating the experimental results obtained directly, or indirectly (i.e. via direct support) with remote sensing techniques.

2. PLUMES: DIFFUSION, AVERAGING AND OTHER TOPICS

2.1 Some preliminary concepts

Traditionally the study of plumes was one of the first areas in which remote sensing correlation spectrophotometers were applied. Most applications have been made in the plume location mode, i.e. looking vertically upwards from ground based platforms.

Initially, and subject to the conditions discussed in chapter IV, it will be assumed that in this mode the instrument output is proportional to:

$$\int_0^{z=\infty} \chi(x, y, z, t)\, dz = cL \qquad (1)$$

where $\chi(x, y, z, t)$ is the concentration field of the target gas above the sensor ($\mu g\ m^{-3}$), z is the effective height of the atmospheric scattering layer used as the source of radiation, and cL is the overhead burden or optical thickness of the target gas expressed in ppm-m ($\mu g\ m^{-2}$) or μm of the target gas at STP, respectively.

To gain some initial insight into what can be expected from the sensors, equation (1) can be evaluated for any of the available simple plume diffusion models. These give the concentration of the target gas in a steady state (average) plume as:

$$\chi = (\bar{Q}/\bar{u})\, F(x, y)\, G(x, z) \tag{2}$$

where $(\bar{Q}/\bar{u})$ is the ratio of source strength to the "average" or advecting wind speed (kg. m^{-1}), F(x, y), G(x, z) are the horizontal and vertical diffusive functions (ref. 2) which, as equation (2) indicates, are considered decoupled in the Gaussian type models. The x coordinate is along the "axis" of the plume model, y is in the horizontal plane and z is taken vertically upwards. The origin of coordinates is at the base of the stack.

The most usual forms for the diffusive functions are:

$$F(x, y) = \frac{1}{\sqrt{2\pi}\,\sigma_y(x)} \exp\{-\tfrac{1}{2}(\frac{y}{\sigma_y(x)})^2\} \quad (m^{-1}) \tag{3}$$

$$G(x, z) = \frac{1}{\sqrt{2\pi}\,\sigma_z(x)} \left[\exp\{-\tfrac{1}{2}(\frac{z-H}{\sigma_z(x)})^2\} + \exp\{-\tfrac{1}{2}(\frac{z+H}{\sigma_z(x)})^2\}\right] \quad (m^{-1}) \tag{4}$$

where σ_y, σ_z are the horizontal and vertical standard deviations of the concentration distribution in the (average) plume cross-sections at each x, assumed (obviously) Gaussian, and H is the effective height of emission. The vertical diffusive function used assumes, among other things, perfect reflection of the target gas at the ground (ref. 14).

Using equations (1), (2), (3) and (4), we find that the output of the instrument at any location of the x, y plane is given by:

$$V(xy) = K(\bar{Q}/\bar{u})\, F(x, y) \tag{5}$$

where V is the output voltage and K (in V $kg^{-1}m^2$, or V/ppm-m) is a constant of proportionality determined by instrumental calibration.

The diffusive functions considered, σ_y and σ_z values and, correspondingly, the Gaussian plume model purport to be "averages", of a normally meandering plume (ref. 15). Initially a linear array of sensors could be deployed along $x = x_o$, $z = 0$, and the plot of their outputs, averaged during a comparable period, would yield

$$\langle V(x_o, y_i)\rangle = K(\bar{Q}/\bar{u})\, F(x_o, y_i) \tag{6}$$

This is a scaled version of the lateral diffusive function of the plume at the distance x_o along its "axis". From this distribution the inverse problem could be solved to obtain $\sigma_y(x_o)$ and, similarly, $\sigma_y(x)$ at any other distance.

An underlying hypothesis in all Gaussian plume models is the steadiness of the atmospheric diffusion process, which implies that diffusion takes place in a field of stationary and homogeneous turbulence. In a real atmosphere, the closest that we can get to being stationary is if the "mean wind" is constant with time. This is never the case for long but one could expect it to be a "reasonable" approximation at certain times and for short periods. A more detailed discussion of these processes is given by Csanady (ref. 12) and Pasquill (ref. 9), and we will not dwell on them here. However, we will initially assume that the diffusion process can be considered stationary for limited periods and that we can take advantage of the ergodic property of this kind of process, i.e. time averages equal ensemble averages. This would allow us to consider, instead of an array of sensors, one instrument mounted on a vehicle which can travel (quickly) back and forth under the plume along the route $x = x_o$, $z = 0$. If the crossings under the plume are completed in a time period, short, compared with the specific averaging times of the model, or, in our case, compared with the meandering "period" of the plume, the ensemble average of all the output traces obtained over the route should coincide with the time-and-vertically integrated plume seen by the stationary array of sensors, i.e. with equation (6).

A final point concerns the integration of equation (6), or, conversely, of the calibrated average profile obtained; multiplying it by the "average" wind and integrating yields:

$$K\bar{u} \int_{-\infty}^{\infty} \langle V(x_o, y)dy \rangle = \bar{Q} \qquad (kg.\,s^{-1}) \qquad (7)$$

This is the target gas mass emission at the source.

A typical installation for this kind of plume studies is shown in Figures 1 and 2. The COSPEC telescope looks out of the window of the car and vertically via a 45° mirror adapter. A point sampler to define the area of plume impingement on the ground, a dual pen chart recorder or equivalent data logger and a power inverter complete the installation. In most cases, a speed proportional counter is used to drive the recorder proportionally to the distance travelled or to trigger the data logger at fixed (distance) intervals to expedite data reduction and processing procedures. The overall measurement concept is shown in Figure 3.

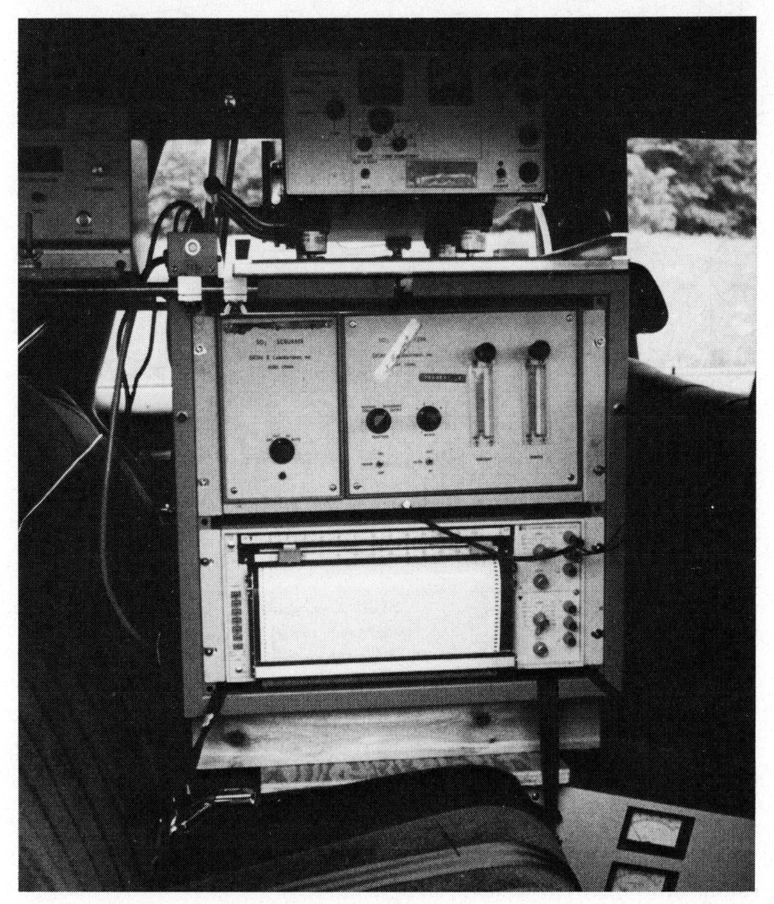

Fig. 1 Typical installation for fast traversing under plumes and perimeter surveys, outside view.

Fig. 2 Inside view of COSPEC installation, COSPEC, ground SO_2 monitor and recorder.

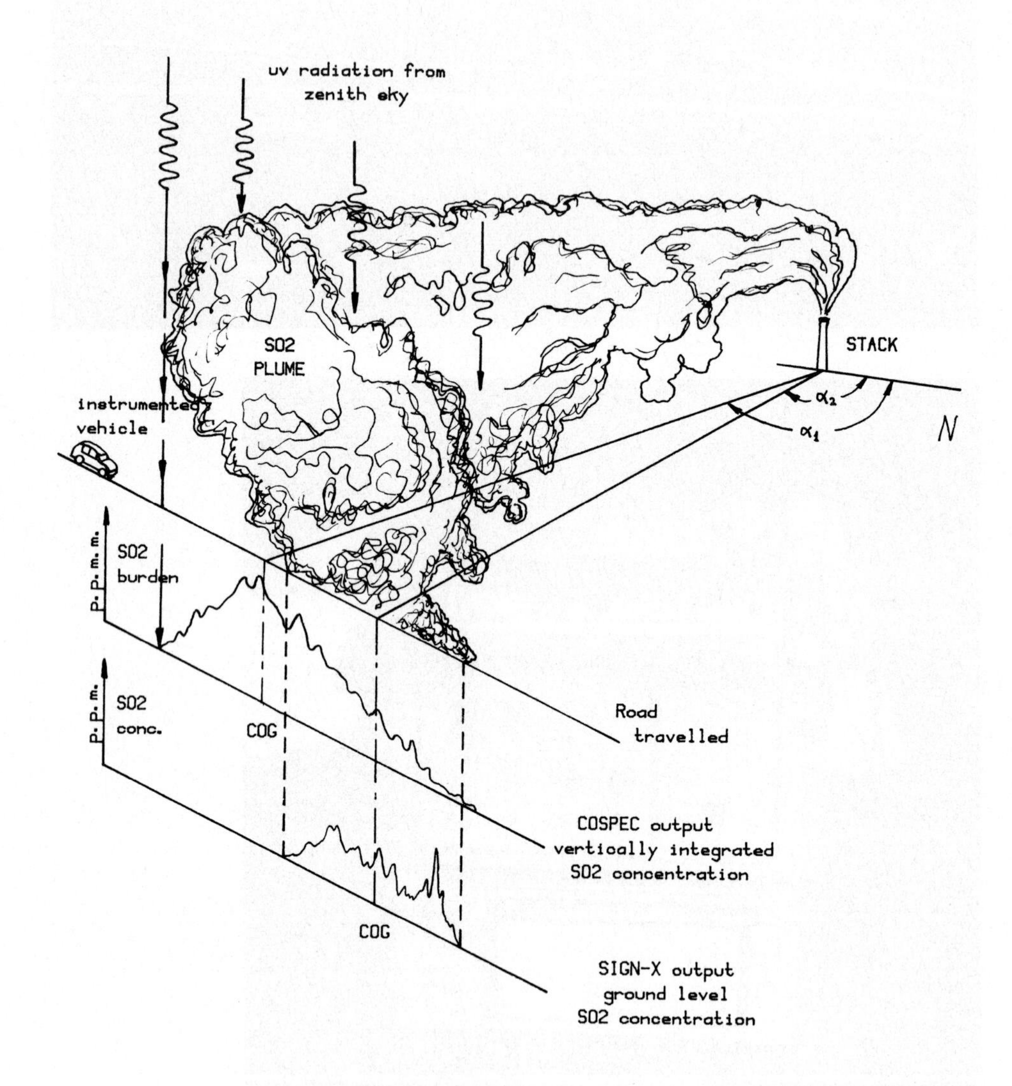

Fig. 3 Conceptual measurement. The COSPEC remote sensor locates the total plume while the ground level monitor outlines the ground impingement area.

One has only to find routes perpendicular to the plume "axis" and to find the "average" wind speed to obtain the lateral dispersion parameters and the mass emission rate of the target gas. This all sounds tremendously attractive but, like the Gaussian plume model, it is deceptively simple and, worst of all, unreal. Some remote sensing instruments are still promoted on the basis of this simplistic approach. Unfortunately, there are more than a few flies in the ointment which we will try to discover now.

2.2 Plume profiling

Before we continue, however, we should introduce some of the terminology used in the rest of this chapter.

- A traverse is defined as the physical path followed by the instrumented vehicle across, or under, the plume. The term Transect is used with the same meaning.
- A plume profile is the data record obtained along the traverse by a vertically integrating remote sensor.
- A transection is a family of traverses, e.g. at varying height and at equal distances from the source, as would be obtained with an instrumented aircraft. The term cross-section is reserved for a transection that is both vertical and perpendicular to the plume centreline or, alternatively, a transection that can be considered equally representative.

In the case of a ground based sensor looking upwards, a transection can also be defined as the surface generated by the line-of-sight of the instrument (ideally, the vertical) as it travels along a certain ground path. This defines a generalized cylindrical surface which may be composed of plane sections, or curved sections, or combinations of these.

In a real atmosphere, one of the first problems confronting the user of an upward looking remote sensor (or any other) is to determine how well the collected data profiles are representative of vertically integrated plane cuts perpendicular to the plume. This problem can be divided into two parts:

a) to find a plume "axis" or a centreline along which one can determine distances from the source, and the condition of perpendicularity of the traverse;
b) to decide, on the basis of the geometry of the traverses relative to the plume, how representative his profile is of a perpendicular cut of the plume. This problem arises because, unlike the ideal situation considered previously, the overall plume shape and its geometrical relationship to the

traverse are only known *a posteriori*.

Aircraft traverses of a plume, for example, are aligned on the basis of either:

a) visual observations of the plume itself, or

b) wind soundings.

In the first case, visual contact with the plume may be difficult if the plume is clean of particulates and its water vapour content has evaporated. In the second case, the wind profile not only changes with height but may vary significantly with distance from the source on account of topographical effects, etc. In either case, it is difficult to define whether the intended traverses are perpendicular to the *instantaneous* or to the *average* plume. Further complications arise due to the plume meander during the time of the measurements and to the non-simultaneity of the traverses at the various heights required to obtain a complete transection. These leave the data as a "curious composite of successive measurements"(ref. 16) and still with an orientation problem.

Still another facet of this problem is that plume "axis" or plume "centreline" are usually vaguely used terms with various interpretations. In our context "axis" will be used exclusively with reference to theoretical or model plumes which are usually assumed:

a) to have (on the average) a normal or Gaussian concentration distribution;

b) to move *ever onwards* in a straight line.

From these it follows that the locus of the points of maximum concentrations coincides with the centre of mass of the cross-section, and it is also the "axis" of the (model) plume.

Real plumes rarely follow a straight path, and their concentration distributions are seldom symmetrical with respect to any point whether considered instantaneously or on the average. In this situation, the concept of "centreline" based on the locus of the centres of mass of each cross-section is more appropriate; it can be applied to both "instantaneous" and "average" plumes as well as uniform or bifurcating plumes (ref. 17). It also better lends itself to theoretical treatment of the data (refs. 9, 12), particularly of the kind which is obtained with a vertically integrating remote sensor (ref. 18).

Operationally, the data gathering procedures consist of:

1) determining the approximate plume direction on the basis of visual observations of a wind sounding or of a perimetre survey around the source;

2) selecting on a regional map a number of possible routes crossing under the suspected plume path move to the area and locate the plume. In gene-

ral, teams closer to the source should communicate their positions and/or any observed plume movements to those operating at larger distances from the source;

3) selecting an origin of traverses and initiating the profiling back and forth under the plume as rapidly as possible.

In its traverse along the route, the sensor encounters a (real) plume that changes in time and space, so that the profile is really non-instantaneous. In general one is forced to assume:

- a frozen plume during the time required to complete the traverse;
- an instantaneous (or very fast) instrumental response to the spatial (or temporal) changes in the overhead burden as seen by the instrument from the moving vehicle.

These assumptions may be justified (initially) when:

a) the atmospheric stability conditions are between very stable to neutral;
b) the time constant of the instrument is short, e.g. one second;
c) each run is completed in a time period short compared with the meandering of the plume.

Unstable atmospheric conditions with vigorous meandering, looping and plume breakup cause significant changes in the plume position and concentration distribution before a single traverse can be completed. In this situation a single profile cannot be considered too representative of the "instantaneous" plume. However, if several traverses are made in succession, the profiles can be combined in two different ways to give:

1) a mean profile of the instantaneous plume averaged in time, i.e. a fixed-distance Lagrangian average, and
2) a mean profile of the plume averaged in time and cross-plume direction, i.e. a fixed-distance Eulerian average.

These are illustrated in Figure 4 and will be discussed later.

The general data processing procedure is to average all the profiles obtained, on the road selected, during a specified period of time. This period is usually determined by the subsequent use of the data, e.g. validation of diffusion models, comparison with ground level concentrations, calculation of mass fluxes, etc. In general, one half to one hour's worth of profiles are averaged. In the author's experience this may include up to 15 profiles for traverses close to the stack (in an uncluttered road) to just two or three profiles at distances of 30 to 50 km from the stack. This is fortunate, in a sense,

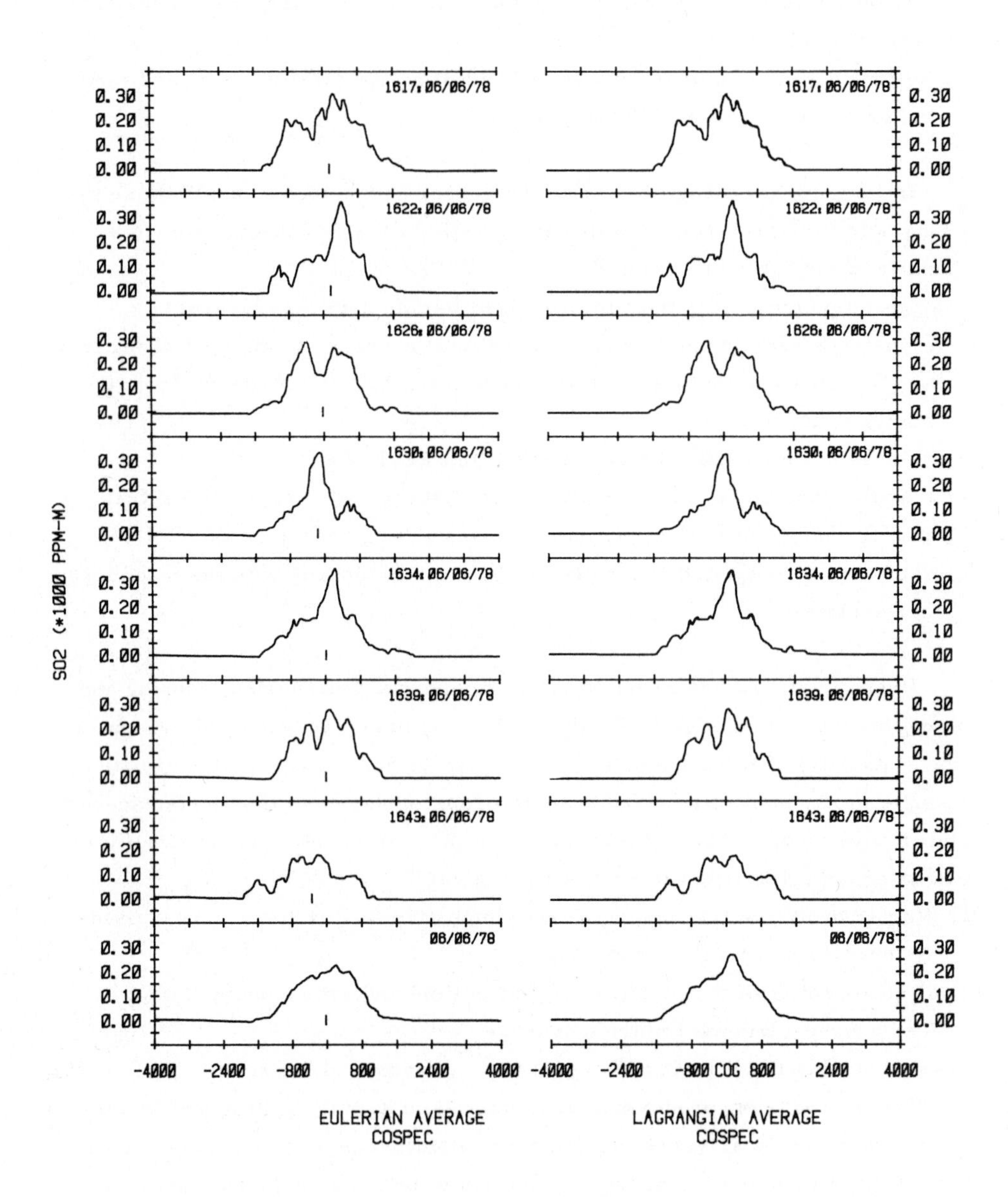

Fig. 4 Averaging concepts. The Eulerian profile (lower left) obtained by averaging, as they occur, the individual profiles. The Lagrangian is obtained after re-centering and averaging. The centres of gravity of the individual profiles are shown (at left) by a short vertical line. Profiles are shown as seen from the source looking down wind.

because at short distances from the stack the plume is being acted upon by small scale (higher frequency) turbulence and its characterization requires a large number of profiles to obtain representative averages. At the larger distances, the plume is rather passive and its (larger) shape does not change as rapidly, so that a lesser number of profiles is adequate.

The space and time road-averaged-profile has now the characteristics of the road followed, which may not be straight or perpendicular to the centreline of the average plume. This is determined by its position over the map, that reported by other teams closer to the source, and, finally, by the location of the source itself. Certain criteria (ref. 19) can be followed to determine whether the average profile is representative of a perpendicular cut to the centreline, and a projecting plane is determined as the effective plane of the traverse. The individual profiles are projected onto this plane and those projections are subsequently used for all the data processing that follows. The projected profiles and all averages derived from them can be considered representative of "instantaneous" or "average" plume cuts if the road followed meets certain angular constraints with the experimentally determined centreline (refs. 18, 19, 20). It should be quickly emphasized that the criteria in those references are just some of a possible few, and that this area of research is fully open.

As shown in Figure 4, the Eulerian profile is obtained by averaging, as they occur, all the profiles of the set. The second average, the Lagrangian, is also obtained by recentering the projections of the individual profiles with respect to their individual centres of gravity and averaging. The first average represents the plume that affects the ground, and the second, the plume one would see if one could ride the plume centre as it meanders over the road traversed. It retains the intrinsic shape of the instantaneous plume, e.g. skewness, bifurcation, etc.

Under very stable situations, with little plume meander, the two averages converge. This is shown in Figures 5 and 6 for a power plant plume in the very stable region of a lake breeze circulation cell (ref. 20).

After all of this, one of the most obvious features that a COSPEC user may notice is that no matter how perpendicular to the plume centreline his traverses are, the plume averages, the Lagrangian in particular, are almost always skewed. This is further enhanced if the ground level concentrations measured along with the COSPEC are also plotted. Finally, this becomes even more obvious for plumes from tall stacks, as shown in the example of Figure 7. It can be seen that not only the individual profiles are skewed but

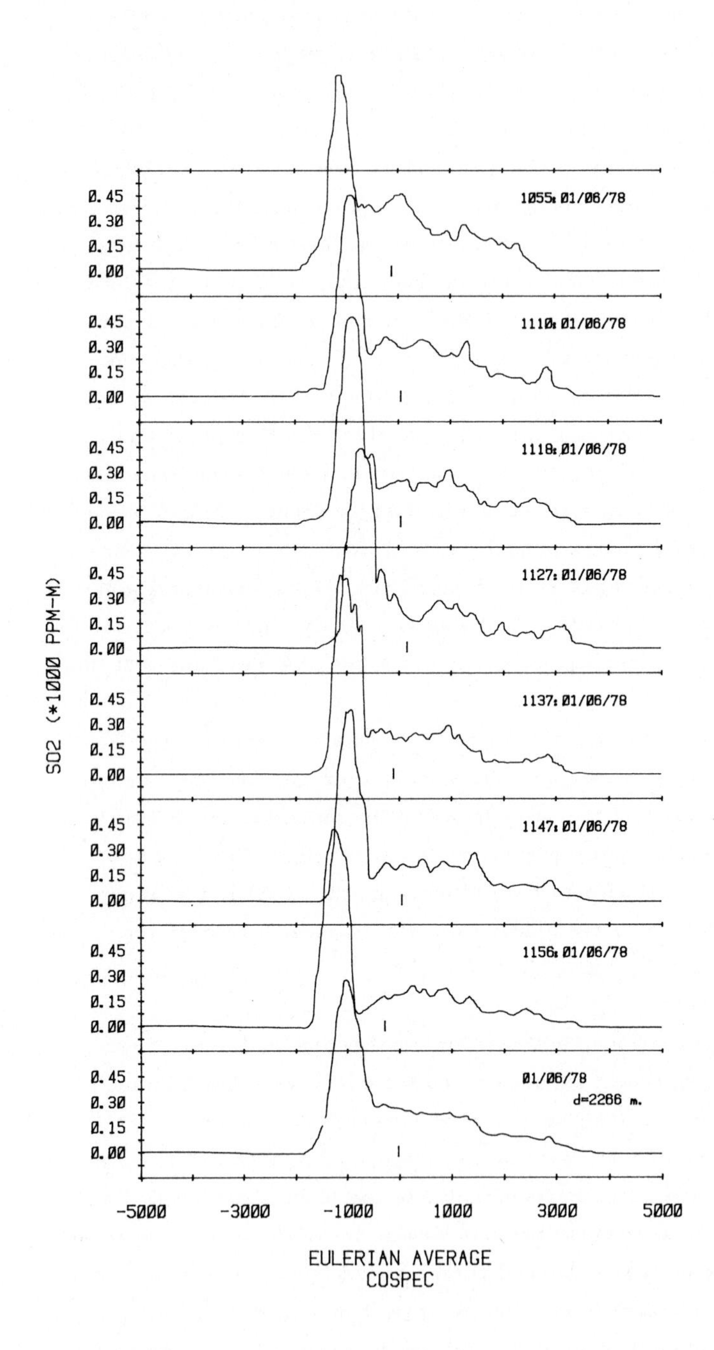

Fig. 5 Eulerian average of plume(s) in very stable flow. Distance to source 2.2 km.

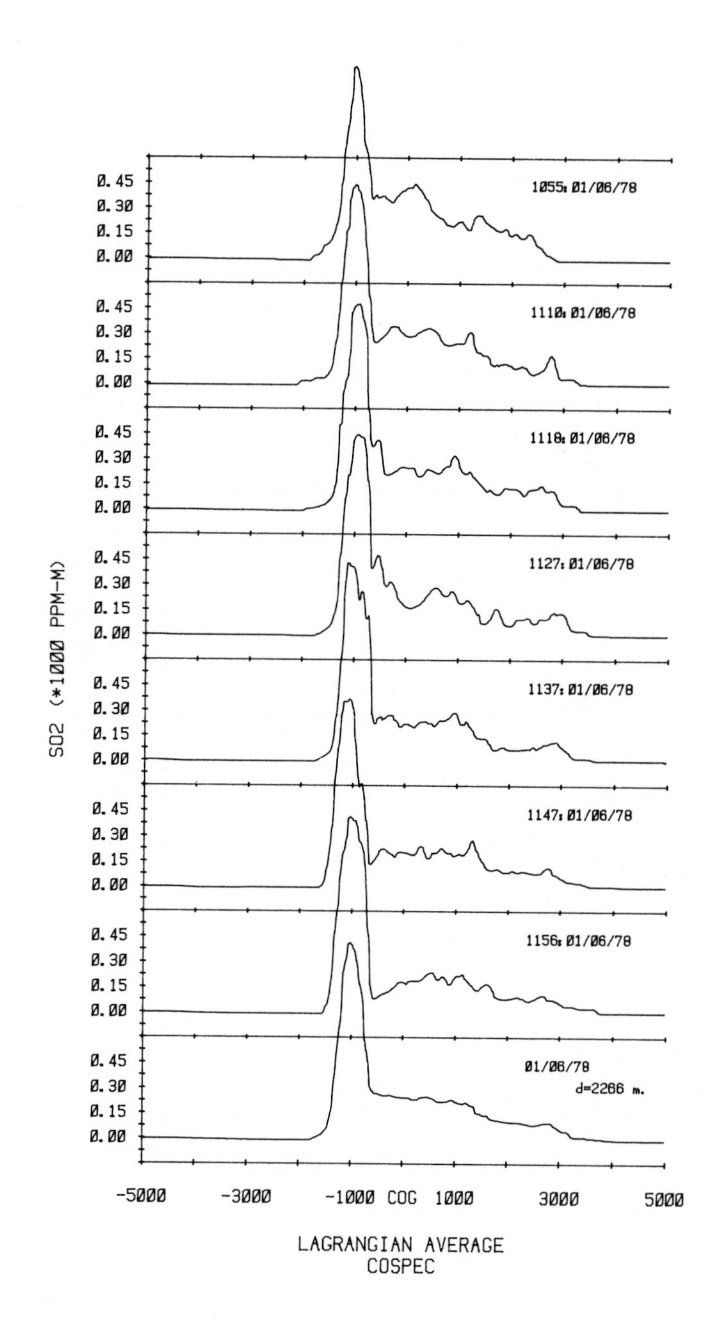

Fig. 6 Lagrangian average of plume(s) in very stable flow. Distance to source 2.2 km.

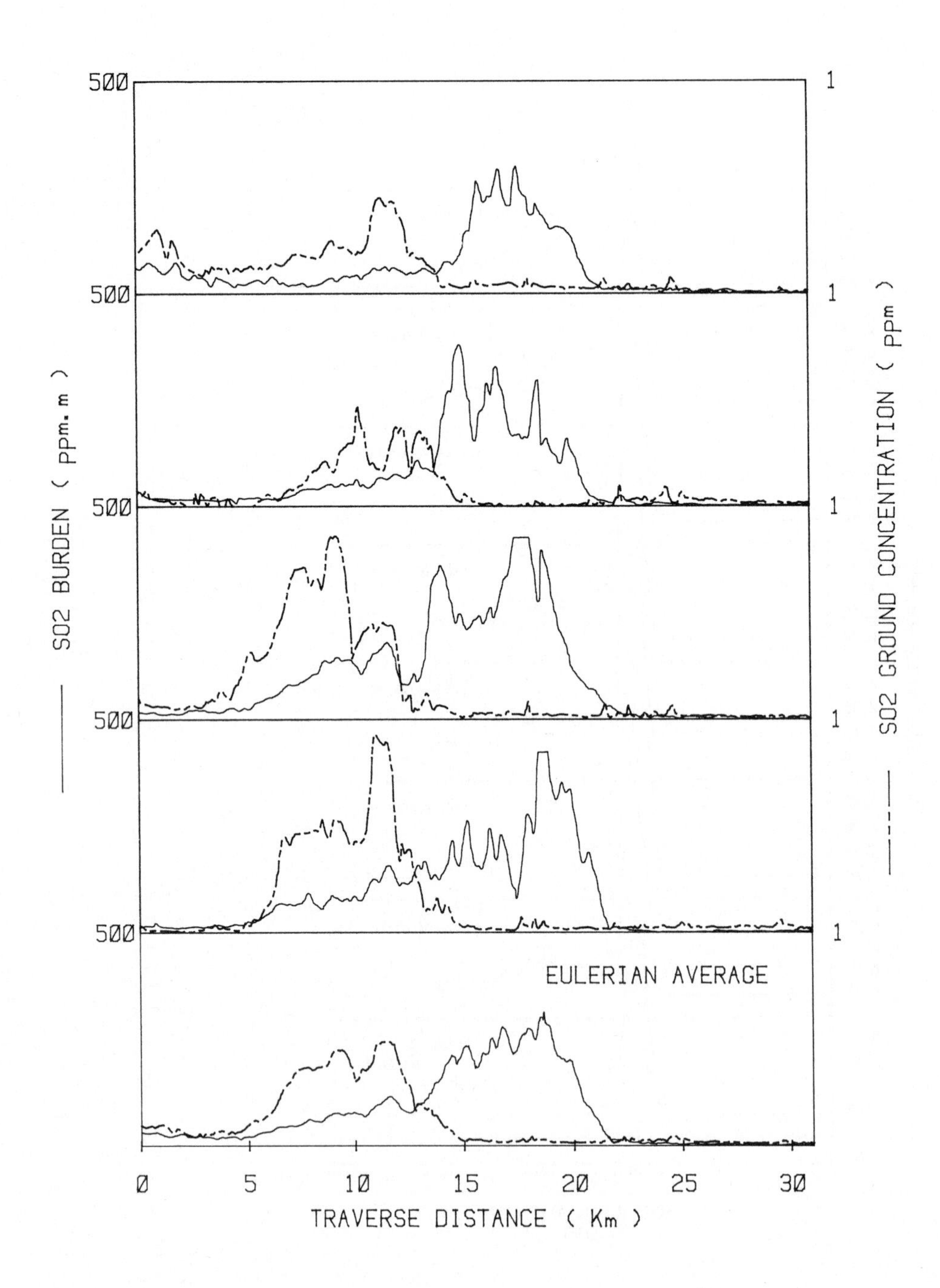

Fig. 7 Plume profiles from a tall stack showing the effects of wind directional shear. Distance to source 16 km.

that the ground impingement occurs over the tail of the COSPEC profile.

This would appear to have a simple explanation when one considers that the wind (normally) changes direction with height (veers to the right) and that the plume cross-section is most likely tilted in space. It would appear that the lowermost part of the plume, further to the left as seen from the source looking downwind, is the first to impinge on the ground (on the average) and results in the observed records. One could also wonder whether a similar situation would occur if the wind backs with height (i. e. turns to the left). This situation can occur with cold air advection, as may be observed during the initial onset of a lake-breeze. Figure 8 shows two Eulerian averages of a plume from a coastal power plant in this situation (upper graph), and after the lake breeze is fully established. In the first case, the skewness is to the right, and the centre of gravity of the ground impingement is to the right of that of the vertically integrated profile. In the lower part, with the lake breeze well established and rotating to the right under Coriollis forces (ref. 21), the plume has reverted to its "normal" mode and the skewness is again to the left.

Another example of a plume skewed to the right is provided by Figures 9 and 10. These show (~ 1/2 hour; 6, 4 and 3 traverse) Lagrangian averages at three distances from the source. The available profiles were selected to offset the travel time of the air parcel. It can be observed that the skewness and other plume features are well conserved up to 36 km for the conditions prevailing at the time. This aspect of plume diffusion has been examined by a number of researchers of which we can mention Pasquill and Csanady (refs. 9, 11). In this diffusion process, the plume is distorted laterally as soon as it grows vertically to the point that it begins to diffuse in layers that move in different directions, i. e. crosswind shear effects. The result of this is a combined diffusion process which yields lateral plume spreads larger than those due solely to turbulent mixing. This process has been described as shear augmented diffusion.

It should be indicated that the effects of wind directional shear on the diffusion of plumes have been observed by various researchers (refs. 9, 11, 22-24), but are still poorly documented by the more conventional techniques. This has been mostly caused by instrumental and sampling difficulties which have plagued instrumented aircraft platforms. These move too quickly through the plume for the response time constant of most conventional sensors, and cannot get too close to the ground. The result is that shear-augmented diffusion has been dismissed as having "negligible effect within the first 10 to

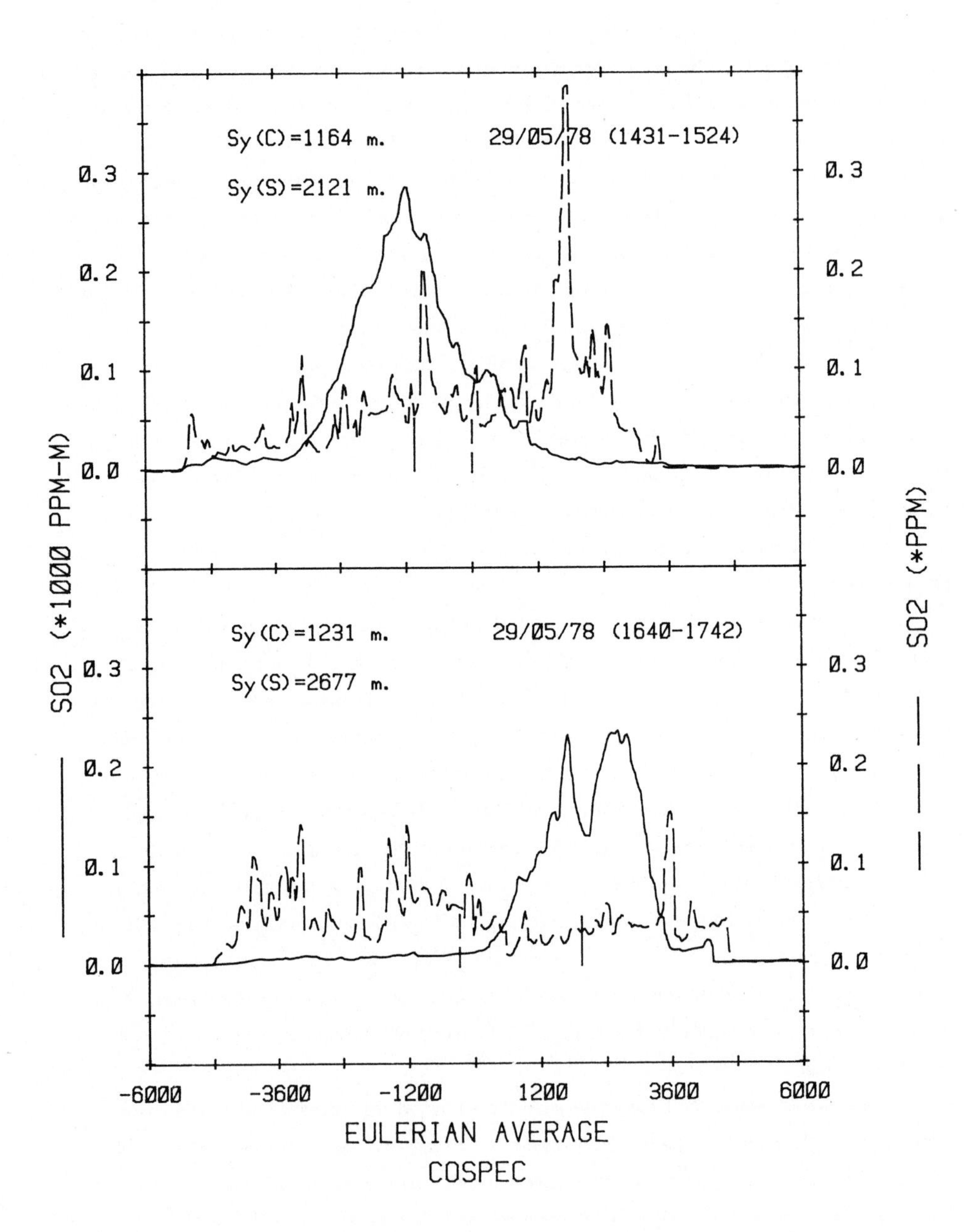

Fig. 8 Eulerian averages of a plume in a convective boundary layer. During the period of the upper graph the wind backed with height throughout the layer. At the time the lower average was obtained (2 hours later) the wind had reverted to veeering with height. Distance to source 11 km.

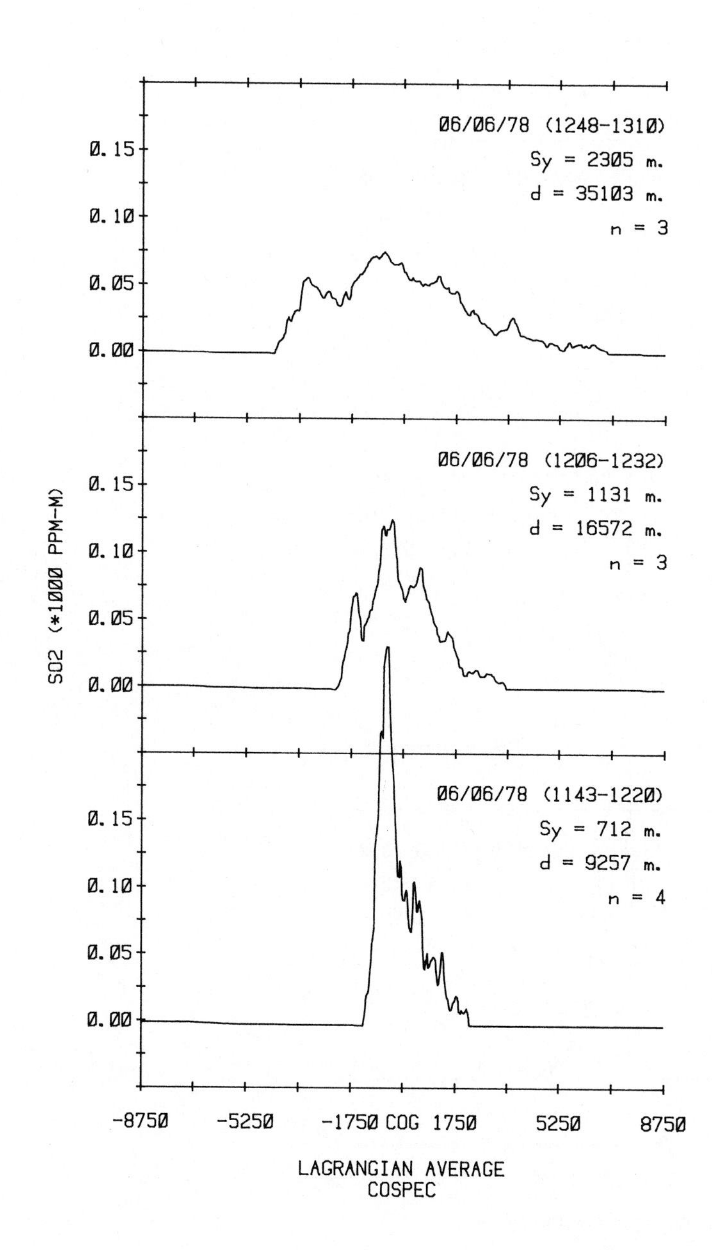

Figs. 9 and 10

Lagrangian plume profiles showing the transport of plume properties downwind. The profiles were selected to offset the time of travel of the air parcel. Distances shown in metres.

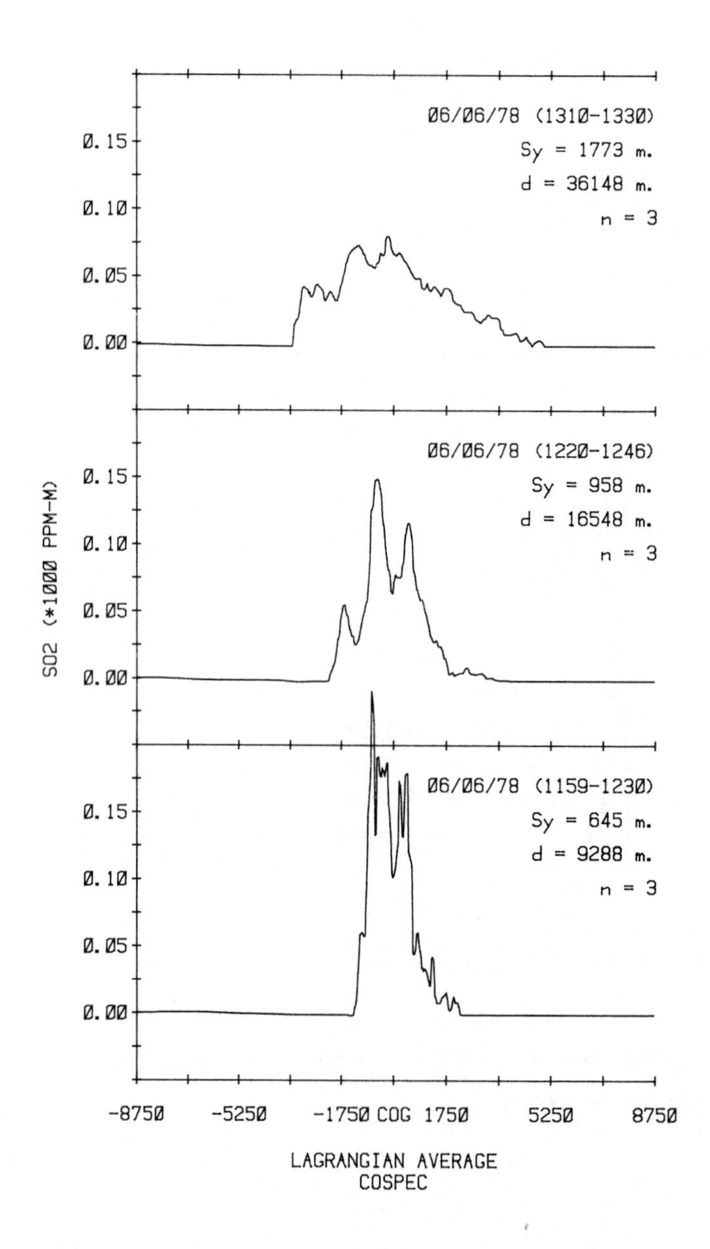

Fig. 10. See page 397 for legend.

12 km of the plume" (Pasquill). After collecting and analysing a few thousand plume profiles collected at distances from a few hundred metres to a few hundred km*), the author would not agree with that statement and believes that this is one of the most obvious breakdowns of the Gaussian plume models. It is even more important since plumes from very tall stacks behave in this manner, and Gaussian plume models are still used (and even legislated) in an attempt to predict the intensity and location of ground concentrations.

Just to illustrate that things are neither so bad nor so clear cut, Figure 11 shows two Eulerian averages, each of seven profiles at ~ 16 km from the source, obtained under conditions close to ideal for description by the Gaussian model, i.e. flat terrain, light cloudy skies and persistent winds. The averages show a smooth profile resembling a normal distribution both for the vertically integrating COSPEC and for the ground concentration sensor. The positions of their respective centres of gravity, however, still reflect a certain degree of skewness.

To end this section, it is interesting to compare the profiles in Figure 11 with those in Figure 8, obtained under strong convective conditions in a lake breeze induced fumigation. The COSPEC profiles are rather smooth in both cases, as can be expected from the vertical averaging properties of this sensor. The ground level concentration averages, however, show how spotty the convective fumigation process can be. This is a situation where the assumption of ergodicity can be seriously questioned. Other assumptions which must be looked at very carefully are those pertaining to

a) a uniformly mixed concentration distribution within a convectively mixed boundary layer, in particular for tall stack plumes and for a few km from the source, and
b) that there are no wind-shear effects within a convectively mixed boundary layer.

Both of these are still widely used by boundary layer modellers.

With this presentation the author has intended to show how remote sensors can support and/or complement conventional techniques to document some hitherto poorly documented phenomena, in the hope that existing theories in air pollution and boundary layer diffusion may become more coherent with observed phenomena.

---

*) with most of them in the 1 - 35 km range.

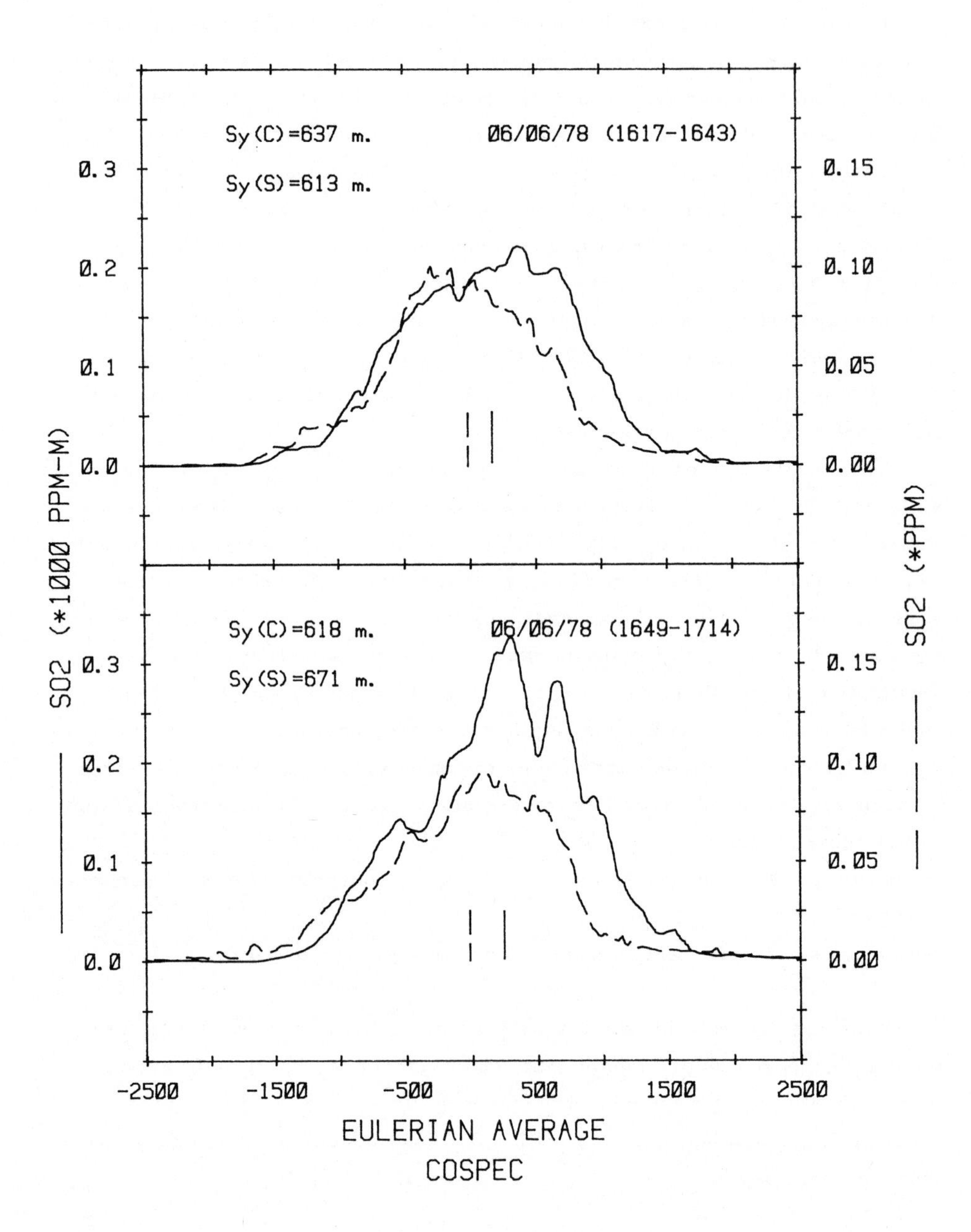

Fig. 11 Eulerian plume profiles under neutral conditions at 16 km from the source.

3. MASS FLUXES

The problem of calculating mass fluxes from COSPEC profiles can be re-examined now in view of what we have seen about plume behaviour.

In principle the main problems are: the three-dimensional nature of the plume and the wind field and that the turbulence is rarely stationary. We can assume that a wind sounding or a suitable average of various soundings is representative of a quasi-stationary wind field during the time of the measurements, so that, for the moment, the three-dimensional characteristics of the wind and the plume are the main sources of concern.

Initially, one can think of two situations for which a "mean wind speed" can be found. The first one arises when the plume can be assumed to be uniformly diffused throughout the vertical. This may occur (at least on the average) in a "well mixed layer" at sufficient distance from the stack. After the previous section, the reader may be aware that the author is not convinced that this ever quite occurs.

The mass flux in this case would become

$$\bar{Q} = \oint \int_0^L \frac{V(s)}{L} \vec{u}(s,z) \cdot \vec{n}(s,z) ds dz \qquad (kg.s^{-1}) \qquad (8)$$

where

- $V(s)$ is the reduced data profile in $\mu g.m^{-2}$ expressed as a function of the distance $\underline{s}$ along the ground path followed by the instrumented vehicle;
- L is the depth of the mixed layer. If the measured target gas burden is assumed to be uniformly mixed, the value V_s/L represents the average concentration throughout the layer;
- $\vec{u}(s,z)$ is the (average) vector field describing the wind at any point in the transection surface;
- $\vec{n}(s,n)$ is the unit vector at point (x,z) perpendicular to the cylindrical surface of the transection.

We can first notice that:

a) the unit vector has only x and y components, and
b) that it is the same along the vertical (viewing) line over any point of the path. In this situation $\bar{u}(s,z).\bar{n}(s) = u_n(s,z)$, i.e. the horizontal components to the wind at any height.

Equation (8) can be written as

$$Q = \oint \left\{ V(s) \frac{1}{L} \int_0^L u_n(s,z)dz \right\} ds \tag{9}$$

$$= \oint V(s) < u_n(s) > ds \tag{10}$$

The function $<u(s)>$ is then the "mean value of the wind" above point s which is effective in transporting target gas across the transection. It can be obtained from a wind and temperature sounding by calculating the average value of the horizontal wind component over the depth of the mixed layer (inferred, in turn, from the temperature sounding). If the wind profile is considered representative of the wind field over the whole traverse, the mass flux can be obtained by calculating (10) along the traverse. The same procedure can be generalized for a traverse within a convergence-free wind field.

The second situation for which a mass flow can be calculated occurs when a plume is contained in a layer with uniform wind speed and direction. This will actually occur close to the source and before the plume attains any substantial depth. It may also occur at further distances from the source if the plume becomes of the ribbon type, i.e. trapped within a stable layer and having very limited vertical and lateral diffusion.

In fact, the information obtained from the analysis of the Eulerian and Lagrangian averages can be of use during the process of calculating stack emissions. The most important factor appears to be the shape of the Lagrangian profile. It should be symmetrical or very nearly so. In the case of a bimodal (bifurcating plume) profile and particularly for plumes from tall stacks, an added condition would be that both edges of the profile should be steep.

Symmetry of the Lagrangian average indicates that the relative diffusion is dominated by turbulence. Asymmetry, long tail ends to either side or, in general, a skewed profile indicate that the diffusion is affected by wind directional shear as well as turbulence. In this situation different layers of the plume move and diffuse in varying directions and a representative wind speed, or plume speed, may be difficult to determine. If other independent means are available to determine the plume boundaries, e.g. Lidar, layer averaging can be used again to determine the mass flux (ref. 24).

Two other characteristics of the profiles give information about the behaviour of the wind during the measurement period. Meandering is an indication of either:

a) fluctuations of the wind direction,

b) intermittent penetration of the plume into layers with different wind direction, or

c) both processes.

Billowing of the plume can be noticed by a family of profiles which change their areas significantly during the measuring period. It can be the result of:

a) fluctuations in the wind speed,
b) intermittent penetration (or fluctuating plume rise) into layers with different wind speed,

or a combination of both. When the area of each individual profile varies greatly, and wind fluctuations are suspect, the use of the harmonic mean of the areas of the profiles, rather than the arithmetic mean, previously considered, appears to result in more accurate estimates of the flux (refs. 24, 25).

The results of an extensive study conducted by Sperling (ref. 26) have indicated that the accuracy of the remotely sensed vs the in stack measurements are within ± 24%, when averaging profiles for a 20 minute period. This author, however, utilized all the available profiles at varying distances from the stack and without any preselection based on the characteristics of the averages. A more selective processing of the data by Hoff, with more comprehensive wind information (ref. 24) and the use of the harmonic means, yields estimates within 7% of the fluxes computed at the source.

As a final note we should indicate that most of these studies were conducted around isolated power plants. The situation is quite different in industrialized areas, mostly because of superposition by other sources (see next section). Mesometeorological effects such as return flows, particularly if these are limited physically within valleys, should be treated with extreme caution. Figures 5 and 6 are an example of a freakish occurrence which confirms that the data profiles should be examined carefully before attempting to compute mass fluxes. They occurred when a stable (old) plume which had drifted over the lake with the land breeze was returned directly over the power plant the following morning at the onstart of the lake breeze to join with the new plume travelling inland (refs. 21, 24).

4. PERIMETER SURVEYS AND MESO-METEOROLOGICAL EFFECTS

In any particular situation one can expect the air movements to be dominated by either macro-scale (strong gradient) winds or by locally induced circulations of the meso-meteorological type under weak gradient conditions. In the case of industrialized areas with complex topography one can expect strong channelling and aerodynamic effects such as entrainment and downdrafts in the wakes and cavities of hills and ridges. These effects can also vary substantially with the relative orientation between the topographical features and the general flow, as the low level winds are forced to conform to the prevailing relief.

Mesometeorological phenomena are usually associated with weak gradients or anticyclonic conditions when "good weather" can prevail. The atmospheric circulation can be driven by the diurnal cycle of solar heating and can give rise to some well known situations of which we can mention:

a) In industrialized valleys - Night entrapment in a stable layer in the valley bottom to be followed by intensive fumigations the following morning. All this can take place simultaneously with drainage, down-valley flow, at night and up-valley flow during the afternoon.

b) In industrialized coastal areas - Offshore flows with the night land breeze can carry the pollutants over the sea (or a large lake) to be returned the following day with the sea breeze, although not necessarily over the emission area.

c) In industrialized areas with combined valley-sea interactions, where both of the effects described above can occur simultaneously and reinforce each other.

With this, we want to indicate that the establishment of cause-effect relationships between the emissions and the ground level concentrations is no simple matter, and that the application of any one of the prevailing air pollution models would be doomed to certain failure. On the other hand, if the behaviour of the air flow in any particular airshed can be classified into several clearly identified "patterns", these could be studied to determine which model is the most suitable to represent them.

It is in this characterization of air pollution patterns that remote sensors can provide a significant amount of support to the air pollution modelling meteorologist. Unfortunately, we cannot presently dwell on this, and we will end this section with a few examples drawn from the author's work in Bilbao in Northern Spain.

The area is located along an estuary that runs nearly 16 km from the centre of the city to the sea, and is aligned in a SE-NW direction. Two mountain ranges run parallel to the waterway, the one in the SW reaches 700 m, and the other 300 m. The location and general topography is rather complex with a number of smaller hills and valleys and a very irregular mix of industry and population centres. Because of its location, air flow within the Bilbao area is strongly influenced by the local topography under all meteorological conditions. This is particularly important under anticyclonic conditions with clear skies when atmospheric motions in the area are almost exclusively dominated by the meso-meteorological phenomena of local origin. These effects have resulted in some serious pollution episodes (ref. 27).

In order to characterize the air movements in the area, two mobile laboratories were instrumented in light vehicles to allow for fast coverage of the routes within the airshed. Figure 12 illustrates how a return flow can be characterized.

Graph a over the left side of the river shows, among other things, the plume of a power plant, marked PP as it drifts towards the mouth of the river in a stable stratified flow during the first hours of the morning. Time 0715-0830 (GMT).

Graph b shows the onset of the morning fumigation at the same time that the polluted airmass stagnated over the mouth of the estuary re-enters with the sea breeze. Time 0840-10. 10 (GMT).

Graph c : The sea breeze is now well established and has pushed the polluted airmass up valley. The coastal area is now clean while the area of Bilbao proper (Feria, Deusto) is still under heavy fumigation. Time 10. 30-11. 00 (GMT).

Graph d: Further intensification of the sea-breeze pushes the pollutants inland and leaves only the effects of the local sources to be noticed. Time 11. 05-14. 11 (GMT).

Graph e: Illustrates the same situation for the industries in the left side of the river. Three large plumes can be seen crossing over at least two routes each. The plume of the power plant (PP) can now be detected in a direction nearly 180° from its morning direction and shows the effect of a small hill upwind by impinging on the ground at a short distance from the stack.

Figure 13 is another example of a plume in drainage stable flow similar to Figure 12a, and it shows how the power plant plume flows towards the sea.

Figure 14 shows hill effects on the plume of another power plant (CT) as it is forced to rise above a 400 m ridge. Approximately 4 km downwind it has been entrained into the wake of the ridge and impinges on the ground over SS DEL VALLE.

Figure 15 to be compared with the previous figure, it shows the plume of the same power plant (CT) under strong NW winds. Under these conditions strong channelling occurs towards Bilbao, however, because there are no other topographical effects, ground impingement occurs only when the plume has grown in the vertical to reach the ground between Burcena and

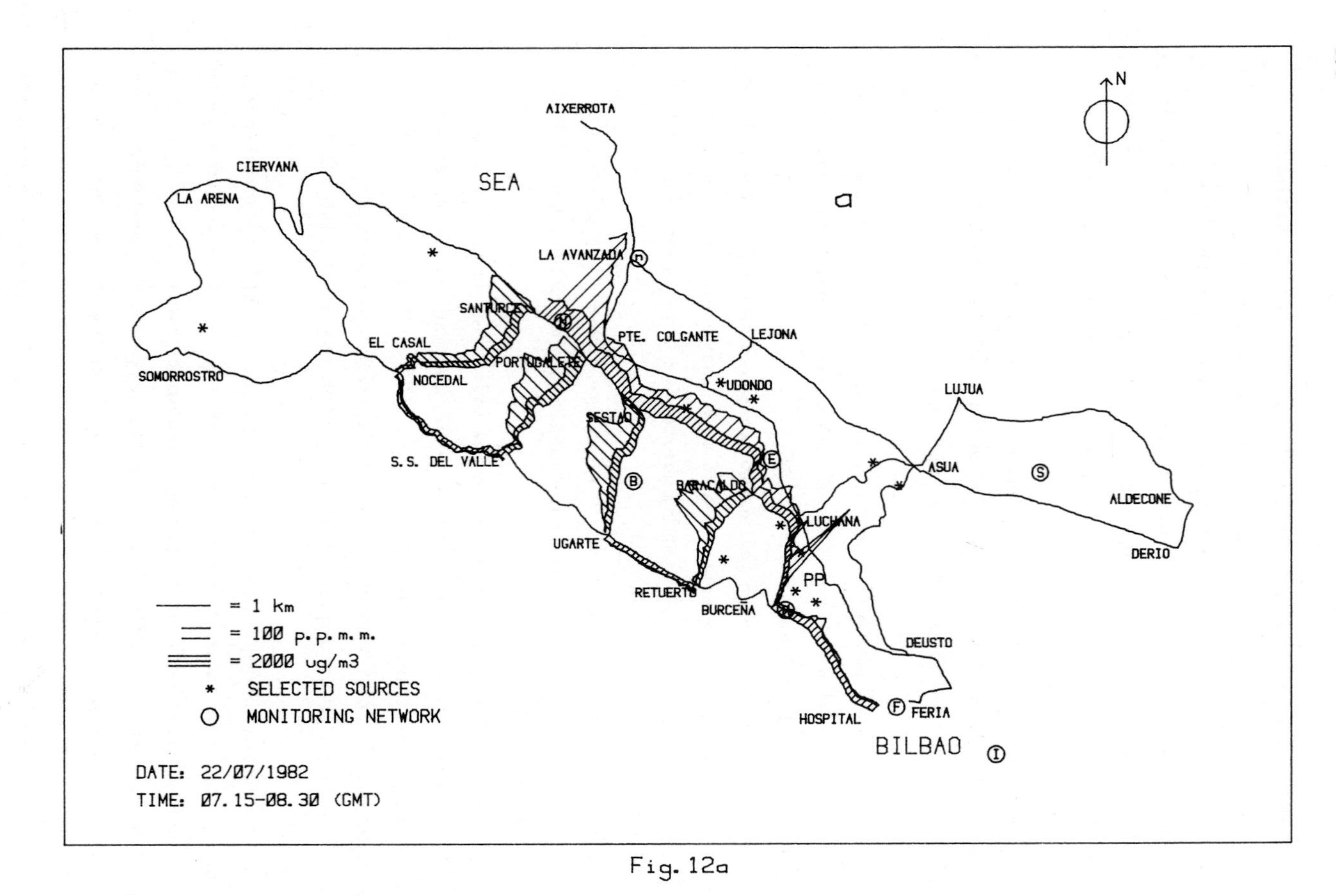
N
AIXERROTA
CIERVANA
LA ARENA
SEA
a
LA AVANZADA
SANTURCE
EL CASAL
SOMORROSTRO
PORTUGALETE
NOCEDAL
PTE. COLGANTE
LEJONA
UDONDO
LUJUA
SESTAO
S.S. DEL VALLE
ASUA
BARACALDO
ALDECONE
LUCHANA
UGARTE
DERIO
PP
RETUERTO
BURCEÑA
DEUSTO
HOSPITAL
FERIA
BILBAO
= 1 Km
= 100 p.p.m.m.
= 2000 ug/m3
* SELECTED SOURCES
MONITORING NETWORK
DATE: 22/07/1982
TIME: 07.15-08.30 (GMT)

Fig. 12a

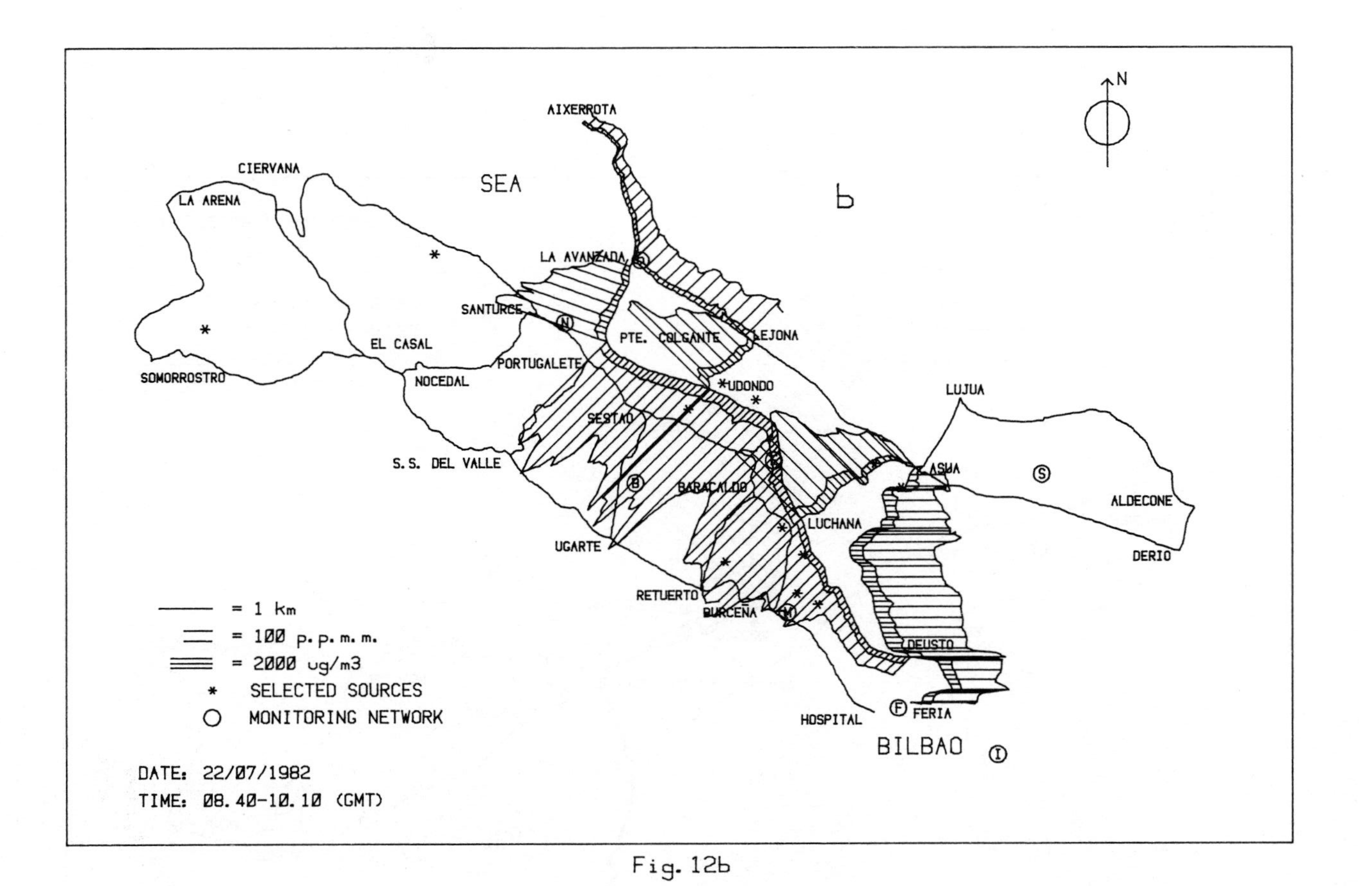

N
AIXERROTA
CIERVANA
SEA
b
LA ARENA
LA AVANZADA
SANTURCE
PTE. COLGANTE
LEJONA
EL CASAL
SOMORROSTRO
PORTUGALETE
NOCEDAL
UDONDO
LUJUA
SESTAO
S.S. DEL VALLE
ASUA
BARACALDO
ALDECONE
LUCHANA
UGARTE
DERIO
RETUERTO
BURCEÑA
DEUSTO
HOSPITAL
FERIA
BILBAO
= 1 km
= 100 p.p.m.m.
= 2000 ug/m3
SELECTED SOURCES
MONITORING NETWORK
DATE: 22/07/1982
TIME: 08.40-10.10 (GMT)

Fig. 12b

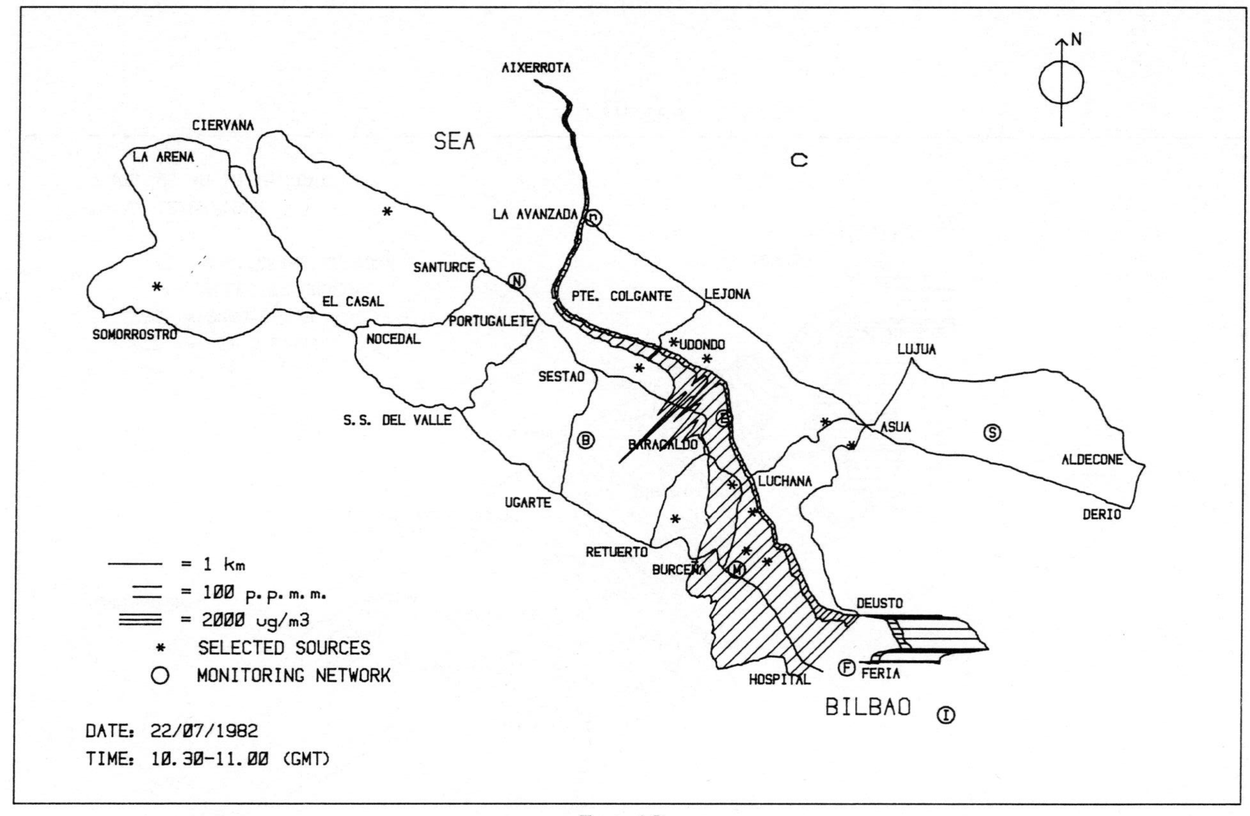
N
AIXERROTA
SEA
C
CIERVANA
LA ARENA
LA AVANZADA
SANTURCE
EL CASAL
SOMORROSTRO
NOCEDAL
PORTUGALETE
PTE. COLGANTE
LEJONA
UDONDO
SESTAO
S. S. DEL VALLE
BARACALDO
LUJUA
ASUA
ALDECONE
DERIO
LUCHANA
UGARTE
RETUERTO
BURCEÑA
DEUSTO
HOSPITAL
FERIA
BILBAO
= 1 km
= 100 p. p. m. m.
= 2000 ug/m3
SELECTED SOURCES
MONITORING NETWORK
DATE: 22/07/1982
TIME: 10.30-11.00 (GMT)

Fig. 12c

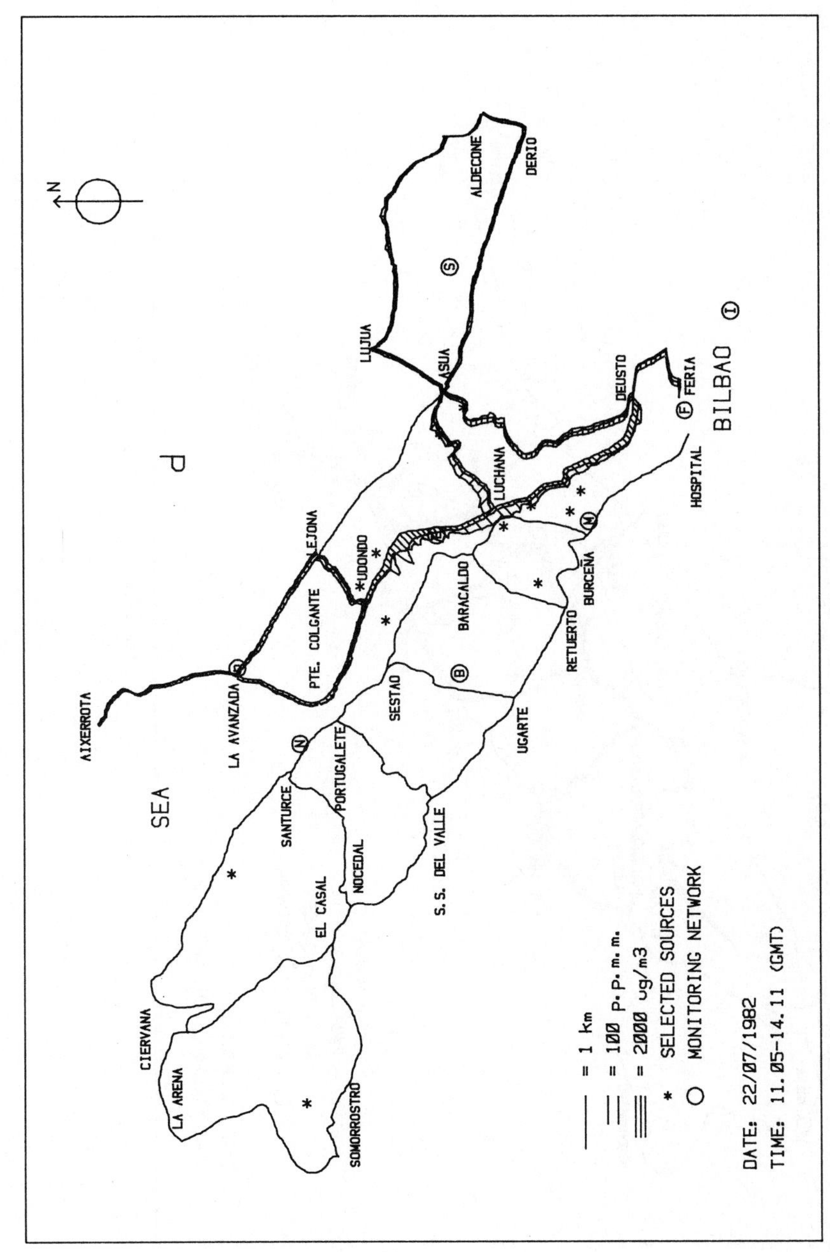
N
P
SEA
AIXERROTA
LA AVANZADA
PTE. COLGANTE
LEJONA
UDONDO
LUJUA
ASUA
ALDECONE
DERIO
DEUSTO
FERIA
BILBAO
LUCHANA
HOSPITAL
BURCEÑA
BARACALDO
RETUERTO
SESTAO
UGARTE
PORTUGALETE
SANTURCE
NOCEDAL
S.S. DEL VALLE
EL CASAL
CIERVANA
LA ARENA
SOMORROSTRO
= 1 Km
= 100 p.p.m.m.
= 2000 ug/m3
SELECTED SOURCES
MONITORING NETWORK
DATE: 22/07/1982
TIME: 11.05-14.11 (GMT)

Fig. 12d

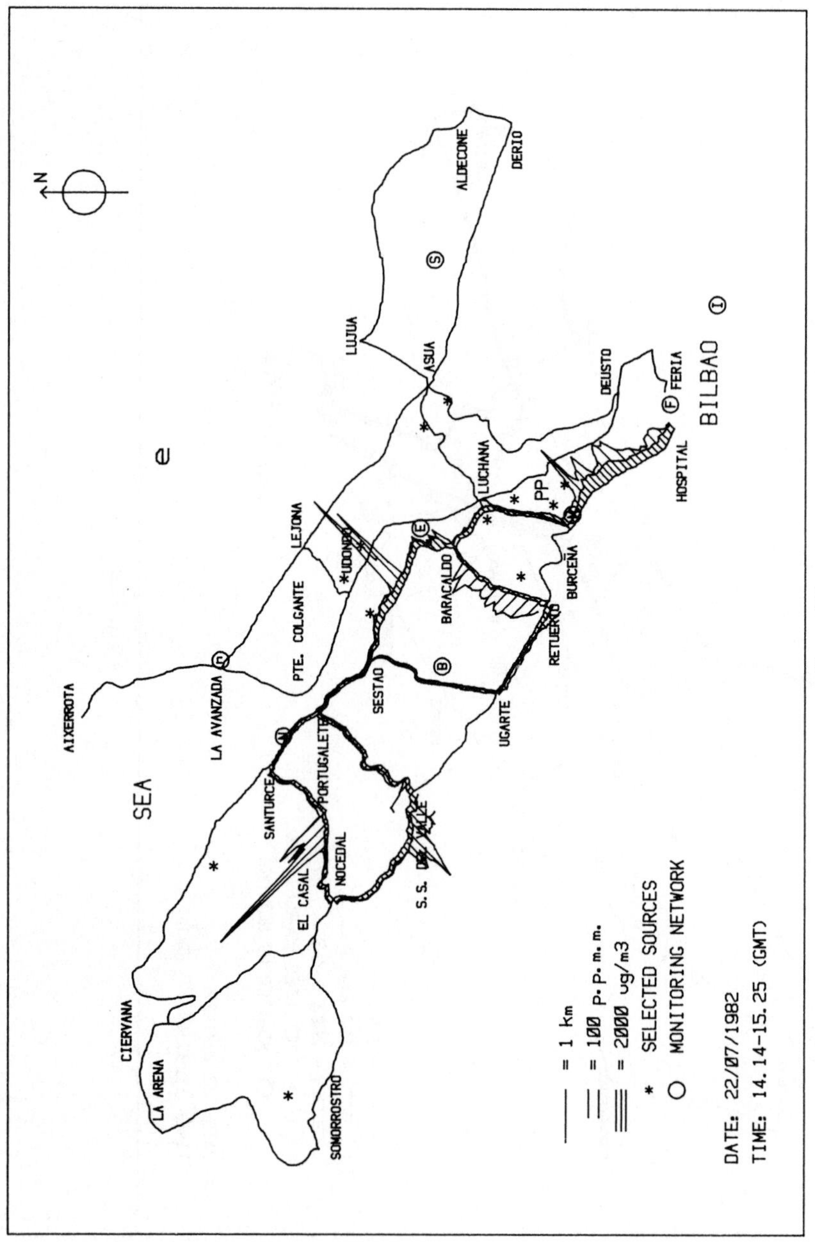
N
e
SEA
AIXERROTA
LA AVANZADA
PTE. COLGANTE
LEJONA
UDONDO
LUJUA
ASUA
ALDECONE
DERIO
DEUSTO
FERIA
BILBAO
HOSPITAL
LUCHANA
PP
BARACALDO
BURCEÑA
RETUERTO
SESTAO
UGARTE
PORTUGALETE
SANTURCE
NOCEDAL
EL CASAL
S.S. DEL VALLE
CIERVANA
LA ARENA
SOMORROSTRO
= 1 Km
= 100 p.p.m.m.
= 2000 ug/m3
SELECTED SOURCES
MONITORING NETWORK
DATE: 22/07/1982
TIME: 14.14-15.25 (GMT)

Fig. 12e

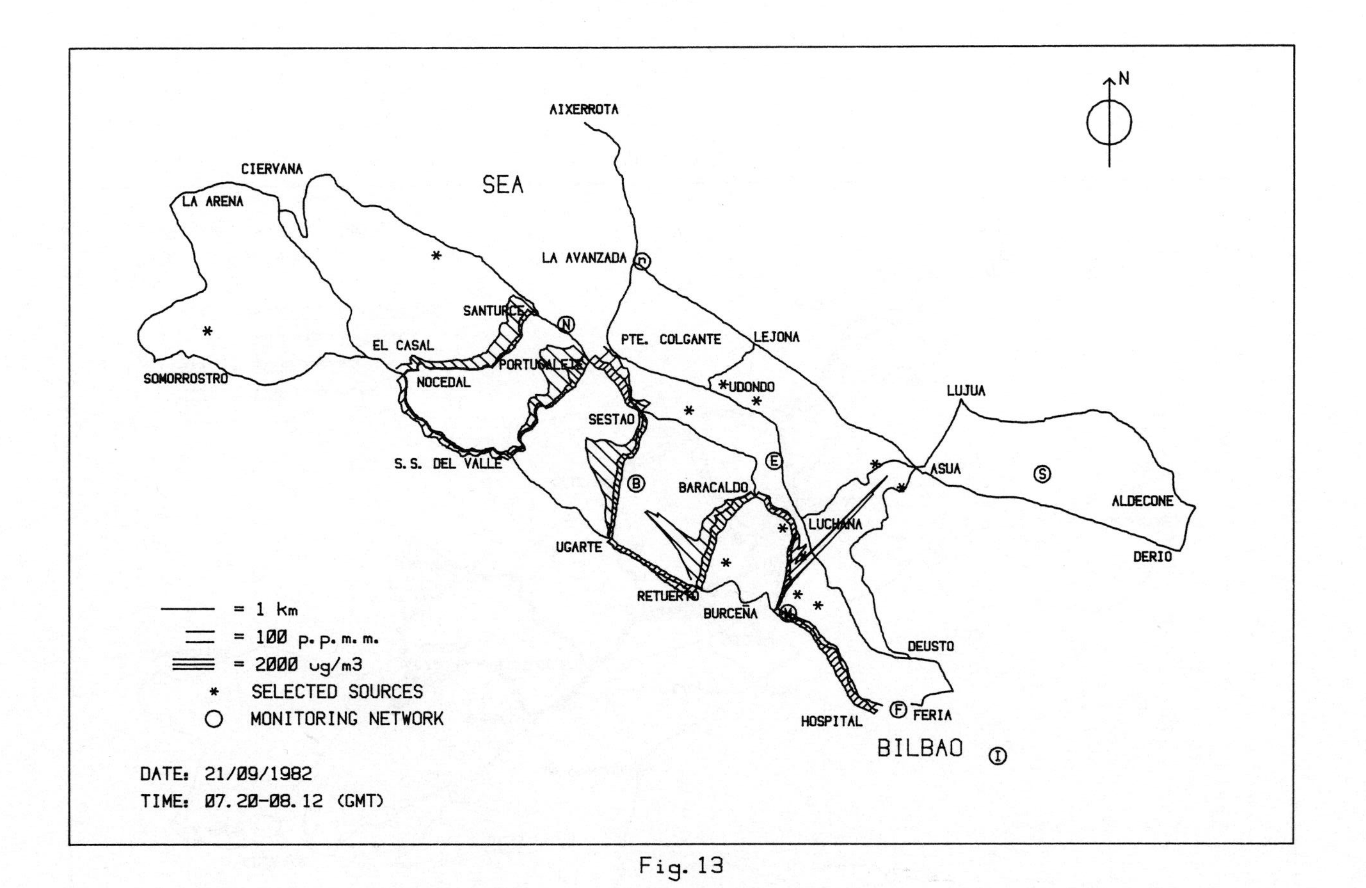
N
AIXERROTA
CIERVANA
LA ARENA
SEA
LA AVANZADA
SANTURCE
EL CASAL
NOCEDAL
PORTUGALETE
SOMORROSTRO
PTE. COLGANTE
LEJONA
UDONDO
SESTAO
S.S. DEL VALLE
LUJUA
ASUA
ALDECONE
DERIO
BARACALDO
LUCHANA
UGARTE
RETUERTO
BURCEÑA
DEUSTO
HOSPITAL
FERIA
BILBAO
= 1 km
= 100 p.p.m.m.
= 2000 ug/m3
* SELECTED SOURCES
MONITORING NETWORK
DATE: 21/09/1982
TIME: 07.20-08.12 (GMT)

Fig. 13

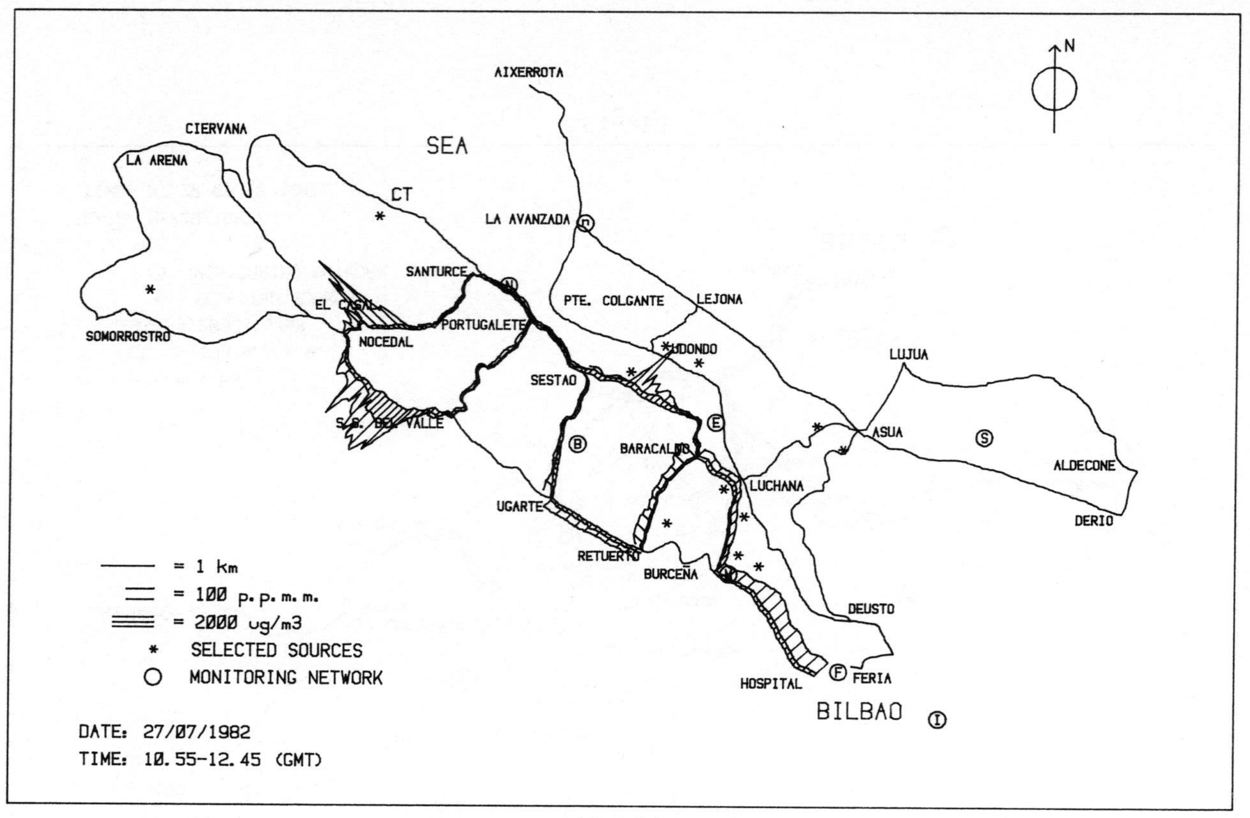
N
AIXERROTA
CIERVANA
LA ARENA
SEA
CT
LA AVANZADA
SANTURCE
PTE. COLGANTE
LEJONA
EL CASAL
SOMORROSTRO
PORTUGALETE
NOCEDAL
LUJUA
SESTAO
ASUA
BARACALDO
ALDECONE
LUCHANA
UGARTE
DERIO
RETUERTO
BURCEÑA
DEUSTO
HOSPITAL
FERIA
BILBAO
= 1 km
= 100 p.p.m.m.
= 2000 ug/m3
SELECTED SOURCES
MONITORING NETWORK
DATE: 27/07/1982
TIME: 10.55-12.45 (GMT)

Fig. 14

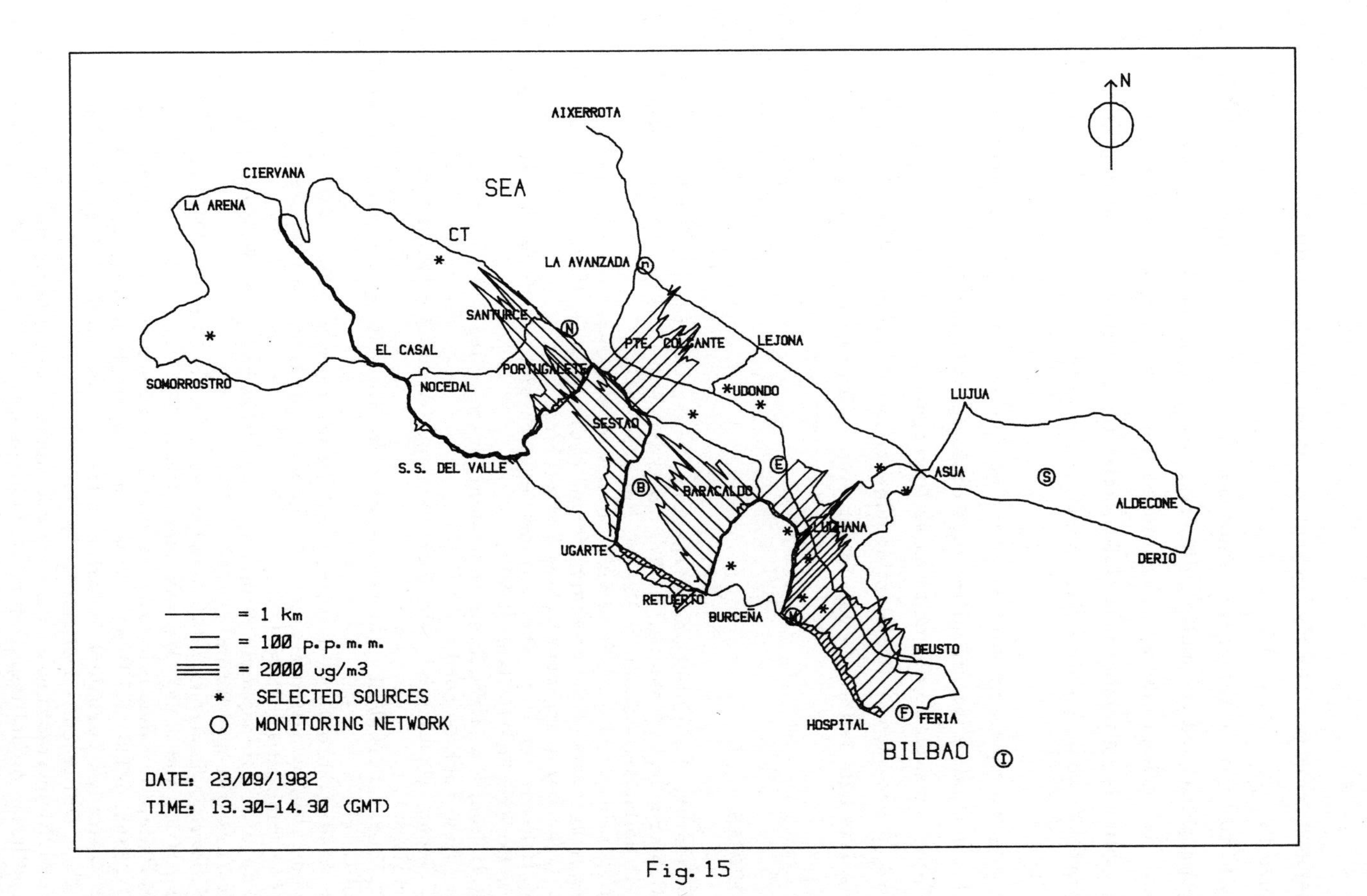
N
AIXERROTA
CIERVANA
SEA
LA ARENA
CT
LA AVANZADA
SANTURCE
EL CASAL
LEJONA
SOMORROSTRO
NOCEDAL
UDONDO
LUJUA
SESTAO
S.S. DEL VALLE
ASUA
BARACALDO
ALDECONE
UGARTE
DERIO
BURCEÑA
DEUSTO
HOSPITAL
FERIA
BILBAO
= 1 km
= 100 p.p.m.m.
= 2000 ug/m3
SELECTED SOURCES
MONITORING NETWORK
DATE: 23/09/1982
TIME: 13.30-14.30 (GMT)

Fig. 15

Hospital. This is a clearly distinct situation from that with N winds of figure 14.

Figures 12 to 15 were selected to show how drainage and re-entry flows, topographical and channelling effects can be documented for the same area. Referring to the section in mass flows we can now see that attempting to estimate mass flows during the conditions prevailing in Figure 12b would yield totally anomalous results.

Finally, we should re-emphasize the need to analyse remotely sensed data in the context of the prevailing meteorological phenomena. Figure 12b is, perhaps, unusable to calculate mass fluxes, but as part of the set in Figure 12, it provides an excellent documentation of a return flow. The only really useless data is the one you do not have or (worse) the one you cannot interpret.

REFERENCES

1 R.S. Scorer, Air Pollution, Pergamon Press, Oxford, UK (1968).
2 A.C. Stern, H.C. Wohlers, R.W. Boubel and W.P. Lowry, Fundamentals of air pollution, Academic Press, New York (1973).
3 A.N. Strahler and A.H. Strahler, Environmental geoscience (Part I, Energy systems of the atmosphere and hydrosphere), Hamilton Publishing Co. (John Wiley and Sons), New York (1973).
4 J.V. Iribarne and H.R. Cho, Atmospheric physics. D. Reidel Publishing Co., Dordrecht - Holland (1980).
5 J.R. Holton, An introduction to dynamic meteorology. Academic Press Inc., New York (second Ed.) (1979).
6 R.E. Munn, Descriptive micrometeorology. Academic Press Inc., New York (1966).
7 J.O. Hinze, Turbulence. McGraw-Hill Book Co., New York (second edition) (1974).
8 D.H. Slade (Editor), Meteorology and atomic energy, 1968. National Technical Information Service, US Dep. of Commerce. Springfield Va, 22161. Publication TID-24190.
9 F. Pasquill, Atmospheric diffusion, Ellis Horwood (John Wiley and Sons), New York (second edition) (1974).
10 R.S. Scarer, Environmental aerodynamics, Ellis Horwood Publishers (John Wiley and Sons), West Sussex, UK (1978).
11 E.J. Plate, Aerodynamic characteristics of atmospheric boundary layers. TID-25465, NTIS, US Dep. of Commerce, Springfield Va, 22151 (1974).
12 G.T. Csanady, Turbulent diffusion in the environment. D. Reidel Publishing Co., Dordrecht - Holland (1973).
13 F.T.M. Nieuwstadt and H. van Dop (Editors). Atmospheric turbulence and air pollution modelling. D. Reidel Publishing Co., Dordrecht - Holland (1982).
14 D.B. Turner, Workbook of atmospheric dispersion estimates, Document 5503-0015, US Gov. Printing Office, Washington DC 20402.
15 F.A. Guifford, Turbulent diffusion-typing schemes: A review, Nuclear Safety 17, 68-86 (1976).

16 R. M. Brown, L. A. Cohen and M. E. Smith, Diffusion measurements in the 10-100 km range. J. Appl. Meteorol. 11, 323-334 (1972).

17 F. Fanaki, Experimental observations of a bifurcated buoyant plume, Boundary-Layer Meteorol. 9, 479-495 (1976).

18 M. M. Millan, A. J. Gallant and H. E. Turner, The application of correlation spectroscopy to the study of dispersion from tall stacks. Atmos. Environ. 10, 499-511 (1976).

19 M. M. Millan, A note on the geometry of plume diffusion measurements. Atmos. Environ. 10, 655-658 (1976).

20 R. M. Hoff, A. J. Gallant and M. M. Millan, Data processing procedures for the Barringer correlation spectrometer (COSPEC). Report TEC 860, Atmospheric Environment Service, Downsview, Ont., Canada (1978).

21 R. V. Portelli, B. R. Kerman, R. E. Mickle, N. B. Trivett, R. M. Hoff, M. M. Millan, P. Fellin, K. G. Anlauf, H. A. Wiebe, P. K. Mistra, R. Bell and O. T. Melo, The Nanticoke shoreline diffusion experiment, June 1978, I. Experimental Design and Program Overview. II Internal Boundary Layer Structure. III Ground-based Air Quality Measurements. IV Air Chemistry. Atmos. Environ. 16, 413-466 (1982).

22 E. Pooler Jr. and L. E. Niemeyer, Dispersion from tall stacks: An evaluation. Proc. 2nd Int. Clear Air Congress, Academic Press, Oxford (1971).

23 D. L. Randerson, Temporal changes in horizontal diffusion parameters of a single nuclear debris cloud. J. Appl. Meteorol. 11, 670-673 (1977).

24 R. M. Hoff and M. M. Millan, Remote SO_2 mass flux measurements using COSPEC, APCA Journal 31, 381-384 (1981).

25 R. H. Varey, R. M. Ellis, P. M. Hamilton, D. J. Moore and S. Sutton, Plume dispersion and SO_2 flux measurements at Drax power station, England. Correlation Spectroscopy Conference, Toronto, Canada (1977).

26 R. B. Sperling, M. A. Peache and W. M. Vaughan, Accuracy of remotely sensed SO_2 mass emission rates. Report EPA-600/2-79-094, U.S. Environmental Protection Agency. Available from NTIS, Springfield Va. 22161, U.S.A.

27 M. M. Millan, L. A. Alonso, J. A. Legarreta, M. V. Albizu, I. Ureta and C. Egusquiaguirre, A fumigation episode in an industrialized estuary: Bilbao, November 1981, Atmos. Environ., in press (1984).

SUBJECT INDEX